EVOLUTION, CULTURE, AND THE HUMAN MIND

EVOLUTION, CULTURE, AND THE HUMAN MIND

Edited by

Mark Schaller, Ara Norenzayan, Steven J. Heine, Toshio Yamagishi, Tatsuya Kameda

Routledge
Taylor & Francis Group
LONDON AND NEW YORK

First published in 2010 by Psychology press

Published 2016 by Routledge
2 Park Square, Milton Park, Abingdon, Oxfordshire OX14 4RN
711 Third Avenue, New York, NY 10017

First issued in paperback 2015

Routledge is an imprint of the Taylor and Francis Group, an informa business

ISBN 13: 978-1-138-99084-5 (pbk)
ISBN 13: 978-0-8058-5911-9 (hbk)

Library of Congress Cataloging-in-Publication Data

Evolution, culture, and the human mind / editor, Mark Schaller.
p. cm.
Includes bibliographical references and index.
ISBN 978-0-8058-5911-9 (hbk. : alk. paper)
1. Evolutionary psychology. 2. Human evolution. 3. Culture--Psychological aspects. I. Schaller, Mark, 1962-

BF698.95.E953 2010
155.7--dc22 2009026456

Visit the Taylor & Francis Web site at
http://www.taylorandfrancis.com

Contents

SECTION II EVOLUTIONARY BASES OF CULTURAL PHENOMENA

SECTION III EVOLUTIONARY UNIVERSALS AND CROSS-CULTURAL DIFFERENCES

Contributors

Roy F. Baumeister
Department of Psychology
Florida State University

Nicholas A. Bowman
Center for Social Concerns
Notre Dame

Abraham P. Buunk
Department of Psychology
University of Groningen

Avinish Chaturvedi
Department of Business Administration
University of Illinois

Chi-yue Chiu
Nanyang Business School
Nanyang Technological University

Florette Cohen
The College of Staten Island
City University New York

Martin Daly
Department of Psychology, Neuroscience, and Behaviour
McMaster University

Yulia Chentsova Dutton
Department of Psychology
Georgetown University

Steven W. Gangestad
Department of Psychology
University of New Mexico

Jeff Greenberg
Department of Psychology
University of Arizona

Reid Hastie
Graduate School of Business
University of Chicago

Chip Heath
Graduate School of Business
Stanford University

Steven J. Heine
Department of Psychology
University of British Columbia

Joseph Henrich
Departments of Economics and Psychology
University of British Columbia

Tatsuya Kameda
Department of Behavioral Science
Hokkaido University

Douglas T. Kenrick
Department of Psychology
Arizona State University

Young-hoon Kim
Department of Psychology
University of Illinois

Lee A. Kirkpatrick
Department of Psychology
College of William and Mary

Shinobu Kitayama
Department of Psychology
University of Michigan

Damian R. Murray
Department of Psychology
University of British Columbia

Randolph M. Nesse
Departments of Psychology and Psychiatry
University of Michigan

Sarynina Nieuweboer
Department of Psychology
Arizona State University

Ara Norenzayan
Department of Psychology
University of British Columbia

Daniel M. Ogilvie
Department of Psychology
Rutgers University

Yohsuke Ohtsubo
Department of Sociology
Nara University

Tom Pyszczynski
Department of Psychology
University of Colorado, Colorado Springs

Debi Roberson
Department of Psychology
University of Essex

Paul Rozin
Department of Psychology
University of Pennsylvania

Mark Schaller
Department of Psychology
University of British Columbia

Azim F. Shariff
Department of Psychology
University of British Columbia

Sheldon Solomon
Department of Psychology
Skidmore College

Naoto Suzuki
Department of Behavioral Science
Hokkaido University

Masanori Takezawa
Department of Psychology and Society
Tilburg University

Margo Wilson
Department of Psychology, Neuroscience, and Behaviour
McMaster University

Toshio Yamagishi
Department of Behavioral Science
Hokkaido University

1

Introduction

ARA NORENZAYAN, MARK SCHALLER,
and STEVEN J. HEINE

How and why does the human mind work the way it does? Consider two very different perspectives on this important question.

One influential perspective emerges from the study of human evolution: The basic psychological repertoire of the human species consists of adaptations, or their by-products, accumulated over the course of a very long evolutionary history. Key aspects of our emotions are adaptations, as are specific aspects of cognition. Many basic behavior patterns can also be conceptualized as adaptations, such that we respond to environmental cues with behaviors that, in ancestral environments, were associated with incremental advantages in reproductive fitness. In sum, the workings of the human mind—and the resulting psychological phenomena—must be considered the product of evolution.

There is a second influential perspective, provided by the study of culture: Our psychological experiences and responses to the world are fundamentally shaped by cultural learning. Other animal species may show evidence of some of the things that are commonly associated with the concept of culture (socially transmitted traditions, between-group differences that are independent of reproductive events), and there is lively debate over the extent to which chimpanzees, whales, and songbirds might be considered cultural species as well (e.g., Whiten, 2005), but there is no debate about whether humans are a cultural species nor is there any doubt that human life (and human reproductive fitness), compared to that of other species, is fundamentally connected to the complex sets of shared symbols, meanings, rituals, and behavioral norms that make up the cultures that we create, inhabit, and pass on to our descendents. The process of enculturation starts from birth, and the human brain develops in a cultural context. Abundant bodies of evidence document pervasive cross-cultural differences in psychological phenomena and thus attest to the many ways in which even our most basic thoughts, feelings, and behaviors are culturally shaped.

These two perspectives, and the enormous amounts of scientific research that they have generated, compel two fundamental conclusions: The human mind is the product of evolution, and the human mind is shaped by culture. These conclusions are, we think, unassailable.

Given these two conclusions, it might be expected that psychological scientists would have invested considerable energies to understand exactly how these two perspectives fit together in a rigorous and conceptually coherent manner. But, no, that hasn't been the case. For the most part, inquiry in evolutionary psychology has proceeded independent of inquiry in cultural psychology (Norenzayan, 2006). Evolutionary psychological research documents the many specific ways in which evolutionary adaptations appear to govern the operations of the human mind. But because an evolutionary approach compels researchers to focus on species-typical—and thus universal—elements of human cognition, these inquiries only occasionally grapple seriously with questions about human culture and cross-cultural differences. In contrast, research in cultural psychology has focused primarily on the many ways in which psychological phenomena are shaped by different cultural experiences. Because the focus is on cultural variability (rather than pan-human universality), these inquires rarely consider evolutionary processes. The upshot is that, even after several decades of intensive research on human evolutionary universals and on cross-cultural differences, there has been little in the way of rigorous theory and research linking these two perspectives together.

But that is now changing, and that's what this book is about.

THE INTEGRATION OF EVOLUTIONARY AND CULTURAL PSYCHOLOGY AND WHY IT MATTERS

The seed of this book was planted several years ago at a historic gathering that took place at the University of British Columbia in Vancouver, Canada. Supported by generous funding provided primarily by the University of Hokkaido and the American Psychological Association, dozens of scientists (primarily evolutionary psychologists and cultural psychologists, along with a smattering of biologists and anthropologists and other scholars from related disciplines) came together with the explicit goal of forging meaningful integrations between evolutionary and cultural perspectives on the human mind.

Doing so isn't easy. The difficulty arises not because of any inherent incompatibility between these different perspectives but because these two different perspectives typically imply two distinct sets of questions, and these sets of questions don't necessarily overlap in obvious ways. Forging coherent integrations requires scientists to step outside of the comfort zone of the research questions that they are accustomed to addressing with empirical data and instead to ask a broader, more ambitious, and less wieldy set of questions altogether: Exactly which psychological phenomena are universal and which are culturally variable? How does culture itself fit into an evolutionary perspective on human nature? What specific evolutionary pressures gave rise to the human capacity for culture? Did this capacity alter the processes of natural selection itself, and if so, how? What is the cognitive architecture of this capacity, and what are its consequences? How exactly can substantial cross-cultural differences in psychological functioning arise from evolutionary adaptations that are, at some level, universal across human populations? And so on. These questions pose substantial challenges, both conceptually and empirically. But there are also substantial benefits to be gained by rising to these challenges.

First, this integrative program of inquiry may help put an end to the common and troubling misconception that there is some inherent epistemic gulf between evolutionary and cultural perspectives on human psychology. This persistent misconception has its roots in the hoary myth of "nature versus nature"—a false dichotomy that continues to haunt the psychological sciences, to the intellectual detriment of the discipline. It will be valuable to any enthusiast of the psychological sciences to think in deeper, more sophisticated ways about evolutionary and cultural perspectives on the human mind and how they complement each other.

Rigorous psychological research of this sort also stands to make a substantial contribution to broader scientific conversations about evolution and culture. Although it has been largely overlooked within the psychological sciences, the complex relationship between evolution and culture has been an important topic of inquiry in other social and biological sciences (e.g., Richerson & Boyd, 2005). There is considerable scientific interest in adaptations for culture, and their specific cultural consequences, and the ways in which these consequences alter the process of natural selection itself. Psychological scientists are perfectly poised to contribute in novel and exciting ways to these multidisciplinary research programs. After all, evolutionary processes operate on the phenotypic traits expressed by individuals, including individuals' cognitions, decisions, and actions. Moreover, cultural norms (which guide individual behavior) are themselves the collective consequence of individuals' cognitions, decisions, and actions (Schaller & Crandall, 2004). Individuals' cognitions, decisions, and actions are exactly the sorts of things that psychologists are especially adept at studying. In short, as it has become increasingly clear that a complete articulation of the complex relations between evolution and culture demands attention to the human mind, it has also become clear that psychological scientists must contribute more fully to this conversation.

And, of course, this is just flat-out exciting intellectual territory. There is a great deal we do not yet know about how the pieces of the puzzle—evolution, culture, and the human mind—fit together. The topic is fertile ground for novel theories and new empirical discoveries. These discoveries will contribute importantly to the psychological sciences and will be of considerable interest across a broad range of social and biological sciences as well.

OVERVIEW OF THE BOOK

We begin with a set of five chapters that, in various ways, show how cultural and evolutionary perspectives can fit together within the psychological sciences. Rozin explains why there is no incompatibility whatsoever between evolutionary and cultural perspectives on the human mind. Baumeister discusses a variety of ways in which evolutionary processes created the psychological building blocks necessary for human culture to exist. Chiu, Kim, and Chaturvedi summarize the continuing relevance of Donald Campbell's seminal contributions to the simultaneous study of evolution, culture, and cultural evolution. Dutton and Heath address the topic of cultural evolution. They show how selection, transmission, and retention mechanisms can explain why some knowledge structures become and remain culturally popular while others don't. Kirkpatrick draws on recent advances in evolutionary psychology to describe how a focus on psychological adaptations is necessary to forge connections between the mechanisms of genetic evolution and cultural transmission.

Whereas the first set of chapters emphasizes basic processes that are relevant to understanding culture in all its many manifestations, the second section of this book highlights specific ways in which an evolutionary perspective can help us understand particular prototypic aspects of human culture. Gangestad describes how an adaptationist framework can help frame questions about the many specific things that are central to any definition of human culture. This is followed by two chapters that employ evolutionary frameworks to understand human religions and religious beliefs. Solomon, Greenberg, Pyszczynski, Cohen, and Ogilvie locate one particular theory of cultural origins (terror management theory) within an evolutionary perspective and, in so doing, suggest that supernatural beliefs may arise from specific adaptations. Shariff, Norenzayan, and Henrich show how moralizing religions and cooperative tendencies in large groups might have coevolved. Nesse also focuses on human cooperation as a defining feature of human culture and suggests that it may result in part from a specific form of natural selection: social selection. Kameda, Takezawa, Ohtsubo, and Hastie describe an adaptationist perspective on egalitarian beliefs and discuss its implications for social justice and for cultural variability in justice systems.

In the third and final section of this book, we turn our attention to the fact of cultural variability and speculations about the evolutionary roots of cross-cultural differences. Roberson describes a line of research that documents important cross-cultural differences in color vision. These findings underscore the point that even the most ostensibly "pure" psychological phenomena are influenced by culture. Yamagishi and Suzuki summarize an approach to thinking about culture as a self-sustaining system of beliefs and illustrate this approach with many examples that reveal how culture governs individual preferences, decisions, and actions. Kitayama and Bowman draw further attention to one fundamental dimension of cultural difference (the individualism–collectivism dimension) and its psychological consequences and offer one perspective on why these cross-cultural differences might exist. Daly and Wilson focus on another paradigmatic dimension of cultural variability (differences in a "culture of honor") that has important implications for aggression and violence. They reinterpret the cross-cultural evidence within an evolutionary framework. Schaller and Murray describe how different attitudes, values, and behaviors may have different consequences on reproductive fitness depending on the prevalence of pathogens in local ecologies. This evolutionary analysis successfully predicts a wide range of important cross-cultural differences in traits, values, and belief systems (including differences along the individualism–collectivism dimension). Finally, Kenrick, Nieuweboer, and Buunk show how cultural variability in mating systems can reflect deeper evolutionary universals. They use this example to illustrate how an evolutionary analysis is essential to the simultaneous articulation of cultural similarities and cross-cultural differences.

ENVOI

The successes of cultural psychology and evolutionary psychology have fundamentally altered the landscape of the psychological sciences. But these successes create potential traps as well. By amassing large (and largely independent) literatures, and by creating conferences and journals that are specific to their subdisciplines, cultural psychologists and evolutionary psychologists have perhaps found it too easy to ignore each other's work. This book shows a way out of that trap.

Evolutionary and cultural perspectives on the human mind aren't incompatible or irrelevant to each other. Quite the contrary. The origins of human culture (and cross-cultural differences) cannot be fully understood in the absence of evolutionary considerations. Evolutionary psychologists cannot fully explain individual psychological responses without considering the fundamental "cultureness" of human behavior. Genes and culture are mutually necessary for a complete scientific understanding of the human mind.

REFERENCES

Norenzayan, A. (2006). Evolution and transmitted culture. *Psychological Inquiry*, *17*, 123–128.

Richerson, P. J., & Boyd, R. (2005). *Not by genes alone: How culture transformed human evolution*. Chicago: University of Chicago Press.

Schaller, M., & Crandall, C. S. (2004). *The psychological foundations of culture*. Mahwah, NJ: Lawrence Erlbaum.

Whiten, A. (2005). The second inheritance system of chimpanzees and humans. *Nature*, *437*, 52–55.

Section I

How Evolution and Culture Fit Together

2

Towards a Cultural/Evolutionary Psychology

Cooperation and Complementarity

PAUL ROZIN

> We start with the assumption that individual men possess authentic properties distinctive of *Homo sapiens* and that their actions in society alter them in authentically distinctive ways. (Asch, 1952, p. 119)

In about the past two decades, psychology has been blessed by the development of two new subfields, evolutionary and cultural psychology. Each has provided an important perspective and corrective to what has come before. Evolutionary psychology brings to bear on the phenomena of psychology one of the greatest scientific theories of all time. It introduces the important idea of domain specificity into a psychology previously dominated by general process theories, enriches psychological understanding with evolutionary and adaptive explanations, and places human behavior and mind in their natural context. Cultural psychology calls the attention of psychology to one of the most powerful forces, perhaps *the* most powerful force, that shapes human beings, challenges universal principles of psychology from a direction different from evolutionary psychology, and also emphasizes that humans must be studied in context. Both bring important new questions to the forefront of psychology. We should celebrate these accomplishments and recognize that the two together can do much more than either alone, not just because each can add to our understanding but because there is an interaction effect: They can each improve the other.

Consider the following scenarios. There are important problems to be solved. X has some of the tools to solve them, and Y has the materials to apply the tools to. Should X and Y cooperate? Of course.

In more detail, imagine two sets of researchers working within different paradigms. One group, E, has behind it a powerful, well-articulated, and detailed theory. Because, however, (a) the events the theory refers to are primarily in the distant past, (b) the fundamental process studied takes a long time (many generations) to show marked effects, and (c) the past residues (fossils) that are critical data are particularly lacking in the topic of interest (behavior and mind), the gathering of data and the testing and confirmation of hypotheses are extremely difficult. The second group, C, doesn't have a really good theory, but the events it studies are relevant to a variant of the great theory Group E works with. For the case of the phenomena studied by C, the relevant data are very accessible, occur in the present and recent past, provide a superb record of past events, and change very rapidly, in terms of months, years, decades, or at most centuries. Group E has a strength and weakness, and Group C has a complementary weakness and strength. Given that Group E's theory has substantial and easy application to the phenomena of Group C, Group E should be elated to have a much better testing arena. Group C should be delighted to be able to work from a strong theory.

Of course, Group E is evolutionary psychologists, and Group C is cultural psychologists. They should love each other. Instead of fighting over whether 10% or 90% of the variance in human functions is attributable to genes or culture or what percentage of cultural universals can be accounted for in terms of human evolution, they should be celebrating and sharing each other's strengths.

The matter is simple. Cultural evolution, broadly construed, is a principal account for cultural differences and cultural change. Biological evolution and cultural evolution share many important components: both depend essentially on variation, transmission, and natural selection. The differences between biological and cultural evolution are just what allows for a much speeded up evolution, one in which principles of natural selection, the core idea of evolution, can be seen to be at work rampantly, rapidly, and with clear measures. The evolutionary psychologist does not have a good fossil record of behavior or mental events, whereas the cultural psychologist has actual records of these events, in the form of writing for many thousands of years, of intentionally preserved artifacts, and, in most recent times, of photography, sound recording, and video. And with the speed of cultural evolution (look, for example, at fads in first name popularity, or the shift from landline phones to cell phones, or the entry of television into American life), prospective data can actually be collected and detailed archived data can be accessed. For at least decades, national random sample surveys have been carried out, with data available, on beliefs and practices of individuals in many parts of the world. This is the perfect raw material for evolutionary analysis. To add to this opportune cooperation, the forces of biological and cultural evolution interact with each other in recent human history.

The idea that the best arena for evolutionary theory in the domain of human activities is culture is hardly new with me. Donald Campbell (1965) among others pointed this out, and systematic efforts to apply evolutionary ideas to culture have been promoted and elegantly demonstrated by a number of investigators, most notably Durham (1991), Boyd (1985), Richerson and Boyd (2005), Wilson (2002), and Mesoudi, Whiten, and Laland (2004). Richerson and Boyd (2005), in *Not by*

Genes Alone, did a splendid job of arguing the case both for the importance of the principles of biological evolution in understanding cultural evolution and for the fact that cultural evolution is much more important than biological evolution in accounting for what has been happening in the world and mind of humans for the past thousands of years. (A paper by Newson, Richerson, and Boyd [2007] summarized many of the principal points in the Richerson and Boyd book.)

EVOLUTIONARY AND CULTURAL PSYCHOLOGY

All of psychology is about understanding animal and human behavior, mental events, and the worlds in which animals and humans live. For the case of humans, much of this world was actually made by prior humans. The goal of evolutionary psychology is more focused than that of cultural psychology. Evolutionary psychologists are interested in a particular type of understanding, which I have called *evolutionary-adaptive explanation* (Rozin & Schull, 1988). The focus of interest is dual. First is creating the history of any particular manifestation over a time period that extends before the lifetime of the current generation and often goes back thousands of years or more. Second is explaining contemporary behaviors, mental events, or environments in terms of the adaptive (survival) function that promoted their existence. These are extremely important aspects of understanding and are also difficult to accomplish, because they involve study of times long past, for which there is a minimal record. Cultural psychologists are interested in a wider variety of accounts, including describing cultural differences and similarities and explaining immediate and remote causes of these manifestations. The remote cause account in cultural psychology focuses on the period between birth and adulthood, what we can call *developmental explanation* (Rozin & Schull, 1988). There is also, however, a substantial interest in cultural psychology in explaining causes that are more remote than those within the lifetime; that is, evolutionary as well as development causes. And of course, function and adaptation are important aspects of explanation in cultural psychology.

Evolutionary psychologists can harness evolutionary theory, genetics, and extensive knowledge of nonhuman primate behavior and inferred information about early humans. It is incontrovertible that humans are primates and that they have a whole set of adaptations and features that they share with other primates, including, of course, almost all of their genes. But it is just as obvious, at least to me, that most of what humans do cannot be *directly* traced to our primate origins. Our movies, novels, work, eating, and sociality take on very special forms that have built on the primate base but are often very different from their primate precursors. Forks and knives and dinner parties don't have clear forbears. So if we are trying to explain the contemporary human condition, it seems clear to me that most of the variation in humans involves substantial cultural acquisition, within the lifetime. Put another way, culture and the environment (including the effects of cultural evolution) explain most of the variance. But culture is, to a large extent, what is acquired in development and the residues of past lives as expressed in the environment, including the institutions in the environment. The contents of culture have a history, and they have evolved, by the process of cultural evolution, so that cultural evolution plays a central role in understanding culture and hence contemporary human beings.

Evolution and Culture: Nature and Nurture

The opposition between evolutionary psychology and cultural psychology has its parallel in the much older nature–nurture debate. Although many have commented that it is rarely nature *or* nurture, the arguments continue. The nature folks have one big advantage over the nurture folks: They have classical genetics and modern molecular genetics to provide models, units, exemplars, and processes. The nurture folks don't have a very comprehensive and well-documented theory of how experience works, of how the environment changes behavior. This parallels the difference between evolutionary psychology and cultural psychology. In my judgment, in both cases the evolution–nature folk have by far the best theory, but the cultural–nurture folks have a wider range of things that they can in principle explain; that is, they have most of the variance. And in many cases, what they are studying is more susceptible to experiment. The physical and social environments have enormous influence on individuals, but we don't have a great theory to encompass this. What we must remember is that one can separate the quality of theory from the range of domains on which it is applicable. Whatever one's position on the relative importance of nature and nurture (and our irresistible urge to make claims in terms of explaining variance, etc.), it is surely true that the range of phenomena explainable by nurture, "divided by" the quality of our theoretical models, is a much smaller number for nature than for nurture. If one wishes to work only where there is most light (explanatory power), we all have the option of physics, but we must keep an eye on the type of phenomena we are trying to explain. So nature (e.g., behavior genetics) and evolutionary psychology both share important theoretical material and mechanisms and probably can account for much less than half of what we try to understand in psychology. Nurture theorists and cultural psychologists don't have very elegant or comprehensive theories (what are powerful theories of the way the environment affects us: some principles of learning, ideas about social influence?), but they have most of the phenomena.

What is special about the evolutionary and cultural psychology distinction is that, unlike nature and nurture, the evolutionary principles are powerfully important in explaining much of what cultural psychologists are interested in. Cultural evolution is powerful.

Predispositions, Constraints, and Defaults

Both biological and cultural evolution, and human function as well, are molded to a large degree by predispositions and constraints. Reflexes account for little interesting human function. Evolution often acts by inclining an organism to perform in a certain way (predispositions) or excluding certain options (constraints). This is clearly the case for the laws of learning, as, for example, the predisposition to associate tastes with certain gastrointestinal events. Similarly, cultures operate primarily through predispositions and constraints. Most people are capable of understanding how another culture looks at the world, but it is not their inclination to do so, because of socialization. I have identified this in terms of cultures operating by promoting default ways of looking at the world (Rozin, 2003).

One of the big default differences highlighted in cultural psychology is the difference between focusing on harmony and focusing on agency or valence. Thus, in a study originally done by Menon and Shweder (1997) and expanded by me (Rozin, 2003), Americans and Asian Indians were asked what does not belong of the three terms: *anger, shame, happiness*. Almost all Americans say *happiness*, because it alone is positive. Indians are inclined to say *anger*, because, as they explain, shame and happiness are socially constructive and anger is socially destructive. This is a first parsing on these terms. People in either culture can understand the other culture's default parsing of these terms, once it is explained to them. They just don't tend to think that way. The first parsing an individual makes of something is very important, because the usual process of thinking proceeds from that first default rather than from considering alternative framings.

APPLYING BIOLOGICAL EVOLUTION TO THE STUDY OF CULTURE AND CULTURAL EVOLUTION

In this section, I consider some of the basic ideas in biological evolution and how they apply to the study of cultural psychology and, to some degree, psychology in general. Richerson and Boyd (2005) made the most articulated and systematic attempt I know of to show how we can turn principles of evolutionary psychology into the understanding of culture on its own terms. Population thinking, central to thinking about evolution, can certainly be applied to cultures and cultural change. For example, the S-shaped curve describing the growth of an adaptive new variant, in biology or culture (Girifalco, 1991; see also Figure 2.1), has the same mathematical properties because of the way natural selection works and the way that transmission occurs. The Internet can speed up this process enormously, but the S shape remains, with slow adoption at the beginning and end of the penetration of a new variant and faster adoption in the middle period.

The three essentials of biological evolution—variation, transmission, and natural selection—will now each be examined with respect to their relevance to cultural evolution (see Mesoudi et al., 2004, for a similar set of arguments).

Variation

Darwinian evolution depends on variation. Without it, natural selection cannot operate. Heritable variation occurs randomly, principally through mutation and recombination in the biological world. But in the cultural world, variants are intentionally produced. One does not have to wait until a harder material can accidentally be flaked to produce an arrowhead; one can actively look for a harder material and a way to shape it. We do not have to wait for cars to get gradually bigger so that they can haul tree trunks; we can design a bigger car. We don't have to wait for the highly improbable merging of a calculator and a typewriter; we can just do it when

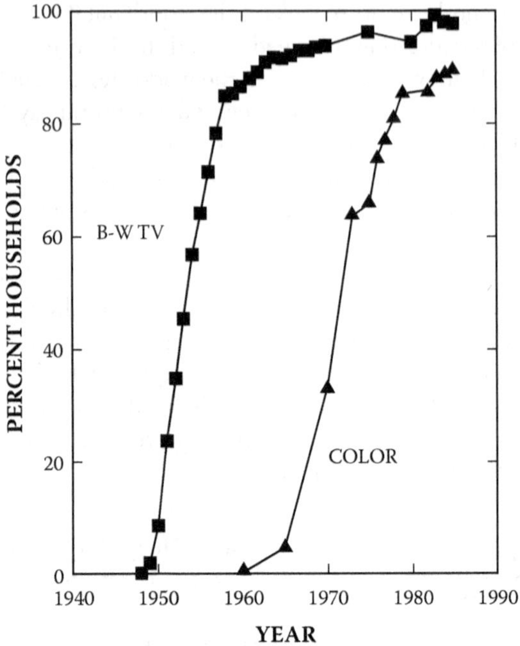

Figure 2.1 Percentage of households adopting black-and-white and color televisions in the United States between 1940 and 1990 (from Girifalco, 1991).

we get the idea and get a computer. If the path to an adaptive new variant involves going through a malfunctioning phase, it is fatal in biological evolution, but human persistence can keep this process going until success is achieved. There are five big advantages in the domain of variation for the cultural evolutionist: One, there are many more variants; two, we have a clear record of them; three, cultural variants can be intentionally produced; fourth, as a result of three, cultural forces allow for the evolution of an adaptation that has to pass through a maladaptive transitional stage; and fifth, groups of people over generations can be involved cumulatively in the production of variants.

Transmission

There are understandable arguments about both the units of transmission and the process of transmission in cultural evolution. The clarity of the genetic unit and genetic assortment is not carried over into cultural evolution. One can, however, understand a process without understanding what the unit is. The best example is Darwin's theory of evolution, expressed in a rich form without knowledge of the units of inheritance (or even that there was a genotype). Dawkins (1976) provocatively suggested the "meme," and arguments abound as to whether this is the only or even an appropriate unit. But what is critical is the process of transmission, as opposed to the unit. Here, those few interested in cultural evolution have ample models, some originating from research in social psychology. There are two types of questions about transmission: The first is the source of transmission, and the second is the process through which transmission occurs.

As to source, Cavalli-Sforza, Feldman, Chen, and Dornbusch (1982) indicated three pathways: vertical (transgenerational, typically parent–child), horizontal (siblings and peers), and oblique (usually one to many, as by teachers or the media). Vertical transmission allows for both genetically determined and acquired features. Somewhat surprisingly, parent–(adult) child resemblance for preferences (e.g., for food or music), which incorporates both genetic and early experience influences, is low (averaging about $r = .15$; Rozin, 1991). Thus, it appears most variation in preferences comes via horizontal or oblique sources. Moral positions (e.g., religion, moral attitudes) show a higher parent–child correlation (Rozin, 1991).

The transmission process for cultural evolution is actually a set of processes. Richerson and Boyd (2005) presented a useful taxonomy (see also Newson et al., 2007). Transfer of information for nonhumans and ancestral humans can occur by observation and by imitation. Explicit teaching seems to be a uniquely human activity. Information transfer in humans was massively increased by a number of uniquely human adaptations or inventions: first language, then narrative, then writing, then photography and video, and now the Internet. A new variant can broadcast to half of the world in hours through personal chain messages or news services on the Internet. Written language and the newer visual and digital media provide a highly accurate means of transmission. People are influenced by norms, and hence the more frequent a variant is, the more likely it is to be adopted. This dynamic generates the ubiquitous S-shaped curve, a feature of transmission and natural selection, in biological evolution and cultural evolution, well illustrated by the acceptance of innovations and new technologies (Girifalco, 1991). Figure 2.1 dramatically displays the rapid S-shaped spread and adoption of both black-and-white and color televisions in the United States. In both cases, the innovation went from being uncommon to very common over about a decade.

The cultural transmission process centrally involves psychology and communication. In fact, an argument can be made that given human cognition and motivations, interpersonal interactions and communication, and differences in ecology and geography across the earth, culture (and culture differences) is virtually an inevitable by-product (Conway & Schaller, 2007). As Conway and Schaller phrase it, "Communication is necessary for culture," and, "Communication is sufficient for culture." Imitation and teaching are deeply psychological processes. The fitness of variants, that is, their success in being communicated and transmitted, depends on both their utility and their "communicability." Recent work (e.g., Heath, Bell, & Sternberg, 2001; Schaller, Conway, & Tanchuk, 2002; summarized in Conway & Schaller, 2007) has emphasized the psychological variables, such as memorability, attention-getting potential, social "interest" (attracting commentary), and emotionality, that help to propel a variant into a cultural norm.

Natural Selection

Natural selection is the key concept in biological evolution, perhaps the most powerful process at work in the living world (Dennett, 1995). It follows incontrovertibly from variation and transmission. It is deeply the same process in biological evolution, cultural evolution, capitalism, and instrumental learning. Lamarck's powerful but incorrect idea about biological transmission of learned acquisitions across generations may have proved untrue, but it is the core of the process of transmission in cultural evolution, and it accounts, along with media such as language, writing, visual media, and digital media, for the cumulative nature of the expansion of cultures. In addition to Lamarck's idea, a powerful form of group selection is at work in cultural evolution. The idea of group selection has been controversial in biological evolution, although it is now resurfacing in biological evolution (Wilson, 2002). There is no doubt, however, that "group selection" operates forcefully at the level of small or large human groups (Richerson & Boyd, 2005; Wilson, 2002).

Inclusive fitness, the coin of the realm for biological evolution, is often replaced by other selection criteria as a result of social consensus; money, influence, power, and prestige can come to dominate procreation in cultural evolution. The low birth rates in the wealthiest countries in the world are ample indicators of this.

What we have, then, in cultural evolution is a rapidly accelerated and expanded version of biological evolution. Intentional variations bridging local minima, rapid and worldwide transmission, and culturally induced strong selection pressures can collapse millennia of biological evolution into years, as shown in Figure 2.1. The power of the principles of evolution is apparent to all in modern developed cultures, as we see the rapid advances of technology. They are so rapid that there is now a major form of vertical transmission from child to parent, as the youngsters help their parents and grandparents navigate the Internet, master their new video recorders and cameras, and digitally manage their finances. A particular feature of biological evolution that blossoms in cultural evolution is preadaptation.

Preadaptation: Important in Biological Evolution, Much More Important in Cultural Evolution

Preadaptation, the use of an entity that evolved for one purpose in another context or system, is a major feature of biological evolution (Bock, 1959; Mayr, 1960; renamed *exaptation* by Gould & Vrba, 1982). Mayr (1960) identified preadaptation (as opposed to mutation) as the principal source of major changes in evolution. The most familiar example in vertebrate evolution is the conversion of a jaw articulation in fish into the middle ear bones of the mammalian ear. In some cases, such as the middle ear bones, the original adaptation is disassembled. In other cases, such as the use of the mouth (evolved for eating and breathing), including the teeth and tongue, for expression of language in humans, the original function is maintained. There are many specific adaptations that would be adaptive in new contexts. Example include the first synapses or neurons (presumably tied to a specific narrow function), lateral inhibition, neural circuits that instantiate diurnal and other rhythms, initially localized neurotransmitters, principles of association (presumably initially evolved in a specific context such as avoiding aversive stimuli or approaching food), chunking in memory, and so on. The problem in biological evolution is creating either a duplication of a circuit or a connection into an existing circuit so that a new module could access an adaptive specialization or module (Rozin, 1976). Even with this enormous problem challenging either the organization of the genome itself and/or the major constraints in development of the nervous system, preadaptation is a very important aspect in biological evolution.

In cultural evolution, the constraints that limit preadaptation in biological evolution are relaxed or even eliminated. An individual human being can see the value of an existing system (whether biologically programmed or socialized through culture) in a new domain and just apply it. Great inventions such as writing and the wheel spread rapidly and widely to many aspects of human life by the process of cultural preadaptation. Preadaptation is a major feature of cultural evolution.

Preadaptation and Virtues and Shortcomings of Carrying Mainline Psychology Over to Evolutionary and Cultural Psychology

There is a special case of preadaptation that applies to the development and evolution of the fields of psychology and its two new offspring, evolutionary and cultural psychology. The model of natural science, developed in the physical sciences, was preadapted for research in biology and psychology. The interplay of theory and data, the idea of testing theories, the idea of careful and objective measurement, and the idea of an experiment all adaptively direct research in the historically new areas of the behavioral sciences. These invaluable features of science were imported into psychology in the first half of the 19th century, and in the past 20 or so years, they have been transferred from mainline psychology to two of its newest offspring, evolutionary and cultural psychology. But, as I have pointed out elsewhere (Rozin, 2001), psychology has adopted only some of the basic features of natural science, and it is only these that have been transferred to cultural psychology and, to a lesser extent, evolutionary psychology.

Progress in physical and biological science, with the scientific method, was generally preceded by careful description of the phenomena of interest. This description often took the form of describing functional relations (as between the volume of a gas and its temperature), and, of course, for Darwin it involved careful description of different species in their natural environments. This piece of basic science has been endowed with negative prestige in modern psychology. Experiment should *follow* careful description in natural settings, and description of major functional relationships. This is particularly clear not only in the development of the psychology of learning in the behavioristic tradition but also in modern social psychology. As Solomon Asch, the seminal figure of the 20th century in social psychology, noted,

> If there must be principles of scientific method, then surely the first to claim our attention is that one should describe phenomena faithfully and allow them to guide the choice of problems and procedures. If social psychology is to make a contribution to human knowledge, if it is to do more than add footnotes to ideas developed in other fields, it must look freely at its phenomena and examine its foundations. (Asch, 1952, p. xv)

Cultural anthropologists have done some of this work for cultural psychologists, in their ethnographies, but these get little attention. Also, the focus of cultural psychology, like the social and cognitive psychology it derives from, has been mental events and has drifted away from behavior (Baumeister, Vohs, & Funder, 2007). Ethnographies have been more behavioral in character. From my perspective, cultural psychology should pay much more attention to behavior, environments, and institutions and not make the same mistakes made by social psychology. Mental and more extended behavioral ethnographies deserve high priority and respect. It is nice to test an idea with sophistication, but it is at least as important that the idea being tested manifests itself in important contexts in real-world settings.

DIFFERENT DOMAINS AND DIFFERENT ROLES FOR CULTURE AND EVOLUTION: FOOD VERSUS SEX

I have focused on the way that evolutionary thinking can be applied to a major issue in cultural psychology, and that is cultural evolution. There are other areas of interaction. One, of course, has to do with cultural and biological cultural coevolution (e.g., Durham, 1991). A second is the idea of domain specificity, which entered late into psychology (e.g., Fodor, 1983; Rozin, 1976; Rozin & Kalat, 1971). It has been embraced by evolutionary psychology, because it seems to be generally characteristic of the way things have evolved (e.g., Tooby & Cosmides, 1992). That is, evolution is likely to begin with adaptation to a specific problem. Although this is sometimes taken to mean that everything is domain specific and modular, there is no reason to believe such a strong claim, but a weaker version surely holds. An adaptation (presumably via preadaptation but possibly by independent reinvention) can go from being very domain limited to available in some domains to generally available and even consciously stated (Rozin, 1976).

The domains of life such as sleeping, eating and food, work, sex, sociality, leisure activities (e.g., the arts, sports), and religion correspond to the focus of description of a species in zoology and of cultures in cultural anthropology. However, they receive almost no attention in social psychology (Rozin, 2006), and this feature of social psychology has been transferred to a large degree to cultural psychology. Cultural psychology is organized like social psychology, in terms of psychological processes. In this section, I propose to indicate how important domains are in both evolutionary and cultural psychology, by contrasting two major domains, sex and food.

Without sex, that is, sexual reproduction, there is no evolution in most species. Without food, there is no opportunity to grow to an age in which sexual reproduction is possible. Sex and food are essentials in evolution, in species survival. Most evolutionary biologists would agree that food, as opposed to sex, is the major force in directing evolutionary change. The reason is that finding, selecting, and capturing food make great demands on a species. It is probably true that the most useful single piece of information about an unknown species, other than its taxonomic status, is what it eats. Sensory systems, motor skills, and many other features of the organism are largely dictated by feeding. Basic changes in food selection and food availability (most notably the development of agriculture and domestication) are foundational in understanding human cultural evolution (Diamond, 1996). The reason for this dominance of food over sex, especially for human evolution, is that food identification, for a generalist animal, is an extraordinarily difficult problem. Food selection in generalists such as humans is what Mayr (1974) called an open system, a system underdetermined by genetic adaptations. This underdetermination results in part because of the enormous variety of forms that nutrients can be packaged in and, simultaneously, the enormous variety of forms that toxins can be packaged in. Mate choice, in contrast, is a much more circumscribed activity, under more genetic control. It is what Mayr (1960) called a closed system.

Evolutionary psychology has paid more attention to sex and mate selection than any other domain of life. The degree of attention to sex is clear in *The Handbook of Evolutionary Psychology* (Buss, 2005), which has six chapters about mating and none about food. Why? The sex bias in evolutionary psychology is understandable, because the primate origins of human sexual behavior are much more direct than the primate origins of human eating. Chimpanzee and human mating are a lot more similar than chimpanzee and human eating.

The understandable focus of evolutionary psychology on sex should be paralleled by a major focus on food in cultural psychology, but it is not. There is one chapter on food, by me, in the new *Handbook of Cultural Psychology* (Kitayama & Cohen, 2007), and this is because of my intervention with the editors to allow me to write such a chapter (they obviously agreed). Briefly, I would like to explain why human relations to food are central to human biological and cultural evolution and coevolution (see Rozin, 2007).

With humans, especially in the approximately 10,000 years since the appearance of agriculture and domestication, food has been transformed from just nutrition and pleasure. Kass (1994) beautifully described this transformation, in European history. Food has become a major mark of civilization, of the distinction between humans and other animals. A simple indication of this is the separation of the verb *to eat* in German into eating by nonhuman animals (*fressen*) and eating by humans (*essen*). In much of the world, we eat at a table, facing each other, using implements, and we eat something that for the most part has been vastly transformed by culinary processes from its natural state. We are offended by the sight of food being masticated in the mouth, yet the civilized human looks right at his or her eating partner while that partner is putting food in his or her mouth and chewing it. And while doing this, we talk to each through the same hole, without displaying the food within (Kass, 1994)—an example of motor virtuosity and high civilization.

In the developed world, the amount spent on food (between 9% and 20% of total income; this contrasts with about 50% for the majority of humanity) is much more than would be necessary to obtain adequate sustenance. By a process of preadaptation, food and eating have adopted many functions besides the original functions of nutrition and oral pleasure. First, food is a source of aesthetic experience, an art form, as in enjoying cuisine and eating out in fancy restaurants. Second, food is a major social instrument, when used as a gift (e.g., on Valentine's Day), as a center of celebrations (e.g., the turkey on Thanksgiving), as a statement of ethnic identity and status, and, most critically, as the frame that much of our social interaction occurs within. The sociality of eating (conviviality) is one of the few things that still brings the American family together on a daily basis. The social functions of food are universal and perhaps less obvious in American culture.

Third, food, because it is intimately involved with (incorporated into) the self and because, by its nature, involves killing life, is fraught with moral significance. In recent American culture, acts such as eating meat or smoking cigarettes have come to be viewed with moral suspicion. In Hindu India, food is an explicitly moral entity, a statement about and a vehicle for manipulating one's moral purity (Appadurai, 1981). The aesthetic, social, and moral functions of food can all be conceived as preadapted on the basic incorporative, nutritional function of food. Fourth, food is one of the major sources of metaphors, as when we describe someone as "bitter" or "sweet" or when we say we are going to get to the "meat" of the argument. Metaphor is quintessentially a preadaptation, the use of a word originally adapted to the food context for explaining or illustrating something else.

The changes in our food world have almost backgrounded the fundamental nutritional role of food. Indeed, an observer of American culture could almost reasonably conclude that eating is a nutritionally maladaptive activity! We have relatively few genetic adaptations to negotiate our complex ancestral food world, but two of them are innate preferences for sweet tastes (well documented; e.g., Steiner, 1979) and fatty textures (not well documented but probably true). Both are indicators of available calories. In the modern developed world, most particularly the United States, calories are treated primarily as a threat to health rather than as a necessity. Our urge for sweets and fats works against our longevity, however adaptive it was in our ancestral environment. So in the modern developed world, some of our ancestral adaptations work against us. Cultural and technological advances have allowed us to create abundant and cheap superfoods. These advances were motivated by our biological predispositions and now have taken on a life of their own. Food is an area that has offered the most striking examples of biological–cultural coevolution (e.g., lactose intolerance; Simoons, 1969, 1970), but the biological evolution just can't keep up with the rate of change of our food world. So although we continue to have sex primarily in the canonical way, our eating has been transformed in its latest instantiation to the consumption of foams and gels in El Bulli, the ultimate in establishing one's food status. Evolutionary adaptations, basically oriented to have us eat (rather than be restrained from eating) and to seek high-calorie foods, are now at the center of understanding humans and food in the developed world, but as maladaptive features.

EPILOGUE

It is time for evolutionary and cultural psychologists to work together and to focus together on how humans function, behaviorally and mentally, in the major domains of life. Evolutionary psychology can be a theoretical foundation for cultural psychology; culture can be the fruit fly for evolutionary psychology as it applies to cultural evolution.

REFERENCES

Appadurai, A. (1981). Gastro-politics in Hindu South Asia. *American Ethnologist, 8*, 494–511.

Asch, S. E. (1952). *Social psychology*. New York: Prentice Hall.

Baumeister, R. F., Vohs, K. D., & Funder, D. C. (2007). Psychology as the science of self-reports and finger movements. *Perspectives on Psychological Science*, *2*, 396–403.

Bock, W. J. (1959). Preadaptation and multiple evolutionary pathways. *Evolution, 13*, 194–211.

Boyd, R. (1985). *Culture and the evolutionary process*. Chicago: University of Chicago Press.

Brown, D. E. (1991). *Human universals*. Philadelphia: Temple University Press.

Buss, D. M. (Ed.). (2005). *The handbook of evolutionary psychology*. New York: Wiley.

Campbell, D. T. (1965). Variation and selective retention in socio-cultural evolution. In H. R. Barringer, G. I. Blanksten, & R. W. Mack (Eds.), *Social change in developing areas: A reinterpretation of evolutionary theory* (pp. 19–49). Cambridge, MA: Schenkman.

Cavalli-Sforza, L. L., Feldman, M. W., Chen, K. H., & Dornbusch, S. M. (1982). Theory and observation in cultural transmission. *Science*, *218*, 19–27.

Conway, L. G., III, & Schaller, M. (2007). How communication shapes culture. In K. Fiedler (Ed.), *Social communication* (pp. 107–127). New York: Psychology Press.

Dawkins, R. (1976). *The selfish gene*. Oxford, UK: Oxford University Press.

Dennett, D. C. (1995). *Darwin's dangerous idea: Evolution and the meaning of life*. New York: Simon & Schuster.

Diamond, J. (1996). *Guns, germs, and steel: The fates of human societies*. New York: Norton.

Durham, W. H. (1991). *Coevolution: Genes, culture and human diversity*. Stanford, CA: Stanford University Press.

Fodor, J. A. (1983). *The modularity of mind*. Cambridge MA: MIT Press.

Girifalco, L. A. (1991). *Dynamics of technological change*. New York: Van Nostrand Reinhold.

Gould, S. J., & Vrba, E. S. (1982). Exaptation: A missing term in the science of form. *Paleobiology*, *8*, 4–15.

Heath, C., Bell, C., & Sternberg, E. (2001). Emotional selection in memes: The case of urban legends. *Journal of Personality and Social Psychology*, *81*, 1028–1041.

Kass, L. (1994). *The hungry soul*. New York: Free Press.

Kitayama, S., & Cohen, D. (Eds.). (2007). *Handbook of cultural psychology*. New York: Guilford.

Lakoff, G., & Johnson, M. (1980). *Metaphors we live by*. Chicago: University of Chicago Press.

Mayr, E. (1960). The emergence of evolutionary novelties. In S. Tax (Ed.), *Evolution after Darwin: Vol. 1; The evolution of life* (pp. 349–380). Chicago: University of Chicago Press.

Mayr, E. (1974). Behavior programs and evolutionary strategies. *American Scientist*, *62*, 650–659.

Menon, U., & Shweder, R. A. (1997). Kali's tongue: Cultural psychology and the power of shame in Orissa, India. In S. Kitayama & H. R. Markus (Eds.), *Emotion and culture: Empirical studies of mutual influence* (pp. 241–282). Washington, DC: American Psychological Association.

Mesoudi, A., Whiten, A., & Laland, K. N. (2004). Perspective: Is human cultural evolution Darwinian? Evidence reviewed from the perspective of the *Origin of Species*. *Evolution*, *58*, 1–11.

Newson, L., Richerson, P. J., & Boyd, R. (2007). Cultural evolution and the shaping of cultural diversity. In S. Kitayama & D. Cohen (Eds.), *Handbook of cultural psychology* (pp. 454–476). New York: Guilford.

Richerson, P. J., & Boyd, R. (2005). *Not by genes alone: How culture transformed human evolution.* Chicago: University of Chicago Press.

Rozin, P. (1976). The evolution of intelligence and access to the cognitive unconscious. In J. A. Sprague & A. N. Epstein (Eds.), *Progress in psychobiology and physiological psychology* (Vol. 6, pp. 245–280). New York: Academic Press.

Rozin, P. (1991). Family resemblance in food and other domains: The family paradox and the role of parental congruence. *Appetite, 16*, 93–102.

Rozin, P. (2001). Social psychology and science: Some lessons from Solomon Asch. *Personality and Social Psychology Review*, 5, 2–14.

Rozin, P. (2003). Five potential principles for understanding cultural differences in relation to individual differences. *Journal of Research in Personality*, 37, 273–283.

Rozin, P. (2006). Domain denigration and process preference in academic psychology. *Perspectives on Psychological Science*, *1*, 365–376.

Rozin, P. (2007). Food and eating. In S. Kitayama & D. Cohen (Eds.), *Handbook of cultural psychology* (pp. 391–416). New York: Guilford.

Rozin, P., & Kalat, J. W. (1971). Specific hungers and poison avoidance as adaptive specializations of learning. *Psychological Review*, 78, 459–486.

Rozin, P., & Schull, J. (1988). The adaptive-evolutionary point of view in experimental psychology. In R. C. Atkinson, R. J. Herrnstein, G. Lindzey, & R. D. Luce (Eds.), *Handbook of experimental psychology* (pp. 503–546). New York: Wiley Interscience.

Schaller, M., Conway, L. G., III, & Tanchuk, T. L. (2002). Selective pressure on the once and future contents of ethnic stereotypes: Effects of the communicability of traits. *Journal of Personality and Social Psychology*, 82, 861–877.

Simoons, F. J. (1969). Primary adult lactose intolerance and the milk drinking habit: A problem and biological and cultural interrelations. I. Review of the medical research. *American Journal of Digestive Diseases*, *14*, 819–836.

Simoons, F. J. (1970). Primary adult lactose intolerance and the milk drinking habit: A problem and biological and cultural interrelations. II. A cultural-historical hypothesis. *American Journal of Digestive Diseases*, *15*, 695–710.

Steiner, J. E. (1979). Human facial expressions in response to taste and smell stimulation. In H. W. Reese & L. P. Lipsitt (Eds.), *Advances in child development and behavior* (Vol. 13, pp. 257–295.) New York: Academic Press.

Tooby, J., & Cosmides, L. (1992). The psychological foundations of culture. In J. Barkow, L. Cosmides, & J. Tooby (Eds.), *The adapted mind: Evolutionary psychology and the generation of culture* (pp. 19–136). New York: Oxford University Press.

Wilson, D. S. (2002). *Darwin's cathedral: Evolution, religion and the nature of society*. Chicago: University of Chicago Press.

3

The Human Mind and the Evolution of Cultural Animals

ROY F. BAUMEISTER

For decades, social scientists have debated the relative influence of nature versus culture on human behavior. By this point, most serious thinkers agree that both nature and culture make important contributions. Even so, and even though it is popular to declare the debate over, disputes continue.

The purpose of this chapter is to provide a summary of the central line of thought from my recent book *The Cultural Animal* (Baumeister, 2005). The project for the book was to construct a working model of the human psyche from a bottom-up approach based on reading of laboratory findings about human behavior. As the model took shape, I gradually had to grapple with the nature–culture questions, and the process led me to conclusions that differ from the prevailing views.

Most participants in the nature–culture debates these days start from the assumption that nature came first and, via evolution, created the foundations of the human being, and then culture was invented and exerted its influence. As Frans de Waal (2001) suggested, nature versus culture isn't a fair fight, because nature is always there. The question should be whether some pattern of behavior is entirely nature or nature plus culture. In other contexts, one frequently encounters questions about how long or short the leash is that nature permits culture to influence. The gist of these approaches is that humans are first and fundamentally (though not necessarily first and foremost) biological beings, and the topic open for debate is how much variation culture can add on top of the biological foundations. This is still a fairly wide-open debate. One might assume that biology contributes little more than a few basics and a capacity for learning, so that culture is responsible for almost all the interesting variations in behavior, or one might conclude that culture is a late-arriving and minor factor. Or anything in between.

In contrast to those views, my hypothesis is that culture influenced evolution. The sequence of nature first then culture may not be correct. Again, this is a speculative extrapolation based on lab studies with modern persons, not on examination of fossils or DNA. The beginnings of culture may have altered the selection factors by changing the terms for survival and reproductive advantage, and so humans were, in a sense, bred to be culture beings. (Hence my title, the "cultural animal.") *Nature made us for culture.*

My approach thus accepts the ineluctable power and centrality of biology and evolution, but it also assigns a new and powerful role to culture. Culture is not a late-arriving influence on human behavior that begins after the biological creature is fully formed. Rather, culture—though perhaps not cultural difference—is in our genes.

Although humans are the quintessential cultural animals, they are not the only creatures to have culture, and in fact this point is crucial to my understanding. de Waal (2001) observed that if one adopts fair but broadly inclusive definitions of culture, then culture has been observed in several dozen different species. He readily acknowledged, however, that these nonhuman cultural activities do not begin to approach the extent and power of human culture. This, again, is precisely the form best suited to my analysis: A pattern that can be observed in rudimentary or peripheral form in other animals evolved to become much more advanced and central in humans. From my perspective, insufficient attention has been paid to one important implication of the observation of nonhuman culture. If the beginnings of culture existed in other species, especially species that existed before humankind evolved, then culture was here on the planet before we were here. That means that culture was part of the environment under which humans evolved. The selection in favor of traits that promoted cultural competence could therefore have been a major theme of human evolution.

WHAT IS CULTURE?

To begin with, let me try to define what I mean by culture. Culture is not easily defined, and although there is certainly some broad agreement among the many writers who use the term (e.g., the authors of chapters in this book), undoubtedly differences of emphasis and connotation are common. The term itself has its roots in the biology of growing (e.g., agriculture) and then in a very different reference to advanced artistic activities such as opera (e.g., a cultured aristocrat). In the twentieth century, anthropologists claimed a sort of primacy in their understanding of the term, but then they fell out in a series of arguments, such as whether ideology or material needs were the central causal powers and whether shared beliefs or shared praxis mattered most. The next time you are having dinner with anthropologists and feel mischievous, ask them how many cultures there are in North America, and if the answer is anything other than two, ask them to name them.

Nonetheless, there are some areas of agreement. Culture refers to things that are learned rather than acquired; anything inborn is not culture by itself. Culture is shared: A single person does not have a culture. Culture is generally understood as having to be transmitted from one generation to another and thereby reproduce itself. Despite the argument between the adherents of belief and those of praxis, to me it seems apparent that both must be included. That is, culture contains common understandings and values, but it also consists of shared ways of doing things.

Putting these together, I understand culture as *an information-based system that enables people to live together in organized fashion and satisfy their basic needs* (Baumeister, 2005). This can benefit from elaboration. Culture is first and foremost a system, in my understanding. Some animals live alone. Some live in social groups, and these groups have some degree of system. Culture is a relatively advanced and complicated way of being social. It is a complex social system with special properties.

These special properties include being based on information. It is no accident that all known human cultures have language, because language is the best if not the indispensable tool for using information—and in particular for using information socially, which is to say enabling more than one creature to use the same information.

The emphasis on living together in organized fashion addresses one basic purpose of culture. As a thought experiment, try to imagine what North America would look like if the culture were suddenly obliterated but all the people remained. That is, take away the language and the social system with its rules and shared understandings. Either people would start to create culture anew quite rapidly, or chaos would ensue. Culture may sometimes impose difficulties on how we relate to each other, but in general its purpose is to facilitate peaceful and harmonious coexistence.

The part about satisfying needs also speaks to a basic purpose of culture. As Marvin Harris (e.g., 1977, 1991) and other cultural materialists emphasized, cultures develop in ways that facilitate providing for the people who belong to them. Cultures that fail to provide food and water for their people are headed for significant change if not rapid demise and replacement by others.

CULTURAL VARIATION VERSUS SIMILARITY

At present, strong and thriving research programs in psychology are devoted to the respective influences of nature and culture. Nature is represented most strongly by evolutionary psychology. Its primary emphasis is on continuities and similarities between humans and animals. One frequent goal of this research is to show how human behavior resembles that of the animals from whom we evolved. To be sure, there are some researchers that focus on species-specific adaptations and what is special about humans, but the main thrust of most work, and certainly most of what gets communicated to researchers outside the area, is that human nature shares many common features with the animals from whom we evolved. Even when it turns its attention to differences, such as those between men and women, it often undertakes to show how human gender differences parallel those in other species.

Meanwhile, culture is represented in psychology by cultural psychology. Its emphasis is overwhelmingly on difference. If you see the word *culture* in the title of a conference talk, book, or article, you can bet that it will be about cultural differences. A decade or two ago, this typically meant that the researchers gave out the same questionnaire in England, France, and Germany or in the United States and then wherever the researcher happened to travel (e.g., Venezuela). More recently, the study of cultural differences has zeroed in on the difference between Westerners and Easterners, such as comparing Canadians and Japanese, or North

Americans and any Asians, or sometimes North American students of European descent and those of Asian descent. Regardless, difference is the stock in trade of cultural studies.

I have no quarrel with these approaches, which continue to produce important and fascinating findings. All I want to suggest is that there is also valuable insight to be gained by switching the target criteria so as to look at the other two possible combinations, specifically evolutionary differences and cross-cultural similarities. That is, it is worth attending not only to evolutionary similarities and continuities but also to differences: What sets humans apart from other creatures? What makes us human? Conversely, what do all (or even most) cultures have in common? Crucially, these may overlap—and hold important keys to understanding human nature.

To contend that nature made us for culture is thus to shift the focus to the common patterns across most cultures generally and to what makes humans different and special. The achievements of human culture are after all quite remarkable in comparison with how most other species live. These achievements probably required some psychological capabilities that are distinctive.

CULTURE AS BIOLOGICAL STRATEGY

All living things need to get certain things from their environment. The most common are air, water, and food. Other needs typically include shelter from the elements, protection from predators, and opportunities to reproduce, which may include caring for their young until the young can take care of themselves. Much of what nature has installed inside creatures is there to enable them to get what they need to live: lungs to breathe air, eyes and noses to find food, mouths to eat, stomachs to digest, reproductive organs, and so forth.

Some creatures use sociality as a strategy for improving their prospects. A lone wolf may not be able to kill a moose or elk, but if the wolves hunt together, they can kill such large game, and so the total amount of food available to them is increased even though the environment is objectively the same. (That is, the moose were there either way, but only the social wolves can hunt them successfully.) Social life thus emerges as a better way for the animals to get what they need to live. The downside is that social beings require more extensive inner structures than do solitary ones. They need to be able to recognize conspecifics, to want to live and work together with them, to participate in group decision making, and to resolve disputes between them. Thus, using sociality as a biological strategy can pay big dividends, but it requires some infrastructure in the form of more advanced psychological capabilities.

Culture is another step. Culture is a better way of being social. It uses systems of information, communication, and organization to enable social interaction to reach a much higher and more advantageous level of complexity. In an effective culture, the whole is more than the sum of its parts. That means that the members of the group live (survive and reproduce) much better within the culture than they would if they had remained solitary. The basis for this is what can be called *system gain*. The system confers advantages.

It must be acknowledged that not everyone always benefits from culture. At the extreme, cultures do execute some of their members occasionally, and so these individuals would arguably have done better to live alone in the woods by themselves. By and large, though, the benefits far outweigh the losses for the vast majority.

What are the biological advantages of culture that are responsible for system gain? Four main ones were outlined by Baumeister (2005). They may not be exhaustive, and they overlap somewhat, but together they make a powerful basis for using culture as a biological strategy. (Indeed, any one of them might have been enough to push natural selection in favor of cultural competencies.) Groups that could use these systems would fare better than groups and individuals who could not. And, more important for an evolutionary argument, individuals who were more capable of functioning in these cultural groups would survive and reproduce better than their less capable peers and rivals.

It is important to make clear that my arguments here are based more on mutual, reciprocal causality than one-way, linear effects. Reciprocal influence has been argued as a useful approach to understanding coevolution of biological and cultural realities (e.g., Boyd & Richerson, 1985). I have said that people created culture because of its material advantages. To do so, they needed the capabilities to function in a cultural environment. As they developed the capabilities, the culture could develop further and provide more of its advantages.

To illustrate, consider language. People clearly need biological capabilities to speak. These "hardware" requirements include the ability to pronounce many different sounds reliably, the ability to hear subtle differences, the memory to store the meanings of many different words, and, perhaps most remarkable, the grammatically competent brain. Neither one-way causal explanation seems plausible. It would be silly to suggest that languages existed there in the environment and people gradually developed the ability to talk. Conversely, it is also implausible to suggest that people developed these advanced language hardware capabilities first, for no apparent reason, and then suddenly one day started creating language. Coevolution seems by far the most plausible scenario. That is, rudimentary communication (if only by grunts and snorts) provided some advantages, and so some creatures got better at grunting and snorting, including differentiated grunts for different meanings. These developed slowly into single words, then more words, then perhaps simple combinations of words. Once groups started to have these simple and preliminary forms of language, individuals who could speak a little better than their peers would likely have reproductive advantages. In that way, culture (in this case, language) became part of the environment and provided factors that would influence natural selection and thereby biological evolution (e.g., Baumeister, Maner, & DeWall, 2006).

Put another way, as long as no language existed in the world, being born with the biological hardware for a gift for gab would not bring any benefit, and being born without it would leave one no worse off than the proverbial fish without a bicycle. But once a community had language, the young man with such a gift might well score better with the ladies than his relatively tongue-tied peers, and so the next generation would show an increase in its average verbal capabilities.

In short, these four advantages are not results of culture so much as essential aspects of it that contribute to producing its benefits. Nonetheless, the benefits created by these four constitute important reasons that our species converged on culture as its biological strategy for facilitating survival and reproduction.

Accumulation of Knowledge

A first advantage of culture is the accumulation of knowledge. In most species, learned information is stored in individual brains, and when the animal dies, the knowledge dies with it. In human societies, knowledge is stored in the group, and so what any member learns can be shared with others. To be sure, some animals learn from each other, but this is far more systematic and extensive in humans. Most important, acquired knowledge can be passed on to the next generation, and so it accumulates.

The accumulation of knowledge makes progress possible in a way that is essentially unknown outside of the human species. It is what some regard as the passing of the torch of progress from nature to culture, with a huge resultant increase in speed of change. That is, evolution via natural selection creates progress, in that creatures gradually adapt better to their environment, and later generations are better than their ancestors. But the process of working via genetic change is quite slow. In contrast, cultural progress can be quite rapid, and it is generally agreed that the speed of change has escalated remarkably over the centuries.

Knowledge accumulation has been one major engine of culture. Many basic aspects of human life are thinkable only on that basis, including cooked food and anything beyond the most basic tools. Tomasello (1999) noted that many primates can come up with creative solutions to problems and even manage (mostly by mimicry) to spread them through a small circle of friends, but then the innovations get lost. The nonhuman primates lack the “ratchet effect” of culture, which retains innovations and passes them along. As a result, each new generation of nonhuman primates starts pretty much from the beginning, whereas each generation of human beings can take up where the previous one left off.

Cooking provides a useful example (e.g., Tannahill, 1988). Cooking food has contributed greatly to human reproduction and survival. Indeed, probably more human beings have lived off rice as their staple than any other food—but rice is inedible without cooking, and so those multitudes would have gone hungry. Yet it is unlikely that cooking was fully invented by a single person. Rather, cooking almost certainly developed over many generations, as the often disastrous results of trial and error enabled knowledge to accumulate about how best to cook various foods.

Language

The second advantage is language. Language exists not within individual brains but in the collective. (You can’t have a language by yourself.) Rudimentary communication is found in many social animals, but most linguists refuse to call it language. Human communication is thus much more powerful than the communication that is found in any other species.

Language brings multiple advantages. Culture is essentially based on information, and language greatly facilitates working with information: storing it, operating on it (e.g., by reasoning), sharing it. The accumulation of knowledge and its transmission to future generations, discussed previously, is greatly facilitated by language. Language also facilitates group projects, insofar as members can communicate with each other more effectively and thereby coordinate their activities, producing better performance. Groups of reclusive monks may take vows of silence, but I doubt hunting or military groups ever do.

Language also improves the power of thought. Undoubtedly some thought occurs without language (see Carruthers, 2002). But I would bet on language to improve the quality of reasoning. Also, crucially, language enables people to share their thoughts, ideas, and reasons with other people and in that process to learn about mistakes. Lone thinkers (even with language) may be prone to error, but by talking through a problem with a friend or colleague, one can reach objectively more correct answers.

One further and powerful advantage of language is that it liberates the person from the immediate stimulus environment. Skinner and the behaviorists were quite right to point out that animal behavior is mainly responses to here-and-now stimuli, informed perhaps by reinforcement history. But human beings can respond to facts that are far removed from the here and now, including in the remote future. We take this for granted, but its adaptive significance could scarcely be overstated. As prime examples, agriculture, schools, religion, and national political government all depend on the ability to conceptualize and respond to factors beyond the immediate stimulus environment. Language enables the mind to represent these distant realities.

Roles and Division of Labor

The third advantage of culture is division of labor. Having different people do different things, especially different parts of a collective task, produces vast and undeniable improvements in the quality of work. The benefits of division of labor are so powerful that even nature began using them long before humans appeared, such as in gendered tasks, but nature's capacity to capitalize on role specialization is fairly limited in comparison with what culture has accomplished. It is no accident that one of the seemingly universal laws of cultural change is steady progress toward ever-greater specialization. This continues even today, not just in the corporate world (imagine a company that refused to rely on division of labor!) and academia but in odd places such as sports teams.

The competitive advantage that comes from division of labor stems from the fact that it is possible to be an expert at one thing but not, usually, at everything. If each person performs all different functions, he or she is not usually able to become highly practiced and skilled at any of them. In contrast, splitting a task into parts and having different people specialize in different parts enables each of them to become highly expert at his or her part. The result is that the task is performed by experts, and because of that the quality of the product or performance rises sharply. For example, most of us live in houses that were built by teams of specialists, including carpenters, plumbers, and electricians. If a single person built the entire house, it would be quite inferior. It would also, ironically, be much more expensive, because the person who built it would have to have a wide array of knowledge and so his or her labor would cost more, hour by hour, than the labor of the specialists. Division of labor enables things to be done faster, cheaper, and better, and so groups that could exploit that principle outperformed and supplanted those that could not.

Exchange Relationships

The fourth advantage of culture lies in a network of exchange relationships, also known as a market economy. Trade is conceivable without division of labor, but it is much more powerful and effective with it. It redistributes resources through the group in an optimal manner. I assume it is uncontroversial to assert that throughout history, places and groups that engaged in economic trade tended to advance and flourish, whereas those that eschewed economic relations stagnated and declined. Trade increases wealth all around. If a simple, early illustration is needed to show how trade can increase benefits, one can imagine a hunting group and a fishing group living in the same general area. If they could occasionally trade meat for fish, then the diets of both would be enriched, and members of both groups would be healthier, hence ultimately better able to reproduce.

If any group is disinclined to appreciate fully the advantages of economic trade, it is, ironically enough, professional psychologists. When Clark and Mills (1979) introduced the profound distinction between communal and exchange relationships, it led to a flurry of research, mainly in the area of close relationships. The general finding was that communal relationships are more advanced and successful than exchange ones. I can well believe that that is correct—but mainly in the sorts of close, intimate relationships that social psychologists study. If instead one were to look at large groups or societies, the reverse is true. Large societies that have remained communal and those that have sought (as in communism) to revert to communal ways of living have been much less successful than those that encouraged exchange relationships.

The fate of the Neanderthals is perhaps one of the best indications of the powerful advantages of culture. The Neanderthals were once thought to be ancestors of *Homo sapiens*, but now it is known that they were a different species and an evolutionary dead end. *Homo sapiens* evolved separately and somehow supplanted them. The Neanderthals occupied the planet for a much longer time than *Homo sapiens*, and in that sense they were highly successful for a very long time, but when they came into direct contact with us, they lost the competition. It is not immediately easy to see why. Their brains were as large as ours, and they were physically a bit larger and stronger, so if the competition had come down to one-on-one combat, they seemingly ought to have prevailed.

What the Neanderthals appear to have lacked is division of labor and economic relations. They may have accumulated knowledge and had language, but their culture lacked some of the crucial powers that ours developed. Individually they would have been a match for *Homo sapiens*, but collectively they had no chance. Recent analyses have suggested that the Neanderthals' lack of economic competence alone could have accounted for their decline and extinction (Horan, Bulte, & Shogren, 2005).

What Price Culture?

The argument thus far has been that what makes us human is culture. Culture can be regarded as a biological strategy that offered clear and powerful advantages to those (our ancestors) who were capable of using it. It is not the only or best strategy, of course, and in the end perhaps the beetles or cockroaches will inherit and dominate the earth. But among the creatures who resorted to sociality as their strategy, culture offered powerful improvements on being merely social and thereby enabled our species to achieve its remarkable success at spreading out across the globe, imposing its will and preferences on the natural environment, creating comforts and pleasures far beyond what most animals could dream of, reproducing into huge numbers that will soon amount to 8 billion individuals, and even increasing its own life span by double or triple by dint of its advances in knowledge about its bodies.

Why haven't other species adopted culture? I mentioned earlier that the beginnings of culture have been observed in dozens of other species, but somehow they never seem to get very far. The answer is that culture requires fairly elaborate and biologically expensive capacities. The evolution from solitary beings to social beings depended on a sharp increase in psychological capabilities. In the same way, the evolution from merely social beings to cultural beings depended on another sharp rise in psychological powers.

In my view, the evolution of cultural capability holds the key to understanding human nature. The traits that are distinctively human, that set us apart from other creatures, that define the parameters of the human condition, are a result of biological selection processes driven by the requirements of culture. The next section will consider some of these.

CULTURE AND HUMAN NATURE

This part of the chapter seeks to draw the link between psychology and the evolutionary scenario I have presented. My purpose behind this project was, again, to be able to describe and explain how the human psyche works. The idea that nature made us for culture is a framework for understanding why it is the way it is.

Background

The human psyche is obviously built on what came before it. It holds much in common with the psychologies of other animals, especially its most recent evolutionary forebears. Like all creatures, humans need oxygen, water, and food, and so their bodies have inner structures that make that possible. The psychological foundations of human motivation are also to be found there, though there is much more.

More relevant and interesting are the carryovers from social animals. Although humans are more than social animals, they are still nonetheless social animals.

There is much for evolutionists to continue to do in terms of mapping how human behavior patterns resemble those of the animals from whom we evolved.

But my focus is on what is different and special about humans. Culture represented a new and better way of being social, and so evolution wrought changes to the primate psyche. To be sure, I am reluctant to posit wholly new traits. Evolution does not easily make something out of nothing or veer off in radically new directions. What I regard as big changes would most plausibly involve taking small, peripheral, incipient traits in social animals and elevating them into large, central, more fully developed traits in the human psyche. After all, this appears to be the picture with culture itself, as noted earlier: Some social animals appear to have achieved the beginnings of culture, but for them these are mostly optional, minor, peripheral, and occasional aspects of life. For humans, culture is central and saturates nearly everything we do. Even many of the most basic animal behaviors, such as eating and sleeping, have been touched and transformed in humans by culture. No other animal eats in restaurants or takes sleeping pills.

Also, not all changes have involved advancing and adding. One of the most interesting adjustments to culture involves aggression. To be sure, in some ways humans are more aggressive than other species, because culture has enabled them to invent whole new categories of aggression, such as war, genocide, and character assassination. Yet in another sense, aggression loses much of its primary function in culture. My sense is that aggression evolved in social animals because disputes are inevitable in social life (e.g., when the group gets food that everybody wants to eat, how is it decided who gets the tastiest and biggest portions?), and aggression was one way of settling them. This includes establishing a hierarchy or pecking order that can resolve disputes, so that it is not necessary to fight over each piece of food every day. Culture, however, offers alternative ways of settling disputes and establishing hierarchies. Hierarchy is established by education, wealth, even political appointment or election. Disputes can be resolved by laws and courts. Aggression can be regarded as an evolutionary hangover (with one big exception, as noted next). Aggression was the social animal's primary way of settling the conflicts of social life, but culture has invented others, and in fact much of the history of human culture has been progressive attempts to curb aggression.

The main exception is of course conflicts between cultures. A functioning culture is usually a system that has consensually accepted ways of resolving conflicts and disputes, but when different cultures clash, the two cultures may not share the same rules and mechanisms for avoiding aggression. Hence even as the slow progress of culture sought to curb and restrain aggression between its members, there were payoffs from cultivating aggressive capabilities toward rival cultures. Perhaps now the movement toward integrated world culture is finally groping for ways to bring even this to an end, but the obstacles are formidable, and the most recently completed century was almost certainly the bloodiest in human history in terms of intercultural violence.

A last general consideration is that the cultural animal argument does not rest on the generally discredited group selection form of evolution, even though the advantages are achieved at and dependent on the group level. Individuals benefited by being part of a successful group. In many respects the interests of the group and the individual coincide.

Human Traits as Adaptations for Cultural Competence

Let me then discuss several prominent human traits in the context of being cultural animals. That these traits exist is hardly news. The point is that their existence makes sense in terms of retrofitting the ape psyche to be capable of creating and sustaining culture. Thus, the goal of this section is to show how understanding humans as cultural animals provides a viable explanatory framework for a broad range of psychological traits.

Where to start? Humankind names itself (*Homo sapiens*) in honor of its own intelligence, and it is fair to regard human intelligence as something quite remarkable. The advantages of intelligence need hardly be recited here. But there are several considerations. All else being equal, more intelligence would seemingly be better for all creatures, so one may pause to wonder why squirrels and bugs never reached Einstein's level. The answer, as Dunbar (1993, 1998) proposed, is that intelligence depends on brainpower, and the brain is a very costly organ in terms of its consumption of caloric energy. Being smart enough to solve sudoku puzzles might be nice for squirrels, but if they couldn't get enough nuts to stoke the fuel needs of the larger brain, they might starve. Hence intelligence increases only in step with its ability to bring palpably large payoffs. For most creatures, those prospects are limited. A larger brain might occasionally enable a creature to outwit a rival or predator, but the increase in daily cost for the larger brain's upkeep could produce a net disadvantage.

All this changes with culture, however. Culture is based on information, and so there is more possibility for intelligence to come into play in acquiring, storing, and working with all this information. The extraordinary capacity of human memory (Landauer, 1986) has given researchers pause, but culture vastly increases the amount of information available to be remembered. (Language alone needs plenty of capacity.)

Another point about intelligence is that it seems to have evolved to deal with specific sorts of problems, despite its reputation as an all-purpose reasoning machine. Cosmides and Tooby (1992; also Cosmides, 1989) demonstrated that modern college students (among others) routinely have trouble getting the correct answer to some standard logic problems phrased in a variety of ways, but when those same problems are phrased in terms of cheating the system, the students solve them quickly and accurately. The implication is that one thrust in the evolution of intelligence involved cheater detectors. This argument is quite conducive to the cultural animal perspective. As I said, economic exchange relationships are crucial to culture (and relatively unknown in other species), and so for people to be successful at them—individually and collectively—it is necessary to watch for, avoid, and punish people who abuse the system (e.g., Fehr & Gächter, 2002; Ouwerkerk, Kerr, Gallucci, & Van Lange, 2005).

Theory of mind is another trait that is vital for culture. After all, culture operates like a collective mind, storing information, imposing order, and the like. For people to participate in it, it is probably essential that they perceive other people to have mentalities similar to their own. No doubt language contributes to this and/or is based on it. At present there are ongoing debates as to whether theory of mind exists in full-blown or watered-down form in the few primates most closely related to humans. The debate tends to fall into the dichotomy of, "See, they are just like us!" versus "Human beings are unique." To me, this is unfortunate because neither view is likely correct. As I said, the most plausible scenario for my theory is that the rudimentary beginnings can be found as peripheral features of primate psyches and that evolution blew these up into powerful and central features of the human psyche. Hence I am hopeful that the resolution will be that our immediate primate ancestors had hints of theory of mind, but it took human evolution for these to develop into the full-fledged common and constant sense that we all belong to a community of creatures with minds that are similar enough to our own to permit extensive communication, mutual understanding, and negotiation.

The human self is quite remarkable and is perhaps another crucial adaptation for culture. Again, it is best to avoid the dichotomy of asking whether animals have selves or not. Most likely, they have rudimentary beginnings of selfhood but nothing like the complex structures humans have. The human self can combine multiple definitions and roles; distinguish or blend public appearances with private, inner structures; blend conceptual information and action; and exert remarkably nuanced levels of control over its own responses.

Vast amounts have been written about the self, including occasional pages by me, and this is no place to revisit or reconsider that literature (e.g., see Baumeister, 1998). For now, it will have to be sufficient to recognize the self as the interface between the biological organism and the cultural system. Because human cultural systems are so much more complex than those of other social animals, the human self requires much greater powers and functions. Unlike the fixed roles of, say, ants, human social systems have permanent, temporary, and novel sets of roles, some appearing just for an ad hoc task group, others held for life. To make culture work successfully, the human self is able to hold multiple roles and move among them. It is, among other things, a tool that enables the physical organism to gain acceptance into social systems and function effectively in them.

Let me give special mention to self-control as one of the functions of the self. The ability of the human self to alter its own states and responses, and to bring them in line with conceptually meaningful standards such as laws, norms, moral ideals, and many more, is vitally conducive to culture. There would be no point in making laws if people were unable to alter their actions (right down to their muscle movements) to conform to them—and without laws, a cultural system would be dead meat, at least in competition with more lawful rival systems.

Self-control is yet another phenomenon that requires sidestepping the false dichotomy of seeing it as all or none in other species. I doubt that the methods yet exist to compare powers of self-control across species. When they do emerge, I predict that we will find that animal species vary in degrees of capacity for self-control but that this capacity will prove far more advanced and central in the human psyche than in our biological ancestors. Most likely some social animals developed an incipient capacity to inhibit some responses, even if just for the sake of not being beaten up by the alpha male. This capacity to learn and execute response inhibitions may have then been progressively strengthened in the course of human evolution. It obviously is still far from fully empowered, and people fail at self-control every day all over the world. In fact, the inadequacy of human self-control could well be the single biggest lack of cultural competence among modern human beings, as indicated by the fact that most widespread social and personal problems have some element of self-control failure in them (e.g., Baumeister, Heatherton, & Tice, 1994; see Baumeister & Vohs, 2004). The glass of human self-control is half full and half empty. Most likely this is because self-control, similar to intelligence, is a biologically costly function (see Gailliot & Baumeister, in press), and so nature can afford to give us only limited amounts. But culture increased the payoff for it and the need for it, and so the humans' glass is far fuller than that of most other creatures.

Another unusual feature of the human mind is the two-chamber, two-process design. Fierce debates continue to rage as to how much of the total work is done by automatic, nonconscious processes and how much by the conscious mind (e.g., Bargh, 1994; Dijksterhuis & Nordgren, 2006; Donald, 2002; Wegner, 2002). My own view is that the automatic system is largely inherited from the control system of our biological ancestors, whereas the relatively new and different conscious system must have its own distinctive powers and functions, most likely in the service of cultural participation.

Much has been made of the mistakes made by conscious processing, as in not being able to realize the inner processes that contribute to some responses (Haidt, 2001; Nisbett & Wilson, 1977; Wegner, 2002). Sometimes people consciously construct explanations for their behaviors that do not correspond to the full actually causal processes that produced them. In my view, this misses the point. Cultural animals do not necessarily need to explain all their true reasons for selecting a particular course of action, but they do more often need to be able to provide a possible explanation for their actions that is consistent with the laws, values, morals, and other rules of their group. For a culture to function effectively, people need to behave according to rules. Insofar as novel situations arise and general, abstract rules need to be applied to them, the social control of behavior depends on deciding whether particular acts are acceptable by these general rules. (For example, it doesn't matter why you actually decided to modify your house in some particular way—only whether the modification is consistent with legal building codes. When the building inspector arrives, your task is not one of introspectively furnishing a valid account of why you built the addition in this manner but only one of explaining it in relation to the law.)

Hence a number of authors have noted that consciousness is especially useful for rule-based forms of thinking (e.g., Sloman, 1996; Smith & DeCoster, 1999, 2000). These may be slow and clumsy but are powerful at the level of culture. Logical reasoning is thus one important form of thinking that enables members of a culture to agree on what course of action is optimal, but agreement depends on their being able to discuss the question on the basis of shared assumptions about what follows from what. Lieberman, Gaunt, Gilbert, and Trope (2002) proposed that the automatic processing system is not good at rule-based logical reasoning, so that this may be something where the evolutionarily newer conscious system serves a distinctive and powerfully valuable function. Consistent with that, a series of experiments by DeWall, Baumeister, and Masicampo (2006) employed a series of manipulations to engage or encumber either the nonconscious automatic system or the conscious processing system, and then had participants attempt to solve logical reasoning problems. The manipulations aimed at the nonconscious system had no effect on reasoning ability (even though they did succeed at altering nonconscious processing in familiar ways). In contrast, those manipulations aimed at the conscious system had substantial and significant effects. These data confirm the view that logical reasoning operations depend on conscious processing (see also de Neys, 2006).

Free will is another term that is controversial, and in fact some scientists regard it as anathema due in part to its seemingly spiritualistic connotations. By free will I mean only a relatively flexible means of behavior control capable of self-regulation, rational intelligent decision making, and controlled processes. Detractors think of free will as a random action generator, but that would have relatively little evolutionary payoff. Indeed, if a mutation had produced in some hominid an enhanced capacity for random action, it is hard to see how that would have afforded sufficient advantages in survival or reproduction as to enable the mutation to survive and spread through the gene pool. In contrast, an enhanced capacity for rational, intelligent choice and self-regulation would likely yield large dividends in fitness, perhaps especially among beings living in a cultural context. Perhaps ironically, free will is needed for enabling the person to obey rules! But a creature who can resist some impulses and instead behave in culturally valued ways is likely to be more successful than its undercontrolled, impulsive rivals, especially in a cultural context. Hence I regard free will in this sense as another important advance in the human psyche associated with the evolution of cultural animals (Baumeister, in press).

The duplex mind has also quite possibly changed the functions of emotion. Conscious human emotion may be different in some ways and serve functions different from the affective responses of other animals. Emotion theory has a long tradition of assuming that emotions are for directly causing behavior, exemplified most commonly by the notion that fear causes fleeing. Extensive studies of human behavior, however, have found relatively meager and erratic evidence of direct links from conscious emotion to behavior (e.g., Schwarz & Clore, 1996; in press). Automatic affective reactions may be fast and efficient enough to guide behavior, but conscious emotion may instead serve complex feedback functions that inform and guide cognitive processing (Baumeister, Vohs, DeWall, & Zhang, in press).

CONCLUSION

The phrase "the social animal" has been used by thinkers from Aristotle to Aronson to describe human beings. It is correct as far as it goes: Human beings are social animals. Yet in my view it does not go far enough. Humans are not the only social animals or even the most social of animals. In contrast, "the cultural animal" captures much more closely the distinctive essence of what makes us human.

REFERENCES

Bargh, J. A. (1994). The four horsemen of automaticity: Awareness, intention, efficiency, and control in social cognition. In R. S. Wyer, Jr., & T. K. Srull (Eds.), *Handbook of social cognition* (pp. 1–40). Hillsdale, NJ: Lawrence Erlbaum.

Baumeister, R. F. (1998). The self. In D. T. Gilbert, S. T. Fiske, & G. Lindzey (Eds.), *Handbook of social psychology* (4th ed., pp. 680–740). New York: McGraw-Hill.

Baumeister, R. F. (2005). *The cultural animal: Human nature, meaning, and social life*. New York: Oxford University Press.

Baumeister, R. F. (in press). Free will in scientific psychology. *Perspectives on Psychological Science*.

Baumeister, R. F., Heatherton, T. F., & Tice, D. M. (1994). *Losing control: How and why people fail at self-regulation*. San Diego, CA: Academic Press.

Baumeister, R. F., Maner, J. K., & DeWall, C. N. (2006). Evoked culture and evoked nature: Co-evolution and the emergence of cultural animals. *Psychological Inquiry*, *17*, 128–130.

Baumeister, R. F., & Vohs, K. D. (Eds.). (2004). *Handbook of self-regulation: Research, theory, and applications*. New York: Guilford.

Baumeister, R. F., Vohs, K. D., DeWall, C. N., & Zhang, L. (in press). How emotion shapes behavior: Feedback, anticipation, and reflection, rather than direct causation. *Personality and Social Psychology Review*.

Boyd, R., & Richerson, P. J. (1985). *Culture and the evolutionary process*. Chicago: University of Chicago Press.

Carruthers, P. (2002). The cognitive functions of language. *Behavioral and Brain Sciences*, *25*, 657–719.

Clark, M. S., & Mills, J. (1979). Interpersonal attraction in exchange and communal relationships. *Journal of Personality and Social Psychology*, *37*, 12–24.

Cosmides, L. (1989). The logic of social exchange: Has natural selection shaped how humans reason? Studies with the Wason selection task. *Cognition*, *31*, 187–276.

Cosmides, L., & Tooby, J. (1992). Cognitive adaptations for social exchange. In J. Barkow, L. Cosmides, & J. Tooby (Eds.), *The adapted mind* (pp. 163–228). New York: Oxford University Press.

De Neys, W. (2006). Dual processing in reasoning: Two systems but one reasoner. *Psychological Science*, *17*, 428–433.

de Waal, F. (2001). *The ape and the sushi master: Cultural reflections of a primatologist*. New York: Basic Books.

DeWall, C. N., Baumeister, R. F., & Masicampo, E. J. (2006). *Evidence that logical reasoning depends on conscious processing*. Manuscript submitted for publication.

Dijksterhuis, A., & Nordgren, L. F. (2006). A theory of unconscious thought. *Perspectives on Psychological Science*, *1*, 95–109.

Donald, M. (2002). *A mind so rare: The evolution of human consciousness*. New York: Norton.

Dunbar, R. I. M. (1993). Coevolution of neocortical size, group size, and language in humans. *Behavioral and Brain Sciences*, *16*, 681–694.

Dunbar, R. I. M. (1998). The social brain hypothesis. *Evolutionary Anthropology*, *6*, 178–190.

Fehr, E., & Gächter, S. (2002). Altruistic punishment in humans. *Nature*, *415*, 137–140.

Gailliot, M. T., & Baumeister, R. F. (in press). The physiology of willpower: Linking blood glucose to self-control. *Personality and Social Psychology Review*.

Haidt, J. (2001). The emotional dog and its rational tail: A social intuitionist approach to moral judgment. *Psychological Review*, *108*, 814–834.

Harris, M. (1977). *Cannibals and kings: The origins of cultures*. New York: Vintage.

Harris, M. (1991). *Cows, pigs, wars, and witches: The riddles of culture.* New York: Vintage.

Horan, R. D., Bulte, E., & Shogren, J. F. (2005). How trade saved humanity from biological exclusion: An economic theory of Neanderthal extinction. *Journal of Economic Behavior and Organization*, *58*, 1–29.

Landauer, T. K. (1986). How much do people remember? Some estimates of the quantity of learned information in long-term memory. *Cognitive Science*, *10*, 477–493.

Lieberman, M. D., Gaunt, R., Gilbert, D. T., & Trope, Y. (2002). Reflection and reflexion: A social cognitive neuroscience approach to attributional inference. In M. Zanna (Ed.), *Advances in experimental social psychology* (Vol. 34, pp. 199–249). New York: Elsevier.

Nisbett, R. E., & Wilson, T. D. (1977). Telling more than we can know: Verbal reports on mental processes. *Psychological Review*, *84*, 231–259.

Ouwerkerk, J. W., Kerr, N. L., Gallucci, M., & Van Lange, P. A. M. (2005). Avoiding the social death penalty: Ostracism and cooperation in social dilemmas. In K. D. Williams, J. P. Forgas, & W. von Hippel (Eds.), *The social outcast: Ostracism, social exclusion, rejection, and bullying*. Sydney Symposium of Social Psychology series (pp. 321-332). New York: Psychology Press.

Schwarz, N., & Clore, G. L. (1996). Feelings and phenomenal experiences. In E. T. Higgins & A. Kruglanski (Eds.), *Social psychology: Handbook of basic principles* (pp. 433–465). New York: Guilford.

Schwarz, N., & Clore, G. L. (in press). Feelings and phenomenal experiences. In E. T. Higgins & A. Kruglanski (Eds.), *Social psychology: Handbook of basic principles* (2nd ed.). New York: Guilford.

Sloman, S. A. (1996). The empirical case for two systems of reasoning. *Psychological Bulletin*, *119*, 3–22.

Smith, E. R., & DeCoster, J. (1999). Associated and rule-based processing: A connectionist interpretation of dual-process models. In S. Chaiken & Y. Trope (Eds.), *Dual-process theories in social psychology* (pp. 323–336). New York: Guilford.

Smith, E. R., & DeCoster, J. (2000). Dual-process models in social and cognitive psychology: Conceptual integration and links to underlying memory systems. *Personality and Social Psychology Review*, *4*, 108–131.

Tannahill, R. (1988). *Food in history*. New York: Random House.

Tomasello, M. (1999). *The cultural origins of human cognition*. Cambridge, MA: Harvard University Press.

Wegner, D. M. (2002). *The illusion of conscious will*. Cambridge, MA: MIT Press.

4

Collective Evolution
Revisiting Donald Campbell's Legacy

CHI-YUE CHIU, YOUNG-HOON KIM, and AVINISH CHATURVEDI

> We have seen that since individuals share common resources, yet independently maximize their outcomes, they tend to get engaged in negative sum competition. Positive sum, cooperative arrangements, which optimize the fitness of the group as a whole, normally get eroded by this competition. Thus, the part–whole competition between groups and individuals tends to be won by the individual level. A social metasystem would overcome this negative sum development by simultaneously differentiating individual's roles (division of labor), and restraining their selfish optimizing (social control), thus creating the conditions for positive sum synergy. However, a necessary condition for such metasystem is to evolve a shift from selection of individuals to selection of groups. (Heylighen & Campbell, 1995, p. 208)

Donald Campbell, master of many disciplines, died in May 1996 at the age of 79. Campbell, who is most well known in psychology as a hardheaded methodologist, is one of the world's most important thinkers in evolutionary philosophy. Inspired by Turchin's (1977, 1981) analysis of metasystem transition and contemporary scholarships on multilevel selection theories (Baum & Singh, 1994; Boyd & Richerson, 1985; Wilson & Sober, 1994) and drawing on the principles of evolutionary epistemology Campbell developed earlier, Heylighen and Campbell (1995) formulated a theory of collective evolution.

Campbell's evolutionary epistemology emphasizes the role of blind variation and selective retention in the evolution of knowledge (see Dutton & Heath, this volume). Subsequently, he introduced the concept of "vicarious selector," which is a control system that anticipates interactions between the system and its environment, with their possible negative effects, and allows the system to compensate anticipatory perturbations in the environment before severe damages are done to the system. He also proposed a hierarchical organization of the vicarious selectors that permits a higher level system to constrain the subsystems (Campbell, 1990).

Campbell's theory of collective evolution is the outcome of applying these evolutionary principles to the evolution of social systems. Because Campbell developed his collective evolution theory in response to some problems with selection at the individual level, to understand the context of his theory, before introducing his principles of collective evolution, we will briefly review what Campbell believed to be the shortcomings of individual selection theories.

PROBLEMS WITH SELECTION AT THE INDIVIDUAL LEVEL

Because of the strong methodological individualism at the base of most social sciences, Heylighen and Campbell (1995) anticipated lukewarm reception to their theory of collective evolution in the social sciences. Methodological individualism is the idea (or dogma) that all social processes can be reduced to the attributes of individual persons (Ho & Chiu, 1998). According to the methodological individualist, social systems have no separate ontological reality. There is a close affinity between the normative (or presently fashionable) research orientation in evolutionary theories and that in economics—they both submit that interacting actors will self-organize into stable and efficient patterns of interaction and that coincidence and order rather than overarching planning accounts for long-term social change. The example of how the laws of demand and supply can efficiently regulate complex economic activities in a free economy gives credence to this idea.

One problem these evolutionary theories face is that maximization of individual fitness and self-organization do not guarantee optimization of the fitness of the group as a whole or efficient management of complex social systems (Heylighen & Campbell, 1995; O'Gorman, Sheldon, & Wilson, 2008; Wilson, 2004). In cooperative interactions, pursuit of individual gains increases the overall benefits of the group. Thus, the group will enjoy maximum benefits when its members maximize their personal gains. In such interactions, self-organization will eventually lead to a stable pattern of interactions. For example, most free market economic transactions are cooperative interactions; without any central regulation, the operation of the invisible hand or market forces will ensure that the market will reach an equilibrium point where market efficiency is optimized and all participating economic actors can maximize their personal utility. In competitive interactions, however, competition for maximum individual benefits may not always lead to an absolute increase in the fitness of the group as a whole. An extreme example is negative sum interactions, where competition for maximum individual benefits decreases the overall fitness of the group. An instance of a negative sum interaction is an arms race. When individual nations compete for military dominance over other nations, each nation will channel resources away from meeting its nationals' fundamental needs, reducing the fit of the individuals in the nation and the nation as a whole. Furthermore, the military buildup will also escalate tension between nations and increase the danger of worldwide destructions when the escalated tension eventually leads to an outbreak of international warfare.

Most human interactions involve mixed motives—individuals are motivated to compete for short-term individual gains despite the incentive to cooperate to increase long-term collective gains (Axelrod, 1984). For example, in "Tragedy of the Commons," it is in the interest of each herdsman to outsmart his competitors by allowing his herds to consume more grass on the common pasture, in spite of the long-term benefits to the group if all herdsmen cooperate to allow the grass to grow back and avoid accelerated depletion of the common resource (Hardin, 1968). In these interactions, competition-induced individual fitness increases more quickly than cooperation-induced group fitness. Furthermore, although everybody profits from cooperation, the noncooperative "free riders" can take advantage of the cooperative synergy produced by the cooperative individuals. Thus, in these interactions, coordination of individual actions for the purpose of maximizing the overall fitness of the group is difficult. A key issue every society needs to resolve is how to move its members away from suboptimal selfish maximization of personal fitness toward the optimal overall fitness of the group as a whole.

COLLECTIVE EVOLUTION: CAMPBELL'S LEGACY

Campbell adopted a nonreductionist stance and asserted that social organizations obey laws that cannot be reduced to the attributes of individual persons. He also maintained that evolution takes place simultaneously at the individual and group levels (Campbell & Gateway, 1994). Applying his evolutionary epistemology described at the beginning of this chapter, Heylighen and Campbell (1995) proposed that individuals do not freely develop cooperative or competitive behaviors by trial and error. Instead, social organization evolves as a vicarious selector, which is a control system that anticipates and regulates possibly lethal effects of natural selection. One such mechanism is culture, or the knowledge or beliefs shared between individuals through communication (see also Boyd & Richerson, 1985; Cheng & Chiu, in press; Kashima, 2007; Lau, Chiu, & Lee, 2001; Lau, Lee, & Chiu, 2004). Thus, contrary to the idea that culture emerges from individual evolution, Heylighen and Campbell (1995) maintained that culture emerges to move evolution out of the potential standstill caused by selfish optimizing.

Collective evolution moves evolution toward the optimal overall fitness for the group by reducing competition through division of labor and restraining selfish optimizing through various social control mechanisms. Division of labor facilitates synergetic cooperation at the group level. Competitive interactions are likely to occur when the coevolving individuals share very similar resources to maximize individual fitness (e.g., all herdsmen use the same common pasture to feed their sheep). Division of labor allows different subgroups of individuals to focus on different subtasks while contributing to the common good. This reduces competition for shared resources and supports internal differentiation of beliefs (or cultural differentiation) within a group.

Social control can be achieved either through internalization of the shared beliefs or by development of external regulatory mechanisms such as the legal system and market regulation. Mutual monitoring of socially unacceptable behavior is a relatively informal but highly effective social control mechanism. Some evolutionary scientists contend that prosocial actions may confer evolutionary benefits to the individuals. For example, Nesse (2007; this volume) argued that natural selection favors sociality: Possession of prosocial traits confers fitness benefits to individuals; those who possess these traits (versus those who do not) have access to more social resources for meeting survival needs and are favored by better partners. Thus, individuals will cooperate when the evolutionary benefits conferred by cooperation outweigh its costs. Even though cooperation may confer some evolutionary benefits, it may also decrease individual fitness in some situations, as the preceding analysis indicates. Accordingly, although maximization of relative fitness of an individual does not preclude cooperation, it also does not guarantee maximization of global fitness. Nevertheless, the fitness advantages associated with being socially responsive provide a genetic basis for the effectiveness of mutual monitoring as a social control mechanism (see also Baumeister, this volume). Because individuals may profit from group solidarity and suffer from ostracism (Williams, 2001), mutual monitoring, which is often accompanied by collective sanction, becomes a widely used mechanism in many societies for holding in check individuals with deviant beliefs (Chao, Zhang, & Chiu, 2008).

THEORETICAL IMPLICATIONS OF CAMPBELL'S LEGACY

Campbell's theory of collective evolution offers a refreshing, dynamic view of how shared cognition evolves. It explicates how culture emerges to address the problem resulting from selfish optimizing. His discussion of how biological evolution acts on the collectively evolved social control mechanisms, which is beyond the scope of the present chapter, also sheds light on the common flaws in capitalist and socialist systems. In this section, we focus on the implications of Campbell's theory of collective evolution for understanding the nature and function of culture, cultural boundaries, and social change.

Nature and Function of Culture

Heylighen and Campbell (1995) proposed a distinction between *culture* and *cultural group*. Although it is convenient to treat two groups as distinct if they do not have direct, physical contacts, in a large, complex society where the possibilities for communication are unlimited, such a static characterization of a group is hardy applicable. According to Heylighen and Campbell, culture is a memotype, or a shared belief that provides a constraint or control on the actions of all individuals having that belief. That is, culture is a shared belief developed for purpose of social regulation. A cultural group, or sociotype, is a collection of individuals who submit themselves to the influence of a shared belief, either because they have internalized the belief or because they submit themselves to its influence out of social, legal, or utilitarian considerations.

Campbell's view has influenced the way we understand culture. Like Heylighen and Campbell, we define culture as a collection of loosely connected ideas that a network of interrelated individuals construct, reproduce, and transform to coordinate the pursuit of goal-directed individual and group actions (Chiu & Hong,

2006, 2007; see also Barth, 2002). We define a cultural group as a collection of individuals who have knowledge of the core beliefs in the cultural tradition and have at least partial knowledge of the distribution of the beliefs in the cultural group (Wan et al., 2007; see also Dutton & Heath, this volume; Kashima, 2007). Because a cultural group is different from a cultural tradition, people in a culture can accept their cultural tradition or comply with but do not internalize it (Chiu & Chen, 2004; Hong, Wan, No, & Chiu, 2007; Wan et al., 2007). Furthermore, instead of treating culture as a determinant of behaviors, we, like Campbell, view culture as a means for constraining and coordinating individual actions (Chao et al., 2008; Cohen, 2001; Fiske, 2000; Kashima, 1999).

Fuzzy Boundaries of Cultures

Heylighen and Campbell (1995) submitted that cultural boundaries are fuzzy. As a belief diffuses through communication from one group to another, it undergoes variations and distortions in the process (Kashima, Woolcock, & Kashima, 2000). Because of such communicative degradation, there are continuous shifts in the reading of the belief. Heylighen and Campbell referred to this phenomenon as "memetic drift," or the spontaneous and continuous divergence of beliefs in a large social system. Because of the continuous nature of these shifts, it is difficult to demarcate where one belief or one subgroup ends and another begins.

Despite this, some sociolinguistic constraints that increase the frequency of in-group communication and reduce the likelihood of diffusion of ideas to neighboring groups can sharpen cultural contrasts (Cheng & Chiu, in press; see Dutton & Heath, this volume). For example, individuals speaking the same language and living in the same geographical location communicate with each other more frequently than individuals speaking different languages and living in different areas. This may explain why relatively deep cultural divides are often found between linguistic and geographical groups (Latané & L'Herrou, 1996; Schwartz, 1992).

Furthermore, external memory devices (books, pictures) or media for storing and transmitting cultural information have relatively high fidelity. Thus using these media for cultural transmission can protect cultural information from communicative degradation and hence may help to sharpen cultural contours. For instance, cultural ideas that are stored in the electronic media are not as vulnerable to communicative degradation and can spread to many individuals quickly, compared to ideas that are encoded in the culture's oral traditions only (Donald, 1993).

Social Change

Like biological systems, a social system also evolves through variation and selection, and continuing development is needed for a social system to maintain its relative fitness. Thus, by necessity, culture must adapt to changes in institutional and historical events. As the institutional and historical conditions supporting a culture change, culture will change accordingly (Yamagishi & Takahashi, 1994). For example, in the 1980s, the Japanese employment system was one of the main factors contributing to the economic success in Japan. Two major characteristics of the Japanese employment system are long-term employment and seniority-based wage. With low job mobility and the absence of formal sanctioning mechanisms in Japanese organizations, social reputation of individuals and mutual monitoring of behaviors become the major social control mechanisms. Under such institutional contexts, the motivation to accommodate the self to the organization culture is particularly strong. Shaped by this institutional context, the Japanese tend to adhere to the value of interdependence, pay attention to context, and prefer conformity to uniqueness (Markus & Kitayama, 1991).

When the institutional context changes, however, such cognitive and behavioral tendencies may become maladaptive and ultimately be replaced by another set of tendencies that is adaptive in the new institutional environment. For example, in many Japanese firms, the middle-age and older age employees receive a salary higher than their productivity justifies, whereas younger employees receive a salary lower than their productivity justifies. For the system to be sustainable, the ratio of middle-age and older employees should be smaller than that of young employees. The system's sustainability also relies on high economic growth and the continued existence of firms. As the baby boomers born just after World War II get older, however, and with the unexpected rapid decline in economic growth and the increased number of firm bankruptcies, the continued survival of the traditional employment system is facing many challenges. Indeed, since the 1980s the average working years in a single company has shortened in the young age group, and the association between seniority and wage has weakened, particularly among university graduate male workers (Hattori & Maeda, 2000).

The potential long-term impact of such institutional changes in Japan on Japanese people's cognitive and behavioral styles is highlighted in a study conducted by Suzuki and Yamagishi (2004). Suzuki and Yamagishi (2004) reported that behavioral variation reminiscent of cultural variation emerged after participants had engaged in different types of experimentally simulated social institutions. For example, Japanese university students, after extensive social interactions in a simulated environment where formal social control mechanisms (e.g., formal judicial systems) dominate, characterize themselves as being more independent and less interdependent, attend more to focal object (versus object–context relation) in the perceptual field, and prefer uniqueness (versus uniformity), compared to those who have engaged in extensive social interactions in a simulated environment where social reputation and mutual monitoring are the dominant social control mechanisms. In short, this study demonstrated how changes in social institutions can alter such seemingly well-learned cognitive styles.

CONCLUSION

Just as there is a collective fear of collective processes in social psychology (Markus & Kitayama, 1994), there also seems to be a collective apprehension of collective evolution in the social sciences. The collective fear of collective evolution may arise partly because people cherish the value of freedom embodied in the principle of self-organization and therefore anxiously resist the idea that individuals are controlled by some collective, impersonal system. As Campbell pointed out, however, the notion of social control does not invalidate the principle of self-organization. For example,

> The spontaneous ordering of a market is the result of complex communications and negotiations that lead to an overall co-ordination and constraint of the different actors' behaviors. Once a perfectly competitive market has reached equilibrium, no participating actor is able to set prices that significantly diverge from those resulting from the supply and demand requirements without being eliminated from the scene ("going broke"). One can debate whether the resulting economic order *controls* the actions of the participating actors, but it certainly *constrains* them. (Heylighen & Campbell, 1995, p. 183)

In this commentary, we reviewed Donald Campbell's theory of collective evolution. This review reveals that collective evolution plays an important part in the evolution of culture and offers penetrating insights into the dynamic relationship between the optimization of individual fitness and the problem of managing complex social systems. Campbell's theory also sheds light on the nature and conception of culture, diffusion of cultural ideas, and social change. Indeed, there is increasing evidence that collective evolution is biologically enabled (Tomasello, 2001; see also Baumeister, this volume). Furthermore, throughout the biocultural history of the humankind, biology and culture have coevolved. For example, the development of some external memory devices had modified the configuration of the human biological and cognitive systems. The invention of written symbols had led to the development of brain areas that specialize in processing pictorial, phonetic, and ideographic symbols (see Chiu, Kim, & Wan, 2008). We hope that this commentary can lower the readers' apprehension of collective evolution and convince them that it is an apprehensible and useful concept for understanding the dynamic relationship between evolution, society, and culture.

REFERENCES

Axelrod, R. (1984). *The evolution of cooperation*. New York: Basic Books.

Barth, F. (2002). An anthropology of knowledge. *Current Anthropology*, *43*, 1–18.

Baum, J. A. C., & Singh, J. V. (1994). *Evolutionary dynamics of organizations*. New York: Oxford University Press.

Boyd, R., & Richerson, P. J. (1985). *Culture and the evolutionary process.* Chicago: Chicago University Press.

Campbell, D. T. (1990). Levels of organization, downward causation, and the selection-theory approach to evolutionary epistemology. In E. Tobach & G. Greenberg (Eds.), *Scientific methodology in the study of mind: Evolutionary epistemology* (pp. 1–17). Hillsdale, NJ: Lawrence Erlbaum.

Campbell, D. T., & Gateway, J. B. (1994). Ambivalently held group-optimizing predispositions. *Behavioral and Brain Sciences*, *17*, 614.

Chao, M. M., Zhang, Z.-X., & Chiu, C.-Y. (2008). Personal and collective culpability judgment: A functional analysis of East Asian–North American differences. *Journal of Cross-Cultural Psychology, 39,* 730–744.

Cheng, S. Y.-Y., & Chiu, C.-Y. (in press). Here is to the crazy ones: A communication perspective to the emergence of a brand culture. In E. Morsella (Ed.), *Expressing oneself/expressing one's self: A festschrift in honor of Robert M. Krauss.* Hillsdale, NJ: Lawrence Erlbaum.

Chiu, C.-Y., & Chen, J. (2004). Symbols and interactions: Application of the CCC model to culture, language, and social identity. In S.-H. Ng, C. Candlin, & C.-Y. Chiu (Eds.), *Language matters: Communication, culture, and social identity* (pp. 155–182). Hong Kong: City University of Hong Kong Press.

Chiu, C.-Y., & Hong, Y. (2006). *The social psychology of culture.* New York: Psychology Press.

Chiu, C.-Y., & Hong, Y.-Y. (2007). Cultural processes: Basic principles. In E. T. Higgins & A. E. Kruglanski (Eds.), *Social psychology: Handbook of basic principles* (pp. 785–809). New York: Guilford.

Chiu, C.-Y., Kim, Y.-H., & Wan, W. N. (2008). Personality: Cross-cultural perspectives. In G. J. Boyle, G. Matthews, & D. H. Salofske (Eds.), *The Sage handbook of personality theory and assessment. Vol. 1: Personality theories and models* (pp. 122–144). London: Sage.

Cohen, D. (2001). Cultural variation: Considerations and implications. *Psychological Bulletin, 127,* 451–471.

Donald, M. (1993). Precis of "Origins of the modern mind: Three stages in the evolution of culture and cognition." *Behavioral and Brain Sciences, 16,* 737–791.

Fiske, A. P. (2000). Complementarity theory: Why human social capacities evolved to require cultural complements. *Personality and Social Psychology Review, 4,* 76–94.

Hardin, G. (1968). The tragedy of the commons. *Science, 162,* 1243–1248.

Hattori, R., & Maeda, E. (2000, January). The Japanese employment system (summary). *Bank of Japan Monthly Bulletin.* Retrieved January 24, 2004 from http://www.boj.or.jp/en/type/ronbun/ron/research/data/ron0001a.pdf

Heylighen, F. (1992). Selfish? Memes and the evolution of cooperation. *Journal of Ideas, 2,* 77–84.

Heylighen, F., & Campbell, D. T. (1995). Selection of organization at the societal level: Obstacles and facilitators of metasystem transitions. *World Futures: The Journal of General Evolution, 45,* 181–212.

Ho, D. Y. F., & Chiu, C.-Y. (1998). Collective representations as a metaconstruct: An analysis based on methodological relationalism. *Culture and Psychology, 4,* 349–369.

Hong, Y.-Y., Wan, C., No, S., & Chiu, C.-Y. (2007). Multicultural identities. In S. Kitayama & D. Cohen (Eds.), *Handbook of cultural psychology.* New York: Guilford.

Kashima, Y. (1999). Culture, groups, and coordination problems. *Psychologische Beiträge, 41,* 237–251.

Kashima, Y. (2007). A social psychology of cultural dynamics: Examining how cultures are formed, maintained, and transformed. *Social and Personality Psychology Compass, 2*(1), 107–120.

Kashima, Y., Woolcock, J., & Kashima, E. (2000). Group impressions as dynamic configurations: The tensor product model of group impression formation and change. *Psychological Review, 107,* 914–942.

Latané, B., & L'Herrou, T. (1996). Spatial clustering in the conformity game: Dynamic social impact in electronic groups. *Journal of Personality and Social Psychology, 70,* 1218–1230.

Lau, I. Y.-M., Chiu, C.-Y., & Lee, S.-L. (2001). Communication and shared reality: Implications for the psychological foundations of culture. *Social Cognition, 19,* 350–371.

Lau, I., Lee, S.-L., & Chiu, C.-Y. (2004). Language, cognition and reality: Constructing shared meanings through communication. In M. Schaller & C. Crandall (Eds.), *The psychological foundations of culture* (pp. 77–100). Mahway, NJ: Lawrence Erlbaum.

Markus, H. R., & Kitayama, S. (1991). Culture and self: Implications for cognition, emotion, and motivation. *Psychological Review*, *98*, 224–253.

Markus, H. R., & Kitayama, S. (1994). A collective fear of the collective: Implications for selves and theories of selves. *Personality and Social Psychology Bulletin*, *20*, 568–579.

Nesse, R. M. (2007). Runaway social selection for displays of partner value and altruism. *Biological Theory*, *2*, 1–13.

O'Gorman, R. O., Sheldon, K. M., & Wilson, D. S. (2008). For the good of the group? Exploring group-level evolutionary adaptations using multilevel selection theory. *Group Dynamics: Theory, Research, and Practice, 12*, 17–26.

Schwartz, S. H. (1992). Universals in the content and structure of values: Theory and empirical tests in 20 countries. In M. Zanna (Ed.), *Advances in experimental social psychology* (Vol. 25, pp. 1–65). New York: Academic Press.

Suzuki, N., & Yamagishi, T. (2004, July). *Socio-institutional basis of holistic cognition*. Paper presented at the Mind, Culture and Evolution Conference, Vancouver, BC.

Tomasello, M. (2001). Cultural transmission: A view from chimpanzees and human infants. *Journal of Cross-Cultural Psychology*, *32*, 135–146.

Turchin, V. (1977). *The phenomenon of science: A cybernetic approach to human evolution*. New York: Columbia University Press.

Turchin, V. (1981). *The inertia of fear and the scientific worldview*. New York: Columbia University Press.

Wan, C., Chiu, C.-Y., Tam, K.-P., Lee, S.-L., Lau, I. Y.-M., & Peng, S.-Q. (2007). Perceived cultural importance and actual self-importance of values in cultural identification. *Journal of Personality and Social Psychology*, *92*, 337–354.

Williams, K. (2001). *Ostracism: The power of silence*. New York: Guilford.

Wilson, D. S. (2004). What is wrong with absolute individual fitness? *Trends in Ecology and Evolution*, *19*, 245–248.

Wilson, D. S., & Sober, E. (1994). Reintroducing group selection to the human behavioral sciences. *Behavioral and Brain Sciences*, *17*, 585–654.

Yamagishi, T., & Takahashi, N. (1994). Evolution of norms without metanorms. In U. Shulz, W. Albers, & U. Mueller (Eds.), *Social dilemmas and cooperation* (pp. 311–326). Berlin: Springer-Verlag.

5

Cultural Evolution

Why Are Some Cultural Variants More Successful Than Others?

YULIA CHENTSOVA DUTTON and CHIP HEATH

What makes some cultural ideas and practices more successful than others? Why do some cultural variants—fairy tales, epics, jokes, urban legends, symbols, or baby names—survive for decades (or centuries), whereas others are forgotten by the time of the next gossip cycle? The etiquette norm that, "It is boorish to wipe one's nose on one's cap or clothing," has persisted since at least 1530, whereas the norm "If given a napkin, put it over either the left shoulder or the left forearm," has not. Why? Why has the European fairy tale of "Little Red Riding Hood" persisted and crossed cultural boundaries but the tale of "Hans My Hedgehog" largely disappeared? In the United States and Great Britain, why did the use of biblical names for boys—such as Nathan and Samuel—decline during the early to middle part of the 20th century and then rise again by the 1980s?

Although it seems obvious that cultural evolution is *evolution*—created in a process of variation and selective retention—social scientists have not historically taken evolutionary logic very seriously. Historically, anthropologists have used the term *evolution* to refer to process, not of variation and selective retention, but of successive development by which cultural institutions, ideas, practices, and behaviors change and become increasingly more complex (Malinowski, 1945; Morgan, 1877; Steward, 1955; Tylor, 1870, 1871; White, 1959).

It is somewhat surprising that a better application of evolutionary ideas to cultural variants comes not from anthropology but from psychology. Almost 50 years ago, psychologist Donald Campbell (1960) identified consistent evolutionary mechanisms of *random variation* and *selective retention* as two forces governing the social and cultural selection of knowledge. Campbell defined *cultural variation* as chance deviations in cultural practices relative to the inherited form and *selective retention* as the process that filters the cultural variants that are successfully transmitted to subsequent generations. Campbell's work generated much interest in the evolution of organizational forms and structure (see Baum & Singh, 1994, for a review), but few psychologists pursued his argument. Recent interest in cultural and evolutionary psychology (Boyd & Richerson, 1985; Sperber, 1996) has sparked a resurgence of research in cultural evolution (Heath, Bell, & Sternberg, 2001; Norenzayan, Schaller, & Heine, 2006). Forty years after publication of Campbell's (1960) article, the assertion that cultural evolution is governed by many of the same principles as evolution of biological species, including variation, competition between variants, and selective retention, remains controversial (see Mesoudi, Whiten, & Laland, 2004). In this chapter, we will revisit Campbell's theory of cultural selection and identify psychological mechanisms that underpin cultural evolution—considering why certain variants are selectively retained and transmitted but also why variation occurs in the first place.

We begin our chapter with selective retention because current research has done most to develop this topic. For example, there is a variety of clever research that has explored how cognitive mechanisms such as memorability lead some cultural variants to be retained (e.g., Barrett & Nyhof, 2001; Norenzayan, Atran, Faulkner, & Schaller, 2006; Rubin, 1995). But although previous work has emphasized cognitive mechanisms, in this chapter we also highlight *emotional* mechanisms of selective retention. Recent work has shown that stories, rumors, urban legends, and factoids are selected not just for their dry, cognitive memorability but also for their ability to deliver an emotional kick (Harber & Cohen, 2005; Heath et al., 2001).

We then discuss part of the process of cultural evolution that has been largely neglected by previous researchers: variation. Variation provides the engine that drives cultural evolution, yet most previous work has focused on low-powered sources of variation such as forgetting. Forgetting produces accidental variation, but much variation in culture is conscious and strategic. By ignoring these conscious, directed, and strategic sources of variation, previous work has dramatically underestimated the size of the engine driving the process of cultural evolution. Instead of a sleepy subcompact, the process of cultural evolution is more likely to be a turbo-charged roadster.

In our discussion, we turn to the question of whether cultural variants are best conceptualized as independent elements or parts of a coherent system. Dawkins (1976), writing after Campbell, proposed the notion of meme, or an independent cultural variant propagating within culture as a Darwinian replicator. One of the major concerns raised in the subsequent debate about memes is whether identifiable elements of culture exist and can be studied (Bloch, 2000; Boyd & Richerson, 2000). Indeed, the first incarnation of the *Journal of Memetics* folded in 2004 in part over a frustration about the lack of empirical submissions. We believe that the meme debate is just the latest example of a broader historical debate in the social sciences between proponents who see "culture as elements" and those who see "culture as a system," and we close the paper by arguing that a "culture as elements" approach is necessary and that the research we review in this paper suggests it can be productively pursued.

SELECTIVE RETENTION

What psychological mechanisms serve to narrow the pool of cultural variants to a few winners that will be remembered and transmitted? In this section we briefly discuss both cognitive and emotional mechanisms of selective retention.

We assume that cultural variants will be selected if they fit stable aspects of the psychological and social environment. As psychologists, we will focus on psychological aspects of the environment such as shared memory mechanisms, schemas, or emotions, but a sociologist might use a similar approach to explore relatively stable aspects of the social environment such as social stratification. Similarly, a political scientist might use this approach to explore institutional structures such as forms of government or long-standing social institutions.

To affect cultural evolution, the psychological and social environments need not be universal across cultures. Fessler and Navarrete took a universal approach to food taboos, arguing that meat becomes the target of taboos across diverse cultures because humans (and other primates) are biologically primed to develop conditioned aversions to meat (Fessler & Navarrete, 2003). Yet, there are also social selection mechanisms for taboos. Aunger (2002) argued that food taboos in Congo were more likely to be accepted and transmitted when the individual introducing the food taboo had high social status and prestige. The ability to identify prestigious individuals in a group and copy their behavior may provide an evolutionary advantage in acquiring culturally valued skills and information (Henrich & Gil-White, 2001). Thus, successful cultural variants may capitalize on universal features of human attention, memory, and social learning and communication systems, but they may also take advantage of attentional strategies and stable social dynamics within a particular culture.

Memorability

Human memory limits which cultural variants can be remembered and transmitted successfully. People are unlikely to retain information that is easily forgotten or misremembered, particularly in cultures relying on an oral tradition. Rubin (1995) provided a brilliant account of how the cognitive structure of memory affects the content of oral traditions such as epic ballads or counting-out rhymes. As one example of his approach, he used work on imagery in cognitive psychology to argue that epic ballads such as the *Iliad* or *Odyssey* tend to focus on concrete, easily visualized actions because people find it easier to remember events that are concrete and easy to visualize. Homer is filled with concrete action, not because the Greeks had trouble with abstraction but because the constraint of human memory makes concrete images more likely to survive generation after generation of oral transmission (see Rubin, 1995, Chapter 3).

Concrete information is selected because of basic cognitive processes that are widely shared across all humans. Selective retention may also rely on a memory infrastructure that is widely shared only within a particular culture. Cultural variants may be memorable when they remind people of culturally common scripts (e.g., going to the dentist or a witch doctor) and schemas (e.g., agreed characteristics of particular social groups). Memory researchers have proposed that evoking scripts or schemas provides a scaffold for learning new information (Bellezza & Bower, 1981). For example, people are more likely to transmit stereotype-consistent information in social conversations (Lyons & Kashima, 2001, 2006) and are more likely to rely on script-consistent features of a story as information moves along a chain of people (Mesoudi & Whiten, 2004). This process may be even more robust when examined over hundreds of years. Rubin (1995) studied oral traditions that have persisted for decades or centuries and highlighted the central role played by widely shared scripts, such as dressing for battle or having a meal, which cue recall and protect narratives against deterioration. If an element of a schema is forgotten, people easily substitute another schema-consistent element, thus preserving the overall meaning of the story and making it more schematic and, thus, more memorable over time.

Schema-consistent information is also favored in scientific communication. Researchers who are summarizing the details of scientific articles often distort the work they cite, shaping their descriptions to be more consistent with scientific schemas (Harris, 1979; Loftus, 1974; Vincente & Brewer, 1993). Harris (1979) surveyed accounts of John Watson's famous study of how infant "Little Albert" learned fear reactions. Accounts tended to distort details of the study in the direction of theoretical predictions of the learning theory. Similarly, Vincente and Brewer (1993) showed how de Groot's famous experiment with chess masters is often reported as having a much more clean and compelling experimental design than it actually did (often featuring a nonexistent control group).

Interruption and Schema Inconsistency: The Role of Surprise

Interestingly, once a cultural variant evokes a schema, people may recall it even better when it judiciously violates that schema. Breaking or interrupting a familiar pattern makes cultural variants more memorable and transmissible. Surprise is an emotion designed to interrupt ongoing information processing, seize attention, and free up cognitive resources to evaluate whether the unexpected event is relevant to ongoing activities (Meyer, Reisenzein, & Schützwohl, 1997). Thus, cultural variants that produce surprise draw on both cognition and emotion.

One way of generating surprise is to include information that is inconsistent with existing schemas or expectations (Neuschatz, Lampinen, Preston, Hawkins, & Toglia, 2002; Pezdek, Whetstone, Reynolds, Askari, & Dougherty, 1989; Rojahn & Pettigrew, 1992; Stangor & McMillan, 1992). Interestingly, including schema-inconsistent details in a narrative makes the schema-consistent information more memorable as well (O'Sullivan & Durso, 1984). Narratives that include counterintuitive elements are more memorable (e.g., the idea of a ghost is counterintuitive because it is a "person" who can pass through walls) (Barrett & Nyhof, 2001; Norenzayan, Atran et al., 2006), and they degrade at lower rates than narratives composed of elements that are intuitive and less surprising (Norenzayan & Atran, 2004). But the relationship is dose dependent, such that narratives are more memorable when they contain a modest sprinkling of counterintuitive features (see Norenzayan & Atran, 2004, for a discussion of this effect). For example, Norenzayan, Atran, et al.

(2006) found that fairy tales including two or three counterintuitive elements were better recalled than stories with too many or too few counterintuitive elements. "Little Red Riding Hood" has two elements—a talking wolf and a miraculous resurrection of the eaten grandmother—and has succeeded better than "Hans My Hedgehog," which has only one element—a talking hedgehog. Thus, a cultural form can become more memorable when it is surprising, schema violating, or counterintuitive, but only when the proportion of surprising content is modest.

Interestingly, although most research has focused on ideas that are memorable because they violate natural properties (e.g., a ghost passing through walls), some cultural variants specialize on creating surprise endogenously—they use the cultural form to first establish and then violate a novel pattern. For example, children's stories across cultures often establish a script by repeating a pattern a couple times that is then unexpectedly contradicted. In the story "Three Billy Goats Gruff," the nasty troll under the bridge allows the first two goats to pass because they are too small to eat, but then he is butted off the bridge when he tries to eat the third, larger goat. When the troll is butted off by the third goat, there is a small jolt of surprise even though we have no preexisting mental schemas that predict how trolls interact with goats. Similarly, jokes often set up a pattern where two different people repeat a particular behavior, and then a third individual violates the established pattern (Loewenstein & Heath, 2005). Evidence shows that interruptions of familiar scripts enhance memory in a Zeigarnik-like effect (Davidson, Larson, Luo, & Burden, 2000), and it is interesting that cultural variants have evolved to take advantage of this interruption mechanism.

Emotions

Surprise is not the only emotion that influences retention and propagation of cultural variants. Emotions provide valuable survival-relevant guidance (Harber & Cohen, 2005), so emotional information tends to capture people's attention (Öhman, Flykt, & Esteves, 2001) and to be easy to remember (see Reisberg & Heuer, 2004, for a review). People also tend to transmit emotional information. Individuals who experience a situation that elicits strong emotions talk about those experiences 90% to 96% of the time (Rimé, Mesquita, Philippot, & Boca, 1991). When an event produces more intense emotions, people talk about it with others more frequently.

Emotions enhance memory for information that is central to the event and diminish memory for peripheral information (Burke, Heuer, & Reisberg, 1992; Christianson & Loftus, 1991). So, we would expect that central elements of emotionally charged stories would be more easily remembered and transmitted than the central elements of emotionally neutral stories. We can also predict that peripheral details of the emotionally charged stories would be easily forgotten and altered, thus generating more cultural variants of the original form.

Perhaps it is somewhat surprising that the most-often researched emotion in studies of cultural transmission is disgust. Disgust is a prototypical emotion with associated specific facial cues (i.e., wrinkled nose) and behavioral program (i.e., a desire to move away and/or spit something out). It has been widely studied because it is relevant for core evolutionary behaviors related to contamination and cleanliness (Rozin, 1996; Rozin, Haidt, & McCauley, 1993), and it tends to be evoked by specific practices with obvious health implications—cutting or piercing the skin, contacting body fluids, or ingesting something that is not food.

Disgust appears to be involved in selecting such diverse cultural variants as food taboos (Fessler & Navarrete, 2003), urban legends (Heath et al., 2001), and etiquette norms (Nichols, 2002), leading variants that evoke high disgust to be retained and transmitted more often. For example, Heath et al. (2001) found that urban legends eliciting stronger disgust were more likely to be passed along than those eliciting less intense disgust. This pattern held even after controlling for the stories' plausibility, informational value, and entertainment value. Interestingly, Oedipal stories involving family incest themes are told across cultures (Johnson & Price-Williams, 1996). Perhaps this indicates that Freud was right, but it could also suggest that stories that involve family incest are a highly disgusting form of urban legend.

Similarly, Nichols (2002) studied the persistence of etiquette norms in Western Europe in the period 1530 to 2002. He found that etiquette norms that evoked some core element of disgust (e.g., "Withdraw when you are going to vomit"; "It is boorish to wipe one's nose on one's cap or clothing") were more likely to survive across the centuries than etiquette norms that did not ("When sitting down at a banquet have both hands on the table, not clasped together, nor on the plate"; "If given a napkin, put it over either the left shoulder or the left forearm").

In summary, emerging research suggests that emotional content offers evolutionary advantages to cultural variants. Although we have summarized several studies that focused on the emotion of disgust, there is evidence that other emotions yield similar advantages—the anger, hope, and fear in wartime rumors (Knapp, 1944); the fear in social panics about "flesh-eating bacteria" or drug epidemics (Glassner, 1999); or the awe people experience when they encounter God or a deep scientific theory (Keltner & Haidt, 2003).

Ease of Communication

In addition to interpersonal mechanisms of selective retention such as memory and emotion, the success of cultural variants also depends on interpersonal dynamics. When cultural variants are easy to communicate, they are more likely to be communicated broadly. For example, Schaller, Conway, and Tanchuk (2002) found that communicable traits—traits judged to be easier to talk about in social conversations—were more likely to persist over a 50-year period in stereotypes of ethnic groups.

We have briefly reviewed several mechanisms that might lead cultural variants to be selectively retained. Successfully tapping the cognitive, emotional, and communication systems of humans allows cultural variants to be noticed, understood, communicated, and retained over time.

VARIATION

Almost any cultural form has variants. Any item of gossip can be told to emphasize the role of different individuals, events, or outcomes. Researchers who are citing scientific papers can describe a sociological theory (Mizruchi & Fein, 1999) or the methods of psychological studies (Harris, 1979; Loftus, 1974; Vincente & Brewer, 1993) in very different ways.

But what drives variation in cultural practices and ideas? The chief answer from previous studies is memorability, or lack thereof. Researchers have conducted a lot of research on what makes some cultural variants more memorable, so it may not be surprising that they tend to hold up forgetting or misremembering as a key source of variation (Allport & Postman, 1947; Campbell, 1997; Rubin, 1995). Campbell (1997) believed that "chance deviations" were an important source of cultural variation, and he argued that variation arose by chance as people forgot elements of the original cultural form and filled in new details to make up for the information they forgot.

But we suggest that variation is likely to be more strategic and intentional than has been highlighted by previous theories of blind cultural evolution. If we are right, the mechanisms generating cultural variation are likely to be much more powerful than has previously been assumed. Next, we summarize a small amount of research that has explored processes of variation that are strategic and intentional.

Novelty Seeking

Humans have evolved to seek novelty and stimulation (Cloninger, 1987). The ability to produce novel cultural variants may serve as a sexually selected indicator of phenotypic and genotypic qualities (Haselton & Miller, 2006; Miller, 1999, 2000). A few researchers have argued that this preference for producing and detecting novel cultural variants is an important force driving cultural innovation.

In the provocative book *The Clockwork Muse*, Martindale (1990) studied how artistic traditions vary over long periods of time; for example, British poetry from 1290 to 1949 or Egyptian paintings from 1501 to 1235 B.C. Martindale's main argument is that successful artists must systematically push the boundaries of their art over time, creating novel variations that surprise and engage their audience. He showed compelling evidence that within particular fields of art, indicators of novelty increase systematically over time; for example, the phrase length of British poems tends to become more variable, and poets increasingly use highly ambiguous words.

Although artists may consistently try to create surprising cultural variants that capture the attention of their audience, Martindale argued that the novelty of art can be alternately carried by *content* (e.g., thoughts expressed in a poem) or *form* (e.g., the poem's verse structure). In periods where form is fixed by convention, novelty must be carried by content. Martindale measured novel content with a somewhat confusing measure of "primordial cognition," which codes for dozens of classes of ideas including emotions terms, concrete imagery, and bodily functions. He provided some evidence that there is a trade-off between the novelty carried by content and form. His measure of primordial content tends to increase over time as artists create content that is increasingly bizarre, but when an artistic field adopts a new form, primordial content drops again because artists are able to create surprise with novel forms rather than novel content.

The push for novelty is not limited to art. In an article called "That's Interesting!" about the success of scientific theories, Murray Davis (1971) argued that scientists tend to favor and discuss scientific theories not so much for their predictive power but for their interest value. He argued that scientific theories are more interesting if they refute assumptions held by their target audiences than if they support these assumptions, so there is a pressure—at least in the social sciences—for novel scientific theories that are counterintuitive to succeed.

One way to generate novelty is to tweak an existing cultural form (Lieberson, 2000). Fashion provides us with endless examples of this approach. Of course, most fashion domains involve constraints on the kinds of variation that can be generated—for example, skirt lengths are bounded within a few feet—so variation in many domains is cyclical: After a style has been out of fashion long enough, it can be reintroduced as fashionable once more.

It may be tempting to attribute cycles of fashion not to a psychological desire for variation but to marketing ploys by fashion manufacturers. Individuals in consumer societies are bombarded with messages inviting them to exchange their clothes, nail polish, and cars for novel and more fashionable choices. Some fashion domains, however—such as the names parents choose for their children—are untouched by marketers and yet show similar cycles. Lieberson (2000) argued that preferences for baby names develop as variations on existing tastes. Every generation of parents prefers modest amounts of novelty in baby names relative to the names of their own generation. Over time, this results in a "ratchet effect," a gradual cyclical shift in the popularity of names. For example, in the United States and Great Britain, the use of biblical names for boys, such as Nathan and Samuel, declined during the early to middle part of the 20th century and then rose again by the 1980s. These shifts were better predicted by motivated novelty seeking than by the religious beliefs or activity of parents. The ratchet effect Lieberson observed resembles those in other fashion domains (e.g., dress length, facial hair), suggesting that gradual innovation on familiar standards is an important mechanism of novelty generation, even in the absence of marketing.

Establishing Social Identity

All cultural contexts are composed of diverse subcultures. Recently, subcultural variation has become a prominent interest of cultural psychologists (Kashima et al., 2004; Kitayama, Ishii, Imada, Takemura, & Ramaswamy, in press; Plaut, Markus, & Lachman, 2002; Snibbe & Markus, 2005; Tsai & Chentsova-Dutton, 2003). Even in seemingly monolithic cultures such as Japan, subcultures can be characterized by very different models of basic social processes such as agency (Kitayama et al., in press).

For a subculture to persist, it must adapt to its larger society while maintaining its distinctiveness (Brewer, 1991; Tajfel & Turner, 1986). Members of subcultures work to achieve distinctiveness by adopting a unique combination of cultural variants such as dress, values, beliefs, and behavioral norms. As a result, the emergence of new social groups serves as a key mechanism of generating new cultural variants. The ever-present tension between subculture and mainstream culture can further stimulate ongoing cultural innovation (Thornton, 1996). People are constantly seeking novelty, and members of the mainstream culture often discover novel variants by poaching the signals of a subculture (e.g., suburban youth may borrow rap music from urban youth, or middle-class European Americans may adopt aspects of Tibetan Buddhist practice). But this poaching process diminishes the capacity for subcultural signals to distinguish a unique social group, and members of the subculture whose signals are poached may react by refining or changing the signals to make their group more distinctive once again (Berger, Heath, & Ho, 2006).

Cultural Exchange

Another novelty-generating mechanism is cultural borrowing or hybridization. Common borders, globalization forces, and migration provide individuals in one culture with access to cultural variants generated in other cultural contexts (Burt, 1997). Hybridization of cultural variants is a fertile ground for innovation. Historical examples of the effects of cultural exchange are pervasive: The Silk Road increased religious diversity in Central and East Asia (Foltz, 1999), trade between Poland and the Western European countries (e.g., Germany and Italy) enhanced the cultural diversity of 16th- and 17th-century Lübeck and Danzig (Cowan, 2003), and communication with the West increased the diversity of musical tastes in 20th-century Japan (Burt, 1997). Barbas (2003) argued that gastronomic variety has increased substantially in the past 20 years in the United States because of the introduction of Chinese restaurants into the United States and the subsequent borrowing of Chinese cooking techniques by mainstream American home cooks.

Some studies have compared the relative effects of cultural exchange versus cultural transmission inside a culture, and their basic conclusion is that the importance of cultural exchange is context dependent (see Collard, Shennan, & Tehrani, 2006, for a review). For example, studies of California and New Guinea tribes suggest that cultural variants such as patterns on textiles are at least as predictable by knowing whether two groups could conduct cultural exchange, as measured by the geographic adjacency of cultural groups, as by knowing whether two groups share a common history as measured by language similarity (Jordan & Shennan, 2003; Moore & Romney, 1994; Welsch, 1996). In contrast, studies of African, Turkmen, and European Neolithic tribes show that cultural exchange is far less important than group history (Collard & Shennan, 2000; Guglielmino, Viganotti, Hewlett, & Cavalli-Sforza, 1995; Tehrani & Collard, 2002). It appears that cultural contact leads to borrowing and exchange under certain circumstances, such as in settings that promote intergroup marriages (Jordan & Shennan, 2003).

Not only does cultural contact lead to an immediate increase in the variation in cultural variants, it may also have important downstream effects on subsequent variation. For example, scientific laboratories that include members from different but overlapping fields benefit from the ability of their members to make use of diverse analogies in interpreting data, thus fostering more innovative subsequent studies (Dunbar, 1995). The effects of cultural contact can be even more long lasting, spanning generations: Simonton (1997) demonstrated that Japan's contact with foreign cultures increased levels of creative achievement by subsequent generations of the Japanese. The peak level of subsequent variation happened not in the generation that experienced the immediate contact but in two generations downstream.

In summary, a number of psychological and sociocultural mechanisms account for the amazing diversity of cultural variants within any given culture. These engines of variation range from very basic, such as forgetfulness or a preference for novelty, to more complex, such as the development and maintenance of social group identities and interaction with ideas and practices of other cultural groups.

CULTURE AS ELEMENTS VERSUS CULTURE AS A SYSTEM

There is a long-standing feud in the social sciences between scholars who see culture as a collection of more or less independent and loosely organized elements and scholars who see culture as a relatively integrated and coherent system. By arguing that there is merit to taking an evolutionary approach to the propagation of cultural variants, we have implicitly been promoting the "culture as elements" view. This is approach is not without controversy, so in this section we defend it.

To understand some potential problems with treating culture as elements it helps to consider the recent debate that has unfolded about the idea of memes (Blackmore, 1999; Bloch, 2000; Boyd & Richerson, 2000; Dawkins, 1976; Hull, 2000; Sperber, 2000). The idea of memes has accumulated lots of baggage, so it is now a less helpful term than it might otherwise have been, but it remains a compact, effective way to refer to an element of culture and to urge scholars to consider the possibility that elements undergo an evolutionary process of variation and selective retention. Therefore, it is worth understanding some of the key problems that it has faced because it has claimed that culture comes in the form of elements.

The idea of memes as cultural elements has been criticized on both empirical and theoretical grounds. On empirical grounds, critics have highlighted the lack of empirical examples of things that look and act like replicable elements of culture (Edmonds, 2002; Gatherer, 2005); indeed the *Journal of Memetics* folded in its original incarnation, in part because people felt there were too many theoretical papers chasing too few empirical examples of memes (Edmonds, 2005). Philosopher of science David Hull, who is not opposed to the meme concept, nonetheless told a conference of meme theorists in 2000 that it was time to stop talking about memes and start studying them (Hull, 2000).

But critics of the meme notion have also raised deeper theoretical issues. Bloch (2000), an anthropologist addressing a conference of meme researchers, noted that the candidates for memes are a "ragbag" of proposals, many less than convincing:

> At first, some seem convincing as discrete units: catchy tunes, folk tales, the taboo on shaving among the Sikhs, Pythagoras' theorem, etc. However, on closer observation, even these more obvious "units" lose their boundaries. … The Sikh taboo is meaningless unless it is seen as part of Sikh religion and identity. (Bloch, 2000, p. 194)

He summarized his argument: "In reality, culture simply does not normally divide up into naturally discernable bits" (Bloch, 2000).

The problems faced by memes are not new. This recent debate about memes echoes a long-standing debate between scholars who have treated culture as a set of elements and those who have treated culture as a system.

The "culture as a system" position that Bloch defends has a long history in anthropology. Bronislaw Malinowski, one of founders of modern social anthropology, argued that culture is not "a loose agglomeration of customs … a heap of anthropological curiosities, but a connected living whole … all its elements are interconnected and each fulfills a specific function in the integral scheme" (1929, p. 864).

Indeed there is evidence that cultures do exhibit fit and consistency. Cultural psychologists show that core cultural ideas and values manifest themselves in such dissimilar cultural products such as Olympic coverage (Markus, Uchida, Omoregie, Townsend, & Kitayama, 2006), advertisements (Aaker, Benet-Martínez, & Garolera, 2001; Cho, Kwon, Gentry, Jun, & Kropp, 1999; Kim & Markus, 1999), aspects of physical environments (Miyamoto, Nisbett, & Masuda, 2006), and music lyrics (Rothbaum & Tsang, 1998; Snibbe & Markus, 2005). Empirical studies within this tradition show that cultural products show distinct differences across cultures that correspond to core cultural ideas and values. These studies demonstrate that when cultural variants are relevant to core cultural ideas, they evolve to be coherent and systematic.

But though there have been long-standing theoretical arguments that culture is a system and some empirical evidence supporting this notion, scholars in a number of areas—in anthropology, sociology, psychology, and memetics—have implicitly or explicitly favored the culture as elements view by treating cultural variants as largely independent (Brumann, 1999; DiMaggio, 1997; Swidler, 1986).

Historically, even the field of anthropology—home base of the culture as system view—has always had scholars who favor an element-based approach. Most notably, in the first half of the 1900s, Franz Boas and his students rejected the idea that culture was a systematic whole, arguing that it was a "loose and accidental assemblage of traits" (Kuper, 1999, p. 68). The Boasians faced an anthropological field that was enamored with approaches it called "evolutionary" but that defined evolution as the process of systematic development whereby cultures progressed through fixed stages, including savagery and barbarism, on their way toward civilization. The Boasians scoffed at the idea of systematic development because many cultural traits were not developed inside a society but rather inherited through diffusion.

The diffusion critique by the Boasians, which explicitly assumes that elements of culture can be exchanged among cultures, was so compelling that evolutionary approaches fell from favor in anthropology. But a negative side effect to this fall was that many scholars abandoned any attempt to discover general mechanisms of cultural change. According to one scholar, the Boasians became convinced that "the facts of social life were so convoluted and so discrepant that finding any clear and concrete cultural laws among them was all but hopeless" (Carneiro, 2003, pp. 238–239).

Although the Boasian critique doomed models of systematic development, it did not necessarily encourage anthropologists to focus on elements of culture. Roy

D'Andrade (1995) described cognitive anthropology as the only subfield of anthropology that has taken seriously the idea that cultural elements could be studied. He quoted a major review in the 1960s by Kroeber and Kluckhorn as saying, "Most anthropologists would agree that no constant elemental units like atoms, cells, or genes have yet been satisfactorily established with culture in general" (p. 247). Kroeber and Kluckhorn conceded that elemental units have been found in the study of language but not in other domains. D'Andrade noted that cognitive anthropologists were comfortable with an elements approach—they considered elements of culture such as schemas, propositions, and theories and asked questions about how widely such elements were shared among people within a culture. But he conceded that the work of cognitive anthropologists "did not even make sense within much of the mainstream framework of cultural anthropology. One has to have a notion of separable units before the study of their distribution has any meaning" (p. 247).

Examples of a culture as elements approach are also found in other social sciences. In sociology, Swidler (1986) proposed that culture is a tool kit of independent cultural variants. She argued, for example, that middle-class Americans work with a number of disconnected, and even inconsistent, models of romantic relationships. For example, there is obviously some disagreement between the idea that love can happen "at first sight" as seen in Hollywood movies and the idea that love is about long-term relationships that require careful nurturing, growth, and tending. In her ethnographic interviews of couples, Swidler found evidence that individuals freely use elements from each of these apparently inconsistent models to suit their reasoning in specific situations (see Quinn, 1996, for an argument from an anthropologist that there is more coherence in relationship models than Swidler claimed). Interestingly, Swidler's finding that individuals rely on inconsistent models of romantic relationships is echoed in other domains such as the work on the persistence of naive theories (Dunbar, Fugelsang, & Stein, in press) and naive geographical models (Friedman & Montello, 2006), despite acquisition of inconsistent concepts.

A similar emphasis on culture as a kit of independent tools can be found in recent dynamic constructivist approaches in cultural psychology (Benet-Martínez, Leu, Lee, & Morris, 2002; Gardner, Gabriel, & Lee, 1999; Hong, Benet-Martínez, Chiu, & Morris, 2003; Hong, Chiu, & Kung, 1997; Hong, Morris, Chiu, & Benet-Martínez, 2000). These studies illustrate that culture is "internalized in smaller pieces" (Hong et al., 2003, p. 454) and that multicultural individuals are able to shift between multiple cultural frames depending on which one is cued by their current situation. Interestingly, even monocultural American individuals shift their self-construals, value endorsements, and social judgments depending on situational cues (Gardner et al., 1999), suggesting that American culture provides distinct models of agency that individuals can invoke in different situations.

As is clear from this review, the debate between culture as a system and culture as elements has a long history in the social sciences. It is therefore with some trepidation that we propose a simple resolution to this protracted debate: The culture as elements proponents are right.

Next, we summarize some compelling reasons to favor an elements approach. At the same time, we don't dismiss the appearance of fit highlighted by the culture as system proponents. If culture diffuses as a set of elements, then "coherence" becomes a puzzle, a phenomenon to be explained, so we also sketch some ideas about when cultural elements will cohere enough to appear systematic.

The major reason to favor an elements approach is that comprehensive coherence is a computational impossibility. Individuals may be driven to seek consistency in their norms, attitudes, and beliefs (Festinger, 1957), but computer scientists have noted that it is computationally impossible for an individual to ensure complete coherence among any reasonable number of elements (e.g., see Perlis, 1997). Checking a sample of, say, a couple of hundred cultural elements for consistency (e.g., "Does this cluster of 150 harmonize with that cluster of 50?") is a computational task that would tie up the thoughts of one individual for more than a lifetime. And culture is made up of thousands of elements, not hundreds.

Of course, evolution toward fit may be a process that unfolds at the cultural level as individuals across a culture jointly seek coherence. But even that cultural process is not sufficient to eliminate incoherence: Which is true? "Out of sight, out of mind" or "absence makes the heart grow fonder"? If "you're never too old to learn," then why is it impossible to "teach an old dog new tricks"?

Anthropologist George Peter Murdock favored a culture as systems view, but he acknowledged that the process of coherence making was not instantaneous:

> A culture is a system in the process of achieving equilibrium by the integration of its elements. Any change disturbs the balance and initiates a process of readjustment … [that may] require years or even generations for their accomplishment. Long before an equilibrium has been attained, other changes have occurred and set in motion new processes of adjustment. … Hence cultures appear forever in a state of flux, always approaching but never achieving integration. (Carneiro, 2003, p. 180)

If cultures are always in a state of flux, then an elements approach is the best place to start.

Computational complexity excludes the possibility of a completely coherent system, but some scholars might still dispute whether elements can be defined and studied productively. Much of the historical debate between elements and systems, as in the case of the recent discussion of memes, has pivoted on the claim that it is difficult to identify elements. Bloch's critique of memes in 2000 echoes the skepticism expressed by Kroeber and Kluckhorn in the 1960s when they claimed that "no elemental units have been established."

Yet establishing elemental units is a solvable problem. In this chapter, we have summarized many rich empirical studies where scholars have studied a wide range of elements: children's names, etiquette norms, fairy tales, jokes, scientific theories (and scientists' accounts of scientific findings), supernatural beings, skirt lengths, and social stereotypes. There are other domains of elements research that we didn't have room to cover in this chapter. Folklorists have been systematically studying elements for over one hundred years—ranging from fairy tales that were collected by the Grimm brothers in 1812 to contemporary legends (Dégh, 1971; Fine, 1980) and folk medical cures (Brown, 1952–1964). Psychologists have studied rumors (Allport & Postman, 1947; Knapp, 1944; Rosnow, 1980), sociologists have studied moral panics (Goode & Ben-Yehuda, 1994), and political scientists have studied bogus public policy factoids (O'Neill, 2006). There is also an extensive literature on the diffusion of cultural practices ranging from whole-language reading instruction to the multidivisional form (see the hundreds of studies reviewed in Rogers's book, the *Diffusion of Innovation*, 1995). The studies we review in this chapter are notable because they have elucidated interesting mechanisms that lead certain cultural elements to be selected, but there should be no serious question about whether elements exist and can be studied.

True, in many cases, the right elements may be tough to pick: What is an "element"? The tale of Snow White? Or the motif of a magic potion that features so prominently in that tale (for this critique, see Bloch, 2000; Chick, 2001)? This question makes empirical research more complicated, but it certainly doesn't make it impossible. Elements can be defined at multiple levels, all of which may be interesting. Norenzayen, Atran et al. (2006) argued that successful folktales depend on an optimal number of counterintuitive motifs, and they find that it was easy enough to tally the relative frequency of elements at two different levels: motifs and tales. Lieberson (2000) in his work on names found cycles of fashion happen at the levels of names (e.g., the rise and fall of David or Nathan as boys' names) and also at the level of sounds within names (e.g., the "ee" sound in Nancy and Kathie), demonstrating that a similar process occurs at two different levels of defining elements. How should researchers define a cultural element? They should define it at a level of analysis that allows them to elucidate an interesting selection mechanism.

But if culture comes in the form of discrete elements, then the presence of coherence in some circumstances becomes a puzzle to be explained. When and how do cultural elements collect in semicoherent or systematic groups? For insight, we might look to psychological research on self-concepts where a similar question about coherence arises. Individuals know many things about their "self"; they have beliefs and preferences and attitudes, some of which may be coherent. Research on self-concept finds that elements of the self-concept are more likely to cohere on dimensions of the self that people rate as important and that they endorse with high certainty. These core aspects of self-concept cohere more closely with other thoughts, beliefs, and attitudes than aspects of the self that are regarded as less important (Markus, 1977; Petersen, Stahlberg, & Dauenheimer, 2000; Segal, Hood, Shaw, & Higgins, 1988).

We propose a similar process that forces cultural elements into greater coherence on dimensions that are at the *cultural core*, that is, when elements relate to important cultural dimensions (Hofstede, 2001; Triandis, 1995), values (Schwartz, 1992, 1994), or core cultural ideas (Fiske, Kitayama, Markus, & Nisbett, 1998). Cross-cultural research suggests that each culture converges on different solutions to fundamental adaptive problems—such as balancing individual autonomy and group needs—and core cultural practices reflect these solutions. As a result, cultural variants that are highly relevant to the cultural core evolve under greater pressures toward consistency. Boyd, Bogerhoff-Mulder, Durham, and Richerson (1997) provided an insightful survey of a range of possibilities for modeling cultural descent ranging from culture as a system to culture as a set of elements. In their survey, they discussed the logical and empirical evidence in anthropology for a core-periphery model. Elements that are relevant to the cultural core are more likely to be selectively retained if they agree with the other elements of the cultural core, and rejected otherwise.

For example, the cultural core in Western Europe and America emphasizes personal freedom and independence, and cultural values celebrate self-assertion, personal success, and self-oriented desires (Schwartz, 1994; Schwartz & Bardi, 2001). In turn this cultural core affects the evolution and fit of numerous institutions and practices (see, e.g., Fiske et al., 1998). The U.S. legal system is based on premises of liberty and free will, and its educational system affords unique personal choices. In European American culture, cultural variants that involve how people think about their independent selves (e.g., parenting advice, kids' books, popular music lyrics, advertisements) are more likely to evolve toward fit with core cultural ideas. Cultural and cross-cultural psychologists have accumulated evidence of coherence between measures of culture and psychological functions (Singelis, Bond, Sharkey, & Lai, 1999; Snibbe & Markus, 2005; Tsai, Levenson, & McCoy, 2006) and consistent patterns across diverse range of phenomena (e.g., Kitayama, Duffy, Kawamura, & Larsen, 2003; Masuda & Nisbett, 2001, 2006) because they tend to do their prospecting for coherence around issues at the cultural core.

European American child-rearing practices emphasize nurturing a child's individuality and independence, reflected in daily practices, such as encouraging children to sleep on their own and to make their own choices of snacks, toys, and clothes. European Americans would be less likely to use and transmit a new practice, say cosleeping in a "family bed," that seems inconsistent with other elements of their cultural core. There is a small group of people in America who advocate cosleeping between parents and children, a practice that is the cultural norm in India (see Shweder, 2003), but the advocates of cosleeping in America have faced an uphill battle. In general, elements that are relevant to the cultural core face an additional layer of culture-based selection on top of the cognitive and emotional selection processes we highlight in this chapter, and this additional layer of culture-based selection may drive elements toward more coherence. But coherence testing is a difficult process, and coherence is more likely to be the exception than the rule.

Indeed, not all customs, practices, ideas, and products are relevant for the cultural core. Counting rhymes, scientific factoids, cold remedies, rumors, or baby names may evolve mostly through selection pressure at the psychological level, not the cultural one. For example, research has shown that in the modern United States and England, fashionable baby names reflect a psychological preference for novelty rather than cultural coherence with the religious beliefs of parents or the idea that names should be inherited from preceding family generations (Lieberson, 2000).

In general, we predict that in areas that are close to the cultural core, cultural elements are more likely to face selection pressure in terms of cultural fit (e.g., elements such as stories or proverbs may be selected based on their consistency with a widespread cultural schema about agency). But in areas that are more peripheral, selection pressures are more likely to feature only basic psychological processes (e.g., stories or proverbs may be selected based on their ability to evoke concrete images or provide an emotional kick).

CONCLUSION

In this chapter we have argued that it is worth viewing cultural evolution as a process of variation and selective retention. This approach has potential lessons for several fields. Our emphasis on variation is important. Previous research has focused on sources of variation, such as forgetting, that are mostly accidental, but much cultural variation is conscious and strategic, and it would be worth understanding more about these strategic, conscious sources of variation. And some fields need to discover variation in the first place. Classic research on diffusion (e.g., summarized by Rogers, 1995) is quite consistent with a culture as elements approach but has glossed over variation. Diffusion studies typically code adoption dichotomously, as present or absent, without acknowledging that people may be adopting variants of a practice.

It's also important to look at selective retention. Folklorists, for example, don't. They are quite aware of variation, and indeed they often define folklore as aspects of culture that "exist in variants." But folklorists tend to be obsessed with the range of variation, listing one of every subgenre of a particular folktale or folk medical cure, without providing any evidence about which variants are more or less common. Without knowing which variants are common, we aren't in a good position to understand mechanisms of retention. Research on mechanisms is rapidly emerging. We know relatively more about cognitive mechanisms than other mechanisms such as emotion, but it is worth knowing more about all. By understanding which elements of culture are more likely to be selected, we can understand more about different cultures and about deep universal human truths.

REFERENCES

Aaker, J., Benet-Martínez, V., & Garolera, J. (2001). Consumption symbols as carriers of culture: A study of Japanese and Spanish brand personality constructs. *Journal of Personality and Social Psychology*, *81*(3), 492–508.

Allport, G. W., & Postman, L. J. (1947). *The psychology of rumor*. New York: Holt, Rinehart & Winston.

Aunger, R. (2002). Exposure versus susceptibility in the epidemiology of everyday beliefs. *Journal of Cognition and Culture*, *2*, 113–154.

Barbas, S. (2003). "I'll take chop suey": Restaurants as agents of culinary and cultural change. *Journal of Popular Culture, 36,* 669–686.

Barrett, J. L., & Nyhof, M. A. (2001). Spreading non-natural concepts: The role of intuitive conceptual structures in memory and transmission of cultural materials. *Journal of Cognition and Culture*, *1*, 69–100.

Baum, J. A. C., & Singh, J. V. (Eds.). (1994). *Evolutionary dynamics of organizations*. New York: Oxford University Press.

Bellezza, E. S., & Bower, G. H. (1981). Person stereotypes and memory for people. *Journal of Personality and Social Psychology*, *41*, 856–865.

Benet-Martínez, V., Leu, J., Lee, F., & Morris, M. (2002). Negotiating biculturalism: Cultural frame-switching in biculturals with "oppositional" vs. "compatible" cultural identities. *Journal of Cross-Cultural Psychology*, 33, 492–516.

Berger, J., Heath, C., & Ho, B. (2006). *Divergence in cultural practices: Tastes as signals of identity*. Manuscript submitted for publication.

Blackmore, S. (1999). *The meme machine*. Oxford, UK: Oxford University Press.

Bloch, M. (2000). A well-disposed social anthropologist's problems with memes. In R. Aunger (Ed.), *Darwinizing culture: The status of memetics as a science* (pp. 189–204). Oxford, UK: Oxford University Press.

Boyd, R., Bogerhoff-Mulder, M., Durham, W. H., & Richerson, P. J. (1997). Are cultural phylogenies possible? In P. Weingart, S. D. Mitchell, P. J. Richerson, & S. Maasen (Eds.), *Human by nature* (pp. 355–386). Mahwah, NJ: Lawrence Erlbaum.

Boyd, R., & Richerson, P. J. (1985). *Culture and the evolutionary process*. Chicago: University of Chicago Press.

Boyd, R., & Richerson, P. J. (2000). Memes: Universal acid or a better mousetrap? In R. Aunger (Ed.), *Darwinizing culture: The status of memetics as a science* (pp. 143–162). Oxford, UK: Oxford University Press.

Brewer, M. B. (1991). The social self: On being the same and different at the same time. *Personality and Social Psychology Bulletin*, *17*, 475–482.

Brown, F. C. (1952–1964). *The Frank C. Brown collection of North Carolina folklore* (Vols. 1–7). Durham, NC: Duke University Press.

Brumann, C. (1999). Writing for culture: Why a successful concept should not be discarded. *Current Anthropology*, *40*, S1–S13.

Burke, A., Heuer, F., & Reisberg, D. (1992). Remembering emotional events. *Memory and Cognition*, *20*, 277–290.

Burt, G. (1997). Cultural convergence in historical cultural space-time. *Journal of Cultural Economics*, *21*, 291–305.

Campbell, D. T. (1960). Blind variation and selective retention in creative thought as in other knowledge processes. *Psychological Review*, *67*, 380–400.

Campbell, D. T. (1997). From evolutionary epistemology via selection theory to a sociology of scientific validity. *Evolution and Cognition*, *3*, 5–38.

Carneiro, R. (2003). *Evolutionism in cultural anthropology: A critical history*. Boulder, CO: Westview.

Chick, G. (2001). Culture-bearing units and the units of culture: An introduction. *Cross-Cultural Research*, *35*, 91–108.

Cho, B., Kwan, U., Gentry, J. W., Jun, S., & Kropp, F. (1999). Cultural values reflected in theme and execution: A comparative study of U.S. and Korean television commercials. *Journal of Advertising*, *28*(4), 59–73.

Christianson, S. A., & Loftus, E. F. (1991). Remembering emotional events: The fate of detail information. *Cognition and Emotion*, *5*, 81–108.

Cloninger, C. R. (1987). A systematic method for clinical description and classification of personality variants. *Archives of General Psychiatry*, *44*, 573–588.

Collard, M., & Shennan, S. J. (2000). Ethnogenesis versus phylogenesis in prehistoric culture change: A case-study using European Neolithic pottery and biological phylogenetic techniques. In C. Renfrew & K. Boyle (Eds.), *Archaeogenetics: DNA and the population prehistory of Europe* (pp. 89–97). Cambridge, UK: McDonald Institute for Archaeological Research.

Collard, M., Shennan, S. J., & Tehrani, J. (2006). Branching, blending, and the evolution of cultural similarities and differences among human populations. *Evolution and Human Behavior*, *27*, 169–184.

Cowan, A. (2003). Cultural traffic in Lübeck and Danzig in the sixteenth and seventeenth centuries. *Scandinavian Journal of History*, *28*, 175–185.

D'Andrade, R. (1995). *The development of cognitive anthropology*. Cambridge, UK: Cambridge University Press.

Davidson, D., Larson, S. L., Luo, Z., & Burden, M. J. (2000). Interruption and bizarreness effects in the recall of script-based text. *Memory*, *8*, 217–234.

Davis, M.S. (1971). That's interesting! Towards a phenomenology of sociology and a sociology of phenomenology. *Philosophy of the Social Sciences*, *1*, 309–344.

Dawkins, R. (1976). *The selfish gene*. Oxford, UK: Oxford University Press.

Dégh, L. (1971). The "belief legend" in modern society: Form, function, and relationship to other genres. In W. D. Hand (Ed.), *American folk legend: A symposium* (pp. 55–68). Berkeley: University of California Press.

DiMaggio, R. J. (1997). Culture and cognition. *Annual Review of Sociology*, *23*, 263–287.

Dunbar, K. (1995). How scientists really reason: Scientific reasoning in real-world laboratories. In R. J. Sternberg & J. Davidson (Eds.), *Mechanisms of insight* (pp. 365–395). Cambridge, MA: MIT Press.

Dunbar, K., Fugelsang, J., & Stein, C. (in press). Do naïve theories ever go away? In M. Lovett & P. Shah (Eds.), *Thinking with data: 33rd Carnegie symposium on cognition.* Mahwah, NJ: Lawrence Erlbaum.

Edmonds, B. (2002). Three challenges for the survival of memetics. *Journal of Memetics*, *6*.

Edmonds, B. (2005). The revealed poverty of the gene-meme analogy: Why memetics per se has failed to produce substantive results. *Journal of Memetics*, *9*.

Fessler, D. M. T., & Navarrete, C. D. (2003). Meat is good to taboo: Dietary proscriptions as a product of the interaction of psychological mechanisms and social processes. *Journal of Cognition and Culture*, *3*, 1–40.

Festinger, L. (1957). *A theory of cognitive dissonance.* Stanford, CA: Stanford University Press.

Fine, G. A. (1980). The Kentucky fried rat: Legends and modern society. *Journal of the Folklore Institute*, *17*, 222–243.

Fiske, A. R., Kitayama, S., Markus, H. R., & Nisbett, R. E. (1998). The social matrix of social psychology. In D. Gilbert, S. Fiske, & G. Lindzey (Eds.), *Handbook of social psychology* (pp. 915–981). New York: McGraw-Hill.

Foltz, R. C. (1999). *Religions of the Silk Road: Overland trade and cultural exchange from antiquity to the fifteenth century.* New York: St. Martin's Press.

Friedman, A., & Montello, D. R. (2006). Global-scale location and distance estimates: Common representations and strategies in absolute and relative judgments. *Journal of Experimental Psychology: Learning, Memory, and Cognition*, *32*, 333–346.

Gardner, W. L., Gabriel, S., & Lee, A. Y. (1999). "I" value freedom but "we" value relationships: Self-construal priming mirrors cultural differences in judgment. *Psychological Science*, *10*, 321–326.

Gatherer, D. (2005). Finding a niche for memetics in the 21st century. *Journal of Memetics*, *9*.

Glassner, B. (1999). *The culture of fear.* New York: Basic Books.

Goode, E., & Ben-Yehuda, N. (1994). *Moral panics: The social construction of deviance.* Oxford, UK: Blackwell.

Guglielmino, C. R., Viganotti, C., Hewlett, B., & Cavalli-Sforza, L. L. (1995). Cultural variation in Africa: Role of mechanisms of transmission and adaptation. *Proceedings of the National Academy of Sciences, USA*, 92, 585–589.

Harber, K., & Cohen, D. (2005). The emotional broadcaster model of social sharing. *Journal of Language and Social Psychology*, *24*, 382–400.

Harris, B. (1979). "Whatever happened to Little Albert?" *American Psychologist*, *34*, 151–160.

Haselton, M. G., & Miller, G. F. (2006). Women's fertility across the cycle increases the short-term attractiveness of creative intelligence compared to wealth. *Human Nature*, *17*, 50–73.

Heath, C., Bell, C., & Sternberg, E. (2001). Emotional selection of memes: The case of urban legends. *Journal of Personality and Social Psychology*, *81*, 1028–1041.

Henrich, J., & Gil-White, F. J. (2001). The evolution of prestige: Freely conferred deference as a mechanism for enhancing the benefits of cultural transmission. *Evolution and Human Behavior*, 22, 165–196.

Hofstede, G. (2001). *Culture's consequences: Comparing values, behaviors, institutions, and organizations across nations*. Thousand Oaks, CA: Sage.

Hong, Y., Benet-Martínez, V., Chiu, C., & Morris, M. (2003). Boundaries of cultural influence: Construct activation as a mechanism for cultural differences in social perception. *Journal of Cross-Cultural Psychology*, *34*, 453–464.

Hong, Y.-Y., Chiu, C.-Y., & Kung, T. M. (1997). Bringing culture out in front: Effects of cultural meaning system activation on social cognition. In K. Leung, Y. Kashima, U. Kim, & S. Yamaguchi (Eds.), *Progress in Asian social psychology* (Vol. 1, pp. 135–146). Singapore: Wiley.

Hong, Y., Morris, M. W., Chiu, C., & Benet-Martínez, V. (2000). Multicultural minds: A dynamic constructivist approach to culture and cognition. *American Psychologist*, *55*, 709–720.

Hull, D. L. (2000). Taking memetics seriously: Memetics will be what we make it. In R. Aunger (Ed.), *Darwinizing culture: The status of memetics as a science* (pp. 43–67). Oxford, UK: Oxford University Press.

Johnson, A. W., & Price-Williams, D. (1996). *Oedipus ubiquitous: The family complex in world folk literature*. Stanford, CA: Stanford University Press.

Jordan, P., & Shennan, S. (2003). Cultural transmission, language, and basketry traditions amongst the California Indians. *Journal of Anthropological Archaeology*, *22*, 42–74.

Kashima, Y., Kokubo, T., Kashima, E. S., Boxall, D., Yamaguchi, S., & MacRae, K. (2004). Culture and self: Are there within-culture differences in self between metropolitan areas and regional cities? *Personality and Social Psychology Bulletin*, *30*, 816–823.

Keltner, D., & Haidt, J. (2003). Approaching awe, a moral, spiritual, and aesthetic emotion. *Cognition and Emotion*, *17*, 297–314.

Kim, H., & Markus, H. (1999). Deviance or uniqueness, harmony or conformity? A cultural analysis. *Journal of Personality and Social Psychology*, *77*, 785–800.

Kitayama, S., Duffy, S., Kawamura, T., & Larsen, J. T. (2003). Perceiving an object and its context in two cultures: A cultural look at New Look. *Psychological Science*, *14*, 201–206.

Kitayama, S., Ishii, K., Imada, T., Takemura, K., & Ramaswamy, J. (in press). Voluntary settlement and the spirit of independence. *Journal of Personality and Social Psychology*.

Knapp, R. H. (1944). A psychology of rumor. *Public Opinion Quarterly*, *8*, 23–37.

Kuper, A. (1999). *Culture: The anthropologist's account*. Cambridge, MA: Harvard University Press.

Lieberson, S. (2000). *A matter of taste: How names, fashion, and culture change*. New Haven, CT: Yale University Press.

Loewenstein, G., & Heath, C. (2005). The repetition-shift plot structure: A cognitive influence on selection in the marketplace of ideas. *Proceedings of the Cognitive Science Society, USA*, 1319–1324.

Loftus, E. F. (1974). On reading the fine print. *Quarterly Journal of Experimental Psychology*, *27*, 324.

Lyons, A., & Kashima, Y. (2001). The reproduction of culture: Communication processes tend to maintain cultural stereotypes. *Social Cognition*, *19*, 372–394.

Lyons, A., & Kashima, Y. (2006). Maintaining stereotypes in communication: Investigating memory biases and coherence-seeking in storytelling. *Asian Journal of Social Psychology*, *9*, 59–71.

Malinowski, B. (1929). Social anthropology. In *Encyclopaedia Britannica* (14th ed., Vol. 20, pp. 862–870). Chicago: Encyclopaedia Britannica.

Malinowski, B. (1945). *The dynamics of culture change*. New Haven, CT: Yale University Press.

Markus, H. (1977). Self-schemata and processing information about the self. *Journal of Personality and Social Psychology*, *35*, 63–78.

Markus, H. R., Uchida, Y., Omoregie, H., Townsend, S., & Kitayama, S. (2006). Going for the gold: Sociocultural models of agency in Japanese and American contexts. *Psychological Science*, *17*, 103–112.

Martindale, C. (1990). *The clockwork muse: The predictability of artistic change*. New York: Basic Books.

Masuda, T., & Nisbett, R. E. (2001). Attending holistically versus analytically: Comparing the context sensitivity of Japanese and Americans. *Journal of Personality and Social Psychology*, *81*, 992–934.

Masuda, T., & Nisbett, R. E. (2006). Culture and change blindness. *Cognitive Science*, *30*, 381–399.

Mesoudi, A., & Whiten, A. (2004). The hierarchical transformation of event knowledge in human cultural transmission. *Journal of Cognition and Culture*, *4*, 1–24.

Mesoudi, A., Whiten, A., & Laland, K. N. (2004). Is human cultural evolution Darwinian? Evidence reviewed from the perspective of *The Origin of Species*. *Evolution*, *58*, 1–11.

Meyer, W.-U., Reisenzein, R., & Schützwohl, A. (1997). Towards a process analysis of emotions: The case of surprise. *Motivation and Emotion*, *21*, 251–274.

Miller, G. F. (1999). Sexual selection for cultural displays. In R. Dunbar, C. Knight, & C. Powers (Eds.), *The evolution of culture* (pp. 71–91). Edinburgh: Edinburgh University Press.

Miller, G. F. (2000). *The mating mind: How sexual choice shaped the evolution of human nature*. New York: Doubleday.

Miyamoto, Y., Nisbett, R. E., & Masuda, T. (2006). Culture and the physical environment: Holistic versus analytic perceptual affordances. *Psychological Science*, *17*(2), 113–119.

Mizruchi, M. S., & Fein, L. C. (1999). The social construction of organizational knowledge: A study of the uses of coercive, mimetic, and normative isomorphism. *Administrative Science Quarterly*, *44*, 653–683.

Moore, C. C., & Romney, A. K. (1994). Material culture, geographic propinquity, and linguistic affiliation on the North Coast of New Guinea: A reanalysis of Welsch, Terrell, and Nadolski (1992). *American Anthropologist*, *96*, 370–396.

Morgan, L. H. (1877). *Ancient society*. New York: Holt.

Neuschatz, J. S., Lampinen, J. M., Preston, E. L., Hawkins, E. R., & Toglia, M. P. (2002). The effect of memory schemata on memory and the phenomenological experience of naturalistic situations. *Applied Cognitive Psychology*, *16*, 687–708.

Nichols, S. (2002). On the genealogy of norms: A case for the role of emotion in cultural evolution. *Philosophy of Science*, *69*, 234–255.

Norenzayan, A., & Atran, S. (2004). Cognitive and emotional processes in the cultural transmission of natural and nonnatural beliefs. In M. Schaller & C. Crandall (Eds.), *The psychological foundations of culture* (pp. 149–169). Hillsdale, NJ: Lawrence Erlbaum.

Norenzayan, A., Atran, S., Faulkner, J., & Schaller, M. (2006). Memory and mystery: The cultural selection of minimally counterintuitive narratives. *Cognitive Science*, *30*, 531–553.

Norenzayan, A., Schaller, M., & Heine, S. (2006). Evolution and culture. In M. Schaller, J. Simpson, & D. Kenrick (Eds.), *Evolution and social psychology* (pp. 343–366). New York: Psychology Press.

Öhman, A., Flykt, A., & Esteves, F. (2001). Emotion drives attention: Detecting the snake in the grass. *Journal of Experimental Psychology: General*, *130*, 466–478.

O'Neill, B. (2006). *Folklore of the elite: Historical myths about public policy*. Unpublished manuscript.

O'Sullivan, C. S., & Durso, F. T. (1984). Effect of schema-incongruent information on memory for stereotypical attributes. *Journal of Personality and Social Psychology*, *47*, 55–70.

Perlis, D. (1997). Sources of, and exploiting, inconsistency: Preliminary report. *Journal of Applied Non Classical Logics*, *7*, 13–24.

Petersen, L.-E., Stahlberg, D., & Dauenheimer, D. (2000). Effects of self-schema elaboration on affective and cognitive reactions to self-relevant information. *Genetic, Social, and General Psychology Monographs*, *126*, 25–42.

Pezdek, K., Whetstone, T., Reynolds, K., Askari, N., & Dougherty, T. (1989). Memory for real-world scenes: The role of consistency with schema expectation. *Journal of Experimental Psychology*, *15*, 587–595.

Plaut, V. C., Markus, H. R., & Lachman, M. E. (2002). Place matters: Consensual features and regional variation in American well-being and self. *Journal of Personality and Social Psychology*, *83*, 160–184.

Quinn, N. (1996). Culture and contradiction: The case of Americans reasoning about marriage. *Ethos*, *24*, 391–425.

Reisberg, D., & Heuer, F. (2004). Memory for emotional events. In D. Reisberg & P. Hertel (Eds.), *Memory and emotion* (pp. 3–41). New York: Oxford University Press.

Rimé, B., Mesquita, B., Philippot, P., & Boca, S. (1991). Beyond the emotional event: Six studies on the social sharing of emotion. *Cognition and Emotion*, 5, 435–465.

Rogers, E. M. (1995). *Diffusion of innovations* (4th ed.). New York: Free Press.

Rojahn, K., & Pettigrew, T. F. (1992). Memory for schema-relevant information: A meta-analytic resolution. *British Journal of Social Psychology*, *31*, 81–109.

Rosnow, R. L. (1980). Psychology of rumor revisited. *Psychological Bulletin*, *87*, 578–591.

Rothbaum, F., & Tsang, B. (1998). Love songs in the United States and China: On the nature of romantic love. *Journal of Cross-Cultural Psychology*, 29, 306–319.

Rozin, P. (1996). Towards a psychology of food and eating: From motivation to module to model to marker, morality, meaning, and metaphor. *Current Directions in Psychological Science*, 5, 18–24.

Rozin, P., Haidt, J., & McCauley, C. R. (1993). Disgust. In M. Lewis & J. M. Havlland (Eds.), *Handbook of emotions* (pp. 575–594). New York: Guilford.

Rubin, D. C. (1995). *Memory in oral traditions: The cognitive psychology of epic, ballads, and counting-out rhymes.* New York: Oxford University Press.

Schaller, M., Conway, L. G., III, & Tanchuk, T. L. (2002). Selective pressures on the once and future contents of ethnic stereotypes: Effects of the communicability of traits. *Journal of Personality and Social Psychology*, 82, 861–877.

Schwartz, S. H. (1992). Universals in the content and structure of values: Theoretical advances and empirical tests in 20 countries. In M. Zanna (Ed.), *Advances in experimental social psychology* (Vol. 25, pp. 1–65). Orlando, FL: Academic Press.

Schwartz, S. H. (1994). Beyond individualism/collectivism: New cultural dimensions of values. In U. Kim, H. C. Triandis, C. Kagitçibasi, S. C. Choi, & G. Yoon (Eds.), *Individualism and collectivism: Theory, method and applications* (pp. 85–119). Thousand Oaks, CA: Sage.

Schwartz, S. H., & Bardi, A. (2001). Value hierarchies across cultures: Taking a similarities perspective. *Journal of Cross-Cultural Psychology*, 32, 268–290.

Segal, Z. V., Hood, J. E., Shaw, B. F., & Higgins, E. T. (1988). A structural analysis of the self-schema construct in major depression. *Cognitive Therapy and Research*, 12, 471–485.

Shweder, R. A. (2003). *Why do men barbecue? Recipes for cultural psychology.* Cambridge, MA: Harvard University Press.

Simonton, D. K. (1997). Foreign influence and national achievement: The impact of open milieus on Japanese civilization. *Journal of Personality and Social Psychology*, 72, 86–94.

Singelis, T. M., Bond, M. H., Sharkey, W. F., & Lai, C. S. Y. (1999). Unpackaging culture's influence on self-esteem and embarrassability: The role of self-construals. *Journal of Cross-Cultural Psychology*, *30*, 315–341.

Snibbe, A., & Markus, H. R. (2005). You can't always get what you want: Social class, agency and choice. *Journal of Personality and Social Psychology*, *88*, 703–720.

Sperber, D. (1996). *Explaining culture: A naturalistic approach.* Oxford, UK: Blackwell.

Sperber, D. (2000). An objection to the memetic approach to culture. In R. Aunger (Ed.), *Darwinizing culture: The status of memetics as a science* (pp. 163–174). Oxford, UK: Oxford University Press.

Stangor, C., & McMillan, D. (1992). Memory for expectancy-congruent and expectancy-incongruent information: A review of the social and social-developmental literatures. *Psychological Bulletin*, *111*, 42–61.

Steward, J. (1955). *Theory of culture change: The methodology of multilinear evolution.* Urbana: University of Illinois Press.

Swidler, A. (1986). Culture in action: Symbols and strategies. *American Sociological Review*, *51*, 273–286.

Tajfel, H., & Turner, J. C. (1986). The social identity theory of inter-group behavior. In S. Worchel & L. W. Austin (Eds.), *Psychology of intergroup relations* (pp. 2–24). Chicago: Nelson-Hall.

Tehrani, J. J., & Collard, M. (2002). Investigating cultural evolution through biological phylogenetic analyses of Turkmen textiles. *Journal of Anthropological Archaeology*, *21*, 443–463.

Thornton, S. (1996). *Club cultures: Music, media and subcultural capital.* Middletown, CT: Wesleyan University Press.

Triandis, H. C. (1995). *Individualism and collectivism.* San Francisco: Westview.

Tsai, J. L., & Chentsova-Dutton, Y. (2003). Variation among European Americans in emotional facial expression. *Journal of Cross-Cultural Psychology*, *34*, 650–657.

Tsai, J. L., Levenson, R. W., & McCoy, K. (2006). Cultural and temperamental variation in emotional response. *Emotion*, *6*, 484–497.

Tylor, E. B. (1870). *Researches into the early history of mankind* (2nd ed.). London: John Murray.

Tylor, E. B. (1871). *Primitive culture.* London: John Murray.

Vincente, K. J., & Brewer, W. F. (1993). Reconstructive remembering of the scientific literature. *Cognition*, *46*, 101–128.

Welsch, R. L. (1996). Language, culture, and data on the North Coast of New Guinea. *Journal of Quantitative Anthropology*, *196*, 209–234.

White, L. (1959). *The evolution of culture.* New York: McGraw-Hill.

6

From Genes to Memes
Psychology at the Nexus

LEE A. KIRKPATRICK

A *ring species* refers to a species that, as it expands its range geographically over time, accumulates sufficient genetic change that individuals at the two ends of the range, should they come together, cannot interbreed. In between, however, neighboring populations all the way around the ring hybridize freely. The phenomenon presents a paradox: Is there one species or two (or more)?

I introduce the ring-species phenomenon as a metaphor to describe the scientific community of evolutionary minded students of *Homo sapiens*, from geneticists and evolutionary biologists at one end to anthropologists and memeticists at the other. Researchers at one end focus primarily on biological evolution and the ways it has shaped the structure and functional organization of the human mind/brain; researchers at the other end focus primarily on a different set of evolutionary processes that give rise to culture. In principle, it seems that there is likely a continuum or series of levels of analysis connecting the two ends of the ring, much as Edward Wilson (1998) described the way in which chemistry is built on (but not reducible to) physics, biology on chemistry, and so forth. When researchers from the two ends of this spectrum come together, however, they often look for the all the world like distinct species, incapable of (intellectually) interbreeding.

Over the years, a handful of intrepid researchers have worked explicitly to tie together the two ends of the ring into various kinds of dual-level evolutionary models (e.g., Blackmore, 1999; Boyd & Richerson, 1985; Cavalli-Sforza & Feldman, 1981; Durham, 1991; Lumsden & Wilson, 1981; Mesoudi, Whiten, & Laland, 2006; Richerson & Boyd, 2005) that explicitly distinguish biological evolution (of *genes*) from cultural or memetic evolution (of *memes* or *cultural variants*, etc.) within a single integrated model. Recent books by Dennett (2006) and Dawkins (2006) proposed explicit genes-plus-memes models of the origins of religion, according to which certain evolved psychological systems enable the emergence of religious beliefs, such as ideas about supernatural agents, and then cultural transmission and social learning fill in the details.

These two-level evolutionary approaches take us one important step in the right direction, but in attempting to join the ends of the ring where they come together, they often give short shrift to the spectrum of continuity or levels of analysis in between. We should know better, but we still fall routinely into the age-old "nature versus nurture" trap, arguing about whether a given phenomenon is a biological adaptation or a cultural adaptation or, relatedly, whether its emergence owes to its being "good for genes" or "good for groups." Moreover, focusing on genes on the one hand and cultural evolution on the other has led us to overlook a crucial player in between: the *individual*. In disciplinary terms, crucial areas of *psychology* have routinely been ignored in the zeal to connect evolutionary biology to anthropology. The purpose of this chapter is to call attention to these intermediate processes in between the ends of the ring and to show how their inclusion is essential to a proper understanding of the entire spectrum.

BETWEEN BIOLOGICAL EVOLUTION AND CULTURAL EVOLUTION

The success of contemporary evolutionary psychology, particularly as a qualitative advance beyond its predecessor sociobiology, is owed in large part to the insight that natural selection produces not behavior per se but specialized systems in the brain designed to interact with the environment to produce behavior (Cosmides & Tooby, 1987). Explaining behavior is not a matter of identifying its purported adaptive value per se but instead a matter of identifying the evolved *cognitive adaptations* or *psychological mechanisms*, and conditions, that produce them. This has since become the primary agenda of evolutionary psychology: identifying the functionally specialized systems that make up our species' evolved psychological architecture and specifying their respective designs. The aim of this chapter is to call attention to the fact that the path from evolution to (cognitive) adaptations, as well as that from adaptations to behavior, represents a distinct level of analysis, characterized by processes qualitatively distinct from the biological and cultural evolutionary processes that bracket them.

FROM GENES TO PSYCHOLOGICAL MECHANISMS: ONTOGENY

Although Dawkins (1976) originally likened genes to architects' blueprints, he later (1986) corrected himself and argued for the better analogy of a *recipe*. Genes represent sets of instructions for building organisms (and their adaptations), not models of the final products. Thus, biological evolution gives rise to a species-typical genome, but much happens in between the genes and the final form of adaptations (including cognitive ones). A spate of recent publications has appeared over the past few years calling attention to the crucial role of the *developmental* processes by which cognitive (and other) adaptations emerge ontogenetically within each individual (e.g., Bjorklund & Blasi, 2005; Bjorklund & Pellegrini, 2002; Boyce & Ellis, 2005; Burgess & MacDonald, 2004; Ellis & Bjorklund, 2005; Geary & Bjorklund, 2000). Space here permits mention of just a few insights from this approach that are particularly relevant to the present discussion.

First, a given genotype (evolved via natural selection) can lead to different phenotypic outcomes—for example, differential organization and/or calibration of cognitive systems—as a function of differences between current and ancestral environments. The environment in which children of modern industrial societies grow up is radically different from that of our ancestors in countless ways. For example, the alarming rates of childhood and adult obesity in the United States today are a result of a mismatch between ancestral and modern diets and physical-activity levels. It remains largely unknown how comparable mismatches with respect to social environment—such as mass media, text messaging and cell phones, and online relationship opportunities—may alter the development of social–psychological mechanisms relative to their "intended" design.

Second, many adaptations look and operate quite differently across successive stages of development, an obvious example being mechanisms related to mating and reproduction. In some cases earlier manifestations represent merely immature forms of adaptations for adulthood, but others are designed to be specific to particular developmental stages. For example, Bjorklund and Blasi (2005) reviewed research suggesting that young children's tendency to overestimate their own abilities and skills may be an adaptation specific to this development stage to encourage exploration and practice. Only later in development does it become more adaptive for self-assessments to become increasingly accurate and realistic.

Third, understanding the ontogenetic processes by which psychological mechanisms develop can provide important insights regarding their operation later in life. A widely accepted hypothesis regarding beliefs in gods and other supernatural agents, for example, is that they involve a theory-of-mind module (ToMM) designed for reasoning about other human beings, with such counterintuitive properties as immortality and omniscience added on. Barrett (2004; Barrett & Richert, 2003) reviewed evidence from developmental research suggesting that during early childhood—the time when people first begin to learn about concepts of God—such properties are actually intuitive and thus easily assimilated. The developmental task is not so much to learn that God has such properties but rather to learn that people do not.

Fourth, ontogenetic development is the context within which individual differences unfold as a function of the interaction between genes and environments. Some such individual differences reflect the *conditional adaptations* that select among alternative developmental pathways based on early experience (Boyce & Ellis, 2005). For example, children from homes characterized by high stress, inadequate resources, and harsh and rejecting parenting reach puberty earlier than their peers (Ellis & Garber, 2000) and tend toward short-term rather than long-term reproductive strategies in adolescence (Belsky, Steinberg, & Draper, 1991). Other individual differences are adaptively neutral or maladaptive (Buss & Greiling, 1999). Although such variability represents random noise from an adaptationist perspective, it is very real and functionally important for the lives of individuals and the intra- and interpersonal processes to be discussed later.

Finally, it is obvious that childhood is a crucial period of life for both individual and social learning; indeed, this is presumably an important reason why our species has evolved a life-history strategy involving such an extended childhood period (Bjorklund & Blasi, 2005). Ontogenetic development thus represents a crucial point at which the ring comes full circle, where cultural evolution feeds most directly back to individual psychology.

FROM PSYCHOLOGICAL MECHANISMS TO BEHAVIOR: COGNITION

Much as the path from genes to psychological mechanisms is complex, with both adaptive and potentially maladaptive outcomes, so too is the path from psychological mechanisms to behavior. Between the evolved psychological mechanisms designed by natural selection and the behavior eventually produced (which, in turn, provides the raw material for memetic or cultural evolutionary processes) lies an enormous universe of processes that occur within individual brains/minds. Numerous theorists have observed that at least some cognitive processes, such as individual (especially operant) learning, are very much Darwinian in nature (e.g., Plotkin, 1993; Skinner, 1966); others have argued that Darwinian principles of selection are ubiquitous in virtually all brain activity (Calvin, 1987; Edelman, 1987). Lipson (2005) demonstrated that robots can be programmed to learn how to navigate their environments using Darwinian algorithms for evolving internal models of self and environment. However conceptualized, it is clear that a tremendous amount of processing and evaluation of information goes on inside the head before any behavior is produced that might serve to transmit some part of that information to anyone else.

In the brain/mind, information perceived by the senses is processed through a series of (largely) hierarchical stages—actually, many such information-processing sequences operating simultaneously in parallel—from basic perceptual process to the eventual selection of behavior. At early stages of perception, feature-detectors in the visual system are sensitive to the presence of vertical versus horizontal versus diagonal lines, specific directions of motion, and so forth. Selected outputs from these systems feed downstream to systems designed to match these stimulus properties against stored representations of various objects. Selected outputs from these object-recognition systems are then sent to higher level systems—for example, facial-recognition systems where appropriate—and so on. In humans, matters are complicated, perhaps by orders of magnitude, by high-level cognitive processes associated with terms such as *metaphorical thinking* and *consciousness*. Only after many such stages of information processing does a particular behavior emerge, itself the result of a selection process in which one option is selected from among alternatives.

The crucial point is this: The selection criterion at none of these levels involves computation of anything like *inclusive fitness*. Our evolved psychological architecture was designed by natural selection strictly according to good-for-genes criteria, but it does not therefore follow that this is the same criterion by which these mechanisms themselves are designed to operate. For a variety of reasons, it is impossible in principle for us to have evolved cognitive valuative mechanisms capable of reliably computing inclusive-fitness effects of prospective behaviors (see, e.g., Cosmides & Tooby, 1987; Tooby & DeVore, 1987). Instead, the ultimate criterion of genetic fitness is implemented in cognitive systems that use other, highly task-specific indices as proxies for inclusive fitness—from the degree to which a stimulus contains horizontal lines to the degree it resembles one's grandmother (for related and more technical discussions, see Tooby, Cosmides, & Barrett, 2005, and Tooby, Cosmides, Sell, Lieberman, & Sznycer, in press).

The eventual outcome of these multiple levels of cognitive processing, *behavior,* involves the selection of one among myriad alternatives. Again, such selection is based on proximal criteria rather than inclusive fitness per se. One obvious such class of such criteria concerns, in effect, what is perceived or anticipated to benefit the self as an individual. Selfish genes do not necessarily build selfish organisms, but to a large extent they do. Prospective foodstuffs are evaluated in terms of what will keep the gene-carrying individual alive and well, whereas other systems implement strategies to prevent the organism from falling off cliffs, being eaten by predators, and so forth. The good-for-the-self criterion is clearly evident at another level of analysis as well: In many cases, actual or prospective behavior is evaluated based on feelings—nifty little design features that motivate certain behaviors rather than others—in the sense that some outcomes produce (or are expected to produce) unpleasant emotional states and others produce positive ones. Such psychological states obviously can directly affect the behavioral choices of no one other than the individual experiencing them.

As Hamilton (1964) demonstrated, another strategy implemented by selfish genes involves building organisms motivated to enhance the welfare of their close kin. Like many species, we possess cognitive systems for distinguishing genetic relatives from nonrelatives and for caring about and being motivated to enhance the welfare of the former (Lieberman, Tooby, & Cosmides, 2007). Tooby and Cosmides (1996) suggested that we similarly value the welfare of close friends; we form *deep-engagement* relationships with people to whom our own welfare is directly or indirectly yoked, whose investment in our welfare renders them valuable to us, and who therefore can be counted on for investment and assistance, if needed.

In the selfish-gene model of natural selection, the individual is relegated to the role of gene "vehicle"—survival machines designed by genes to facilitate their own replication (Dawkins, 1976). At the same time, models of memetics and cultural evolution tend to relegate the role of individuals to a kind of relay station for memes, their brain/mind a mere vehicle parasitized by memes to facilitate their own replication. Both of these perspectives are useful and valid for their own purposes, but they cannot be fruitfully integrated without consideration of the intermediate level of analysis that bridges them. Individuals do not matter to natural selection, nor do they matter to cultures or to memes. They do, however, matter a great deal to individuals, and it is individuals who actually make behavioral choices. This intermediate level of analysis—or, more accurately, these many intermediate levels of analysis—provides an essential bridge from genes and evolved psychology on the one hand and memes and culture on the other.

IMPLICATIONS

At a general level, the main point of this chapter is that properly understanding behavior from an evolutionary perspective is considerably more complicated than identifying the evolved psychological mechanisms characterizing the human species. A proper understanding of the multiple layers of complex, proximal processes between genes and behavior is crucial not only for completeness of our model but because these intervening processes operate according to principles and criteria that differ functionally and qualitatively from those of natural selection. It provides insights particularly into a wide variety of problems in which human behavior defies explanation in terms of the criteria of either biological or cultural evolution. For example, it really is true that people often behave in ways simply because (in proximal terms) it "feels good." Of course, a (biological) evolutionary perspective is indispensable in explaining why certain things feel good and others don't, but it is important to acknowledge that the feel-good effects themselves are functionally important for guiding behavioral choices at a proximal level.

More specific, with respect to the problem with which the chapter began—connecting the biological-evolutionary processes at one end of the ring with the cultural-evolutionary processes at the other—the crucial point is that any theory of cultural evolution must begin with a proper understanding of behavior. Individual behavior is where any model of memetic or cultural evolution must necessarily begin; no meme or cultural variant can be transmitted from one person to another in the absence of observable behavior. Although social transmission may often occur via observation, the most obvious kind of behavior contributing to such transmission is deliberate communication, especially via language. Researchers (Conway & Schaller, 2007; Schaller, 2001) recently have begun to address this crucial link between biological and cultural evolution by focusing specifically on interpersonal communication. They have argued that what people talk about—that is, socially transmit information about—tends to involve information that is evolutionarily relevant: For example, people are (by evolved design) naturally disgusted by certain kinds of things and as a consequence tend to talk about those things, which in turn can lead to the evolution of cultural taboos.

The perspective outlined here has at least two important implications for research on communicative behavior as the crucial link to cultural evolution. First, it raises questions about analyzing the content of communication directly in terms of "evolutionary relevance" and points to more proximal levels of analysis such as perceived relevance to the self, one's family, and one's friends. Second, and perhaps more important, it reminds us that deliberate communication is itself a form of behavior, and as such the motivations behind it should be analyzable in the same ways as other kinds of behavior. It is one thing to think about a given piece of information but quite another to choose to pass that information along to another. For example, to share valuable information to another individual expected to benefit from the information is an act of altruism. Such information might be shared with kin and close friends for their benefit, but in other cases it is likely explained in terms of other motives such as acquiring prestige or competing for mates (e.g., Miller, 2000).

Another implication of the perspective outlined here is that it raises thorny questions about the definitions of *adaptations* versus evolutionary *by-products*. The distinction is clear in simple examples, such as the heat incidentally produced by an incandescent bulb designed for the function of producing light (Buss, Haselton, Shackelford, Bleske, & Wakefield, 1998). If, however, human cognitive adaptations are designed (by natural selection) with the function of spawning multiple levels of evolutionary processes in the form of cognition—which in turn spawn yet another level of evolutionary processes in the form of cultural evolution—where does adaptation end and by-product begin? Each higher level of analysis is surely built on, enabled by, and constrained by the next lower level, but each level functions independently in a manner that is not reducible to the lower level. So, are the products of the higher order processes adaptations, or are they by-products of the first level?

The model outlined here also provides a framework for reexamining the complex question of how cultural variants arise and succeed in ways that appear beneficial to groups. It has been widely acknowledged since Williams (1966) that the conditions under which group selection is a viable force in biological evolution are highly circumscribed, although the same restrictions do not apply with equal force to cultural evolution. One way to frame this question is to ask whether cultural adaptations represent an emergent property of cultural or memetic evolution or whether they arise because our evolved psychology includes criteria for evaluating behavioral outcomes in terms of their potential benefits to group welfare—that is, a good-for-the-group valuative criterion alongside good-for-self, good-for-kin, and good-for-friends criteria in cognitive processing. The issue is much too complex to tackle here; my point is merely that the model clearly distinguishes two alternative levels of analysis at which the question might be addressed (actually three, if we include the alternative of group-selection processes in biological evolution as well; e.g., D. S. Wilson, 1975).

CONCLUSION

The purpose of this brief chapter is to show that the yawning chasm apparent between biological evolution and cultural evolution, and between researchers focused primarily on one of these versus the other, is an illusion. We may look like different species when the ends of the ring come together, but there do exist intermediate levels of analysis connecting them if we follow the ring the long way around. Genes, the units of biological evolution, are information packets containing recipes for building organisms, including cognitive adaptations; the units of cognition are the information packets in the brain being input into, processed by, and output from these cognitive adaptations, eventually giving rise to behavior; and memes, the units of cultural evolution, are the bits of information transmitted from one individual to another via this observable behavior, and they give rise to culture. All three levels are characterized by competition and selection processes: The winners in biological evolution become species-specific adaptations; the winners in cognitive evolution become behavior; and the winners in memetic evolution become culture.

Evolutionary biology has the potential to teach us much about how brains/minds are designed; anthropology and related fields have the potential to teach us much about how social behavior and culture are organized. The psychology of the individual lies at the nexus between.

REFERENCES

Barrett, J. L. (2004). *Why would anyone believe in God?* Walnut Creek, CA: AltaMira Press.

Barrett, J. L., & Richert, R. A. (2003). Anthropomorphism or preparedness? Exploring children's God concepts. *Review of Religious Research*, *44*(3), 300–312.

Belsky, J., Steinberg, L., & Draper, P. (1991). Childhood experience, interpersonal development, and reproductive strategy: An evolutionary theory of socialization. *Child Development*, *62*, 647–670.

Bjorklund, D. F., & Blasi, C. H. (2005). Evolutionary developmental psychology. In D. M. Buss (Ed.), *The handbook of evolutionary psychology* (pp. 828–850). Hoboken, NJ: John Wiley and Sons.

Bjorklund, D. F., & Pellegrini, A. D. (2002). *The origins of human nature: Evolutionary developmental psychology*. Washington, DC: American Psychological Association.

Blackmore, S. J. (1999). *The meme machine*. Oxford, UK: Oxford University Press.

Boyce, W. T., & Ellis, B. J. (2005). Biological sensitivity to context: I. An evolutionary-developmental theory of the origins and functions of stress reactivity. *Development and Psychopathology*, *17*, 271–301.

Boyd, R., & Richerson, P. J. (1985). *Culture and the evolutionary process*. Chicago: University of Chicago Press.

Burgess, R. L., & MacDonald, K. (Eds.). (2004). *Evolutionary perspectives on human development* (2nd ed.). Thousand Oaks, CA: Sage.

Buss, D. M., & Greiling, H. (1999). Adaptive individual differences. *Journal of Personality*, *67*, 209–243.

Buss, D. M., Haselton, M. G., Shackelford, T. K., Bleske, A. L., & Wakefield, J. C. (1998). Adaptations, exaptations, and spandrels. *American Psychologist*, *53*, 533–548.

Calvin, W. H. (1987). The brain as a Darwin machine. *Nature*, *330*, 33–34.

Cavalli-Sforza, L. L., & Feldman, M. W. (1981). *Cultural transmission and evolution: A quantitative approach*. Princeton, NJ: Princeton University Press.

Conway, L. G., III, & Schaller, M. (2007). How communication shapes culture. In K. Fiedler (Ed.), *Social communication* (pp. 107–127). New York: Psychology Press.

Cosmides, L., & Tooby, J. (1987). From evolution to behavior: Evolutionary psychology as the missing link. In J. Dupre (Ed.), *The latest on the best: Essays on evolution and optimality* (pp. 277–306). Cambridge, MA: MIT Press.

Dawkins, R. (1976). *The selfish gene*. New York: Oxford University Press.

Dawkins, R. (1986). *The blind watchmaker*. New York: W. W. Norton.

Dawkins, R. (2006). *The God delusion*. Boston: Houghton Mifflin.

Dennett, D. C. (2006). *Breaking the spell: Religion as a natural phenomenon*. New York: Viking.

Durham, W. H. (1991). *Coevolution: Genes, culture and human diversity*. Stanford, CA: Stanford University Press.

Edelman, G. M. (1987). *Neural Darwinism: The theory of neuronal group selection*. New York: Basic Books.

Ellis, B. J., & Bjorklund, D. F. (Eds.). (2005). *Origins of the social mind: Evolutionary psychology and child development*. New York: Guilford.

Ellis, B. J., & Garber, J. (2000). Psychosocial antecedents of variation in girls' pubertal timing: Maternal depression, stepfather presence, and marital and family stress. *Child Development, 71*, 485–501.

Geary, D. C., & Bjorklund, D. F. (2000). Evolutionary developmental psychology. *Child Development, 71*, 57–65.

Hamilton, W. D. (1964). The evolution of social behavior. *Journal of Theoretical Biology, 7*, 1–52.

Lieberman, D., Tooby, J., & Cosmides, L. (2007). The architecture of human kin detection. *Nature, 445*, 727–731.

Lipson, H. (2005). Evolutionary design and evolutionary robotics. In Y. Bar-Cohen (Ed.), *Biomimetics: Biologically inspired technologies* (pp. 129–155). London: CRC Press.

Lumsden, C. J., & Wilson, E. O. (1981). *Genes, mind, and culture: The coevolutionary process*. Cambridge, MA: Harvard University Press.

Mesoudi, A., Whiten, A., & Laland, K. N. (2006). Towards a unified science of cultural evolution. *Behavioral and Brain Sciences, 29*(4), 329–383.

Miller, G. F. (2000). *The mating mind: How sexual choice shaped the evolution of human nature*. New York: Doubleday.

Plotkin, H. (1993). *Darwin machines and the nature of knowledge*. Cambridge, MA: Harvard University Press.

Richerson, P., & Boyd, R. (2005). *Not by genes alone: How culture transformed human evolution*. Chicago: University of Chicago Press.

Schaller, M. (2001). Unintended influence: Social-evolutionary processes in the construction and change of culturally-shared beliefs. In J. Forgas & K. Williams (Eds.), *Social influence: Direct and indirect processes* (pp. 77–93). Philadelphia: Psychology Press.

Skinner, B. F. (1966). The phylogeny and ontogeny of behavior. *Science, 153*, 1205–1213.

Tooby, J., & Cosmides, L. (1996). Friendship and the banker's paradox: Other pathways to the evolution of adaptations for altruism. *Proceedings of the British Academy, 88*, 119–143.

Tooby, J., Cosmides, L., & Barrett, H. C. (2005). Resolving the debate on innate ideas: Learnability constraints and the evolved interpenetration of motivational and conceptual functions. In P. Carruthers, S. Laurence, & S. Stich (Eds.), *The innate mind: Structure and content* (pp. 305–337). New York: Oxford University Press.

Tooby, J., Cosmides, L., Sell, A., Lieberman, D., & Sznycer, D. (in press). Internal regulatory variables and the design of human motivation: A computational and evolutionary approach. In A. J. Elliot (Ed.), *Handbook of approach and avoidance motivation*. Mahwah, NJ: Lawrence Erlbaum.

Tooby, J., & DeVore, I. (1987). The reconstruction of hominid behavioral evolution through strategic modeling. In W. G. Kinzey (Ed.), *Primate models for the origin of human behavior* (pp. 183–237). New York: SUNY Press.

Williams, G. C. (1966). *Adaptation and natural selection: A critique of some current evolutionary thought*. Princeton, NJ: Princeton University Press.

Wilson, D. S. (1975). A general theory of group selection. *Proceedings of the National Academy of Sciences, 72*, 143–146.

Wilson, E. O. (1998). *Consilience: The unity of knowledge*. New York: A. A. Knopf.

Section *II*

Evolutionary Bases of Cultural Phenomena

7

Exploring the Evolutionary Foundations of Culture
An Adaptationist Framework

STEVEN W. GANGESTAD

In the 1970s, a single sentence uttered on a university campus and containing both the word *evolution* and the word *culture* most likely referred to opposing explanations of the same outcome—for instance, a contrast between a view that a behavior is an outcome of genetic evolution (as a sociobiologist might have claimed) and one that behavior is the outcome of a cultural process. During the 1980s, such a sentence may well have spoken of a particular form of evolution: cultural evolution. According to dual inheritance theory, two forms of evolution affecting behavior can be distinguished: genetic evolution, which consists of changes in gene frequencies within a population across time, and cultural evolution, which consists of changes in culturally transmitted information over time (e.g., Boyd & Richerson, 1985; Dawkins, 1976; Durham, 1991; Lumsden & Wilson, 1981; Richerson & Boyd, 2005). Furthermore, both can be modeled within a population genetic framework (Boyd & Richerson, 1985). Though these forms of evolution do not proceed entirely independently of one another (so that, for instance, changes in culture can affect selection and evolution of gene frequencies), the processes are distinct from one another (see, for instance, Laland & Brown, 2002). And one can argue that one or the other is a more potent cause of behavior within a specific domain.

In the latter half of the 1990s and continuing into the current century, theorists have increasingly turned their attention to another topic, one that embraces simultaneously the importance of both (genetic) evolution and culture. Modern humans are an unusual species in that human groups possess what we refer to as culture. Though other species too may possess elements of culture (e.g., Dindo, Thierry, & Whiten, 2008; Laland & Hoppitt, 2003; Lycett, Collard, & McGrew, 2007; Rendell

& Whitehead, 2001; Sakura, 1989; Whiten et al., 2007), the extent and qualities of human culture may be unique within the biological world. Presumably, genetic evolution occurring within the hominin lineage led to the emergence of traits that led modern humans to produce culture. What precise evolutionary changes led humans to this point? Answers to this question purportedly may tell us much about the nature of human cultural phenomena. But what is more fundamental, they may tell us much about ourselves and the nature of the modern human species.

Another way of putting this is that culture itself is a biological phenomenon. This statement is not meant to be provocative. As the strict meaning of the word *biological* is simply "pertaining to a life-form," the statement merely means that human culture is something that a life-form produces. Recognition of this fact, however, does lead to the question of what evolutionary sequence of change in this life-form over time led it to produce that which we refer to as culture.

My goals for this chapter are fairly modest. I do not answer the question of what evolutionary changes led us to produce culture. I don't know that answer, and those who think they have some good ideas about it might want to describe them in a book, not a chapter. Instead, I merely lay out one framework for approaching the issue. That is, I lay out the sorts of specific questions that one would want to answer to address the larger matter at stake, the evolutionary changes that led to human culture. In so doing, I'm not proposing anything particularly novel. Much of what I have to say is at least implicit (often explicit) in analyses of the evolution of components of culture (e.g., Richerson & Boyd, 2005; Sperber & Hirschfeld, 2004; Tomasello, 2000; Tomasello, Carpenter, Call, Behen, & Moll, 2005; Tooby & Cosmides, 1992). It may nonetheless be useful to offer an account of this framework here.

CULTURE IS NOT ONE THING

What is culture? A common definition is something along the lines of a system of shared beliefs, values, customs, behaviors, and artifacts that the members of a society use to interact with their world and with each other and that are transmitted from generation to generation through learning. Though this definition may imply that culture is "a thing," multiple cultural phenomena can be discriminated. (And, indeed, some definitions of culture emphasize yet other features. In the 1950s, Kroeber and Kluckholn [1952] famously identified 164 different definitions of culture.) Hill (2007) recently proposed that three features characterize human culture. First, culture involves social transmission. Knowledge, skills, beliefs, and practices that one individual holds or emits can be transmitted, via social learning processes (with or without the use of language), to other individuals of a group. Second, culture involves normative prescription: regulation of behavior through social norms. Every known culture has rules that prescribe how people should behave; many are explicitly crystallized in language, and some are tacit. Notable among these norms are rules that pertain to marital and sexual relations and those that regulate aggressive or injurious behavior. Third, culture involves rituals, which are behaviors that, through symbolic representation, serve to reinforce norms.

Hill's list may not be exhaustive; perhaps there exists aspects of human culture not fully captured by the concepts of social transmission, normative prescription, and ritual (see also Boesch & Tomasello, 1998).

That these different aspects of human culture can and must be discriminated is made clear when we ponder the question of whether humans alone in the animal kingdom possess culture (Hill, 2007). The answer appears to depend on what we mean by culture. Some species possess well-developed forms of observational learning and hence the potential for social transmission of skills (e.g., Wrangham, McGrew, de Waal, & Heltne, 1996). When scholars refer to "animal culture" (of the nonhuman variety), they almost always refer to within-group spread of practices or skills through social transmission (e.g., Laland & Hoppitt, 2003; Lycett et al., 2007; Sakura, 1989). Individuals of some species moreover possess notions of fair distribution of resources (on capuchin monkeys, see Brosnan & de Waal, 2003). Nonetheless, no extant nonhuman species appear to possess elaborate forms of norm regulation or conduct of rituals whose function is to reinforce norms (Hill, 2007).

THE EVOLUTION OF CULTURE: HOW WE MIGHT FRAME QUESTIONS

The Evolution of Culture Involved the Evolution of Features That Gave Rise to Cultural Phenomena

Again, the premise that human culture is indeed a phenomenon that itself evolved gives rise to the question of how it evolved: What were the deep time historical events that led to the evolution of human culture? And what might those deep-time historical events tell us about the nature of culture and, more generally, human psychological life? I certainly don't mean to imply that this is the only meaningful question pertaining to evolutionary processes involved in culture. As I've already noted, for instance, the recognition that cultural practices and beliefs can evolve through processes that can be modeled in ways very similar to how population geneticists model genetic evolution has led to important insights into gene-culture coevolution (e.g., Laland, Odling-Smee, & Feldman, 2000).

When we ask about what caused the evolution of human culture—that is, the emergence of cultural phenomena in the hominin lineage—we should try to be as precise as we can be about what we are asking. Culture itself probably did not evolve because of selection for culture per se or even its components. Rather, individuals possess traits that make cultural phenomena possible. A fruitful starting point to addressing the question of what deep-time evolutionary events led to the evolution of culture is to identify those individual traits and then attempt to understand the processes that led to the evolution of those individual traits.

Consider, for instance, social transmission. Social transmission requires forms of social learning on the part of recipients of socially transmitted skills, knowledge, or language. Human social transmission almost certainly involves multiple kinds of learning (e.g., observational learning, verbal learning). What led to the evolution of these forms of learning (some of which may have predated the hominin lineage) and the underlying individual traits that predispose them? Some social transmission may involve recipients' merely copying the effective behavior of others, with no active attempt on the part of the targets copied to teach recipients. But other forms of transmission of information from one person to another obviously do involve active teaching by recipients. What intellectual and motivational features enable active teaching in humans? And what led to the evolution of these features? Similarly, we can ask what features enable norm prescription or rituals in humans (e.g., features in individuals that enable norm setting, features that enable norm following) and then try to identify what led to the evolution of these features (see, for instance, Gintis, 2003).

This approach, then, focuses not on the evolution of culture or even its components. Rather, it attempts to identify the features that make cultural phenomena possible. It then asks what evolutionary processes led those features to evolve.

Two Different Historical Causal Forces

For a trait to newly evolve in a lineage, two different causal forces must be at work (e.g., Thornhill, 2007). First, the trait must originate. Traits typically if not exclusively newly originate because of introduction of novelty within a developmental system (e.g., a genetic mutation), which gives rise to a new outcome. Second, the trait must be maintained once it has originated. Selection is an evolutionary force that can account for the maintenance of a trait once it has originated, though other evolutionary processes can maintain traits once they have originated (e.g., genetic drift, chance variation in which genes persist in a lineage). Moreover, selection cannot account for the origin of a trait in the first place; it can act to maintain a trait only once the trait has newly originated.

An illustration between the causal forces at work in trait origin and trait maintenance is given by mammary glands (see Thornhill, 2007). All extant female mammals possess mammary glands. In some monotremes (e.g., the platypus), mammary glands are structurally quite similar to sweat glands, suggesting that the mammary glands originated through introduction of a novelty in the developmental process giving rise to sweat glands. Once originated, however, mammary glands were maintained in all mammalian lineages. Mammary glands are not identical in all mammals, of course (e.g., the precise constituents of milk delivered through them differs). Within individual mammalian lineages, unique novelties originated and were maintained. Were we able to examine the changes that led to the features of any individual species' mammary glands (e.g., human mammary glands), we would find a number of distinct points of origin along the ancestral lineage that led to that species, each corresponding to the origin of a particular feature (or set of features) that characterize its mammary glands. Similarly, any particular psychological feature that contributes to culture (e.g., capacities for language) may have multiple origin points. Hence, even though nonhuman primates do not possess fully developed human language, they surely do possess some psychological features dating to ancestors shared with humans and that ultimately were key to the capacity for language evolving in humans (e.g., Hauser, Chomsky, & Fitch, 2002).

Adaptations and By-Products

Features that possess any degree of interesting complexity were probably maintained through selection. As complex features typically impose some costs (e.g., in the currency of energy needed to develop it), genetic drift alone is unlikely to be able to maintain it. To say that selection has maintained a trait, however, does not necessarily imply that a trait was maintained because it led to benefits in reproduction and survival of its bearers. Two different kinds of traits may be maintained by selection: *adaptations* and *by-products*.

An adaptation is an individual feature that evolved because it directly led its bearers to possess fitness greater than individuals who lacked the feature. Adaptations have functions. The function of an adaptation is the benefit through which it was selected (Williams, 1966). Passerine bird wings are adaptations for the function of flight. Vertebrate eyes are adaptations for seeing. Placental release of gonadotropins appears to reflect fetal adaptations for the function of increasing the probability of being retained by the mother.

By-products (or incidental effects; also spandrels; Gould & Lewontin, 1979), by contrast, are features that evolved because they co-occurred with other features that provided fitness benefits and hence were selected along with other, directly selected features. That is, by-products tagged along with the evolution of adaptations to which they are tethered. The whiteness of bones, the redness of vertebrate blood, the belly button, and the foveal blind spot are all examples of by-products. By-products have no function, even if they happen to confer reproductive benefits. That is, a function is a benefit that led a trait to evolve. By definition, by-products did not evolve because they provided benefits. Hence, by definition they lack functions.

Both adaptations and by-products evolved through natural selection. But only adaptations were selected *for* their effects; by-products were selected *with* adaptations. Adaptations are said to have been *directly* selected. By-products evolved through *indirect* selection (Thornhill, 1990, 1997; Williams, 1966). Next, I discuss criteria biologists use to distinguish adaptations from by-products.

Complexities in Trait Evolution

The preceding description may be misleading, in that it may imply that traits can unequivocally be sorted into adaptations and by-products. If fact, if we were able to examine the evolutionary changes that ultimately led to the shaping of a trait we now observe, we would often find that matters are not nearly so simple. A feature may have most recently been shaped by selection for a particular function. But some of its aspects may have been selected for another function previously and then retained. Yet others may have been by-products of selection. Consider, for example, bird wings. Bird wings and certain feathers have almost certainly been selected for the function of flight, which permitted bird ancestors to enter an aerial niche giving particular benefits (e.g., lower rates of predation during adult life). But primordial wings and feathers may have once been selected for thermoregulation, only later being modified for the function of flight. Indeed, some feathers, in fact, appear to have retained their original function of thermoregulation. And even many feathers later modified for flight have soft, plumaceous barbs close to the surface that were selected and maintained for thermoregulation. More generally, bird wings and feathers are aggregates of features that have multiple functions (for a fuller discussion of this example, see Andrews, Gangestad, & Matthews, 2002). Despite complexities needed to accurately characterize the nature of individual traits, the distinction between adaptations and by-products is a necessary one within evolutionary biology (e.g., Williams, 1966).

Exaptations and Fortuitous Effects

Some features attain benefits after they have already evolved. That is, such a feature evolved because either it was selected for a function or it was a by-product and then attained a new benefit. Though the newly acquired, fortuitous benefit did not lead the trait to be selected (hence by definition it does not constitute the trait's function), it may help maintain the trait. Gould and Vrba (1982) referred to such features as exaptations. If a trait acquires new benefits and is then modified by selection for better producing these benefits, the trait has undergone a process of secondary adaptation. Bird wings and feathers, once adaptations for thermoregulation, presumably became useful for a weak form of flight and subsequently modified through selection—secondarily adapted—for flight. Some aspects of wings or feathers, however, may never have been modified by selection for flight but were nonetheless useful to it. These aspects remain exaptations.

Again, exaptations acquired benefits fortuitously, not because they were designed through selection for these benefits. As such luck may seem to be a rare event, one might imagine that exaptation is rare. It may be much more common than intuition suggests, however, because of a phenomenon referred to as niche selection. Organisms sometimes thrive by entering new niches, ones not occupied by competitors. Typically, an organism's initial success in a new niche is partly due to exaptation—its happening to possess features that bring benefits in the new niche and thereby allow it to enter that niche (Lewontin, 1983). After all, the organism did not evolve in that new niche and, hence, had not been shaped by certain selection pressures present within it; to be able to succeed in the niche, it had to be fortunate enough to possess traits working well it in. As niche selection is a relatively common event, so too may exaptation and secondary modification be common. Indeed, the evolution of bird wings must have involved niche selection through exaptation: Dinosaurs that happened to possess features preadapted for flight (though not having been shaped by selection for flight) could "fly" into a new niche previously unoccupied, one involving airborne feeding and predator reduction. As noted previously, of course, some bird features were subsequently modified for flight (secondarily adapted for flight), but others probably remain primary exaptations.

Understanding the Evolution of Culture Requires Understanding Features That Gave Rise to Culture as Adaptations, By-Products, and Exaptations

The approach to understanding the evolution of culture I've laid out, once again, focuses not on the evolution of culture per se or even its components. Rather, it attempts to identify the features that make cultural phenomena possible. It then asks what led to the evolution of those features. The notions of adaptation, by-product, and exaptation are the key concepts to be applied to discerning the evolution of features. To ask what led to the evolution of features underlying culture, then, leads to questions of the following sorts: What features that make cultural phenomena possible are adaptations? What are the functions of these adaptations? That is, what benefits led these features to be selected? What are the adaptations *for*? Are the cultural phenomena to which they give rise core to an understanding of these functions? Or are these phenomena by-products of these features? Are some features important to the evolution of culture in fact by-products themselves, not adaptations? If so, what are they by-products of? What adaptations are they tethered to? Are exaptation and secondary modification important to an understanding of how cultural phenomena came to be and, if so, in what ways? Has evolution into the "cultural niche" humans now occupy been facilitated by fortuitous effects on culture not adapted for culture per se but rather exapted to culture? Have some adaptations not merely only adapted *for* aspects of culture but also evolved in response *to* culture? That is, does human culture itself imply selective forces that have, through evolutionary time, effectively shaped human psychological traits (e.g., Richerson & Boyd, 2005)?

These questions have not been definitively answered to date (though, of course, answers have been conjectured). Any adequate understanding of the evolution of human culture, however, must address these questions. Indeed, within the perspective I've proposed, an understanding of the evolutionary process that led to human culture effectively consists of answers to these and related questions.

How Can Adaptations, By-products, and Exaptations Be Identified?

Once again, of the many specifiable traits that individual organisms possess, only a small subset are adaptations—traits that evolved because historically they had effects favored by natural selection. Adaptationism has been described as a meth odology for "carving" the organism into those aspects of its phenotype that have evolved because of net fitness benefits historically and nonfunctional by-products (e.g., Thornhill, 1997). In doing so, the researcher not only understands what aspects of the phenotype are indeed functional but also infers the specific nature of important selective forces that shaped the organism and thereby understands some important evolutionary events that led to the organism we now observe. That is, a researcher not only identifies adaptations but also identifies biological function, what those adaptations are *for*.

Williams (1966), often credited with offering the first systematic statements that gave direction to the modern approach of adaptationism, noted that two criteria are inadequate for claiming that a trait is an adaptation. First, as already noted, it is not sufficient to show that a trait is beneficial currently. Second, it is not sufficient to argue that a trait had past utility. Exaptations have utility but need not have evolved as a result of selection for those beneficial effects. Williams (1966) argued that the biological concept of adaptation is an onerous one and requires stringent standards of evidence. Those standards are captured by the concept of *functional* or *special design*.

Arguments of Design

A trait or constellation of traits exhibits special design for a particular function if it performs a particular function effectively and, furthermore, if it is difficult to imagine another scenario that would have led to the evolution of the trait or constellation of traits. The classic example is the vertebrate eye (see, e.g., Williams, 1992). The eye and its detailed features are effective for seeing. Furthermore, it is difficult to imagine an evolutionary scenario under which the eye would have evolved other than one in which its details were selected for their optical properties and thereby the function of sight. Thus, for instance, it's unimaginable that the eye evolved through pure mutation pressure or random drift. And it is very difficult to fathom that the eye is a nonfunctional by-product of selection. The only plausible evolutionary scenario is one in which features of the eye were favored by selection for the function of sight.

An argument of special design is an argument to the best explanation (see, e.g., Sterelny & Griffiths, 1999). In this form of argument, it is considered reasonable to accept (at least provisionally) one explanation over competing explanations if the preferred explanation explains the facts better than competitors do. The theory that the features of the eye evolved through selection for sight explains the exceptionally good fit between their properties and the function of sight. Any other theory leaves these details completely unexplained.

How Is "Good Design" Assessed?

As noted previously, a special design argument states that a feature or set of features exhibits special design for a particular function because it performs that function proficiently and, furthermore, it is difficult to imagine it arising through an alternative evolutionary process. In Williams's (1966) terms, design is recognize when a feature performs a function with sufficient *specificity*, *precision*, *efficiency*, and *economy* so as to rule out chance. Or, as he later put it, "Adaptation is demonstrated by observed conformity to *a priori* design specifications" (Williams, 1992, p. 40). As implied by this passage, a special design argument has two components: a priori design specifications and an assessment of fit to those specifications.

A priori design specifications and engineering analyses. A special design argument claims that a feature or complex of features performs a particular task well. That claim implies an understanding of what it means to perform the task well. In some instances, it can be useful to have an *engineering analysis* that reveals the kinds of devices that would be good for the function our trait is claimed to exhibit.

Evaluation of wing design illustrates how engineering analysis can shed light on biological function. Bird wings vary in shape and other characteristics—along dimensions of breadth, width, degree of camber, rigidity, and so on. One can do engineering analyses on the characteristics of flight that different kinds of wing designs facilitate (e.g., speed, soaring, hovering, diving, maneuverability, and at what flight speeds). Good designs for particular kinds of flight can then be compared with actual wing designs of different species in light of the flight characteristics their foraging patterns might demand. And, in fact, different species of birds tend to possess wings appropriate to the flight demands of their foraging niche (e.g., Norberg, 2002).

Fit to design specifications. The second component of a special design argument is an evaluation of how well the actual feature or set of features an organism possesses satisfies the a priori specifications of good design. As Williams (1992) noted, "Unfortunately those who wish to ascertain whether some attribute of an organism does or does not conform to design specifications are left largely to their own intuitions, with little help from established methodology" (p. 41). There simply are no formal rules by which to evaluate claims of fit. Ultimately, a special design argument is one about probabilities: "whether a presumed function is served with sufficient precision, economy, efficiency, etc., to rule out pure chance [i.e., any possibility other than adaptation for a particular effect] as an adequate explanation" (Williams, 1966, p. 10, bracketed information added; see also Thornhill, 1990, 1997). But the means by which investigators evaluate the possibility that pure chance is an adequate explanation are informal.

An argument from design need not claim that fit to specification is perfect. Indeed, as Williams (1992) observed, the vertebrate eye is a superb example of a feature that simultaneously exhibits design for function and is "stupidly designed" (p. 73). Were the eye intelligently designed, the retinal layers would not be inverted, with nerves and blood vessels on the inner surface of the eye, in front of the photoreceptors (giving rise to the blind spot). Despite the obvious flaws of the eye's design, it nonetheless contains many telltale signs of having been shaped through selection as an optical device. The probability that it would have the details permitting sight without selection for its optical properties cannot be estimated precisely, but it strikes the intuitions of most biologists to be minute.

The Nature of Psychological Adaptations

I've illustrated design arguments using morphological traits such as eyes and wings. In these instances, engineering analyses on design for sight or flight can be performed, and the resulting specifications of design can be compared with the structure of eyes or wings. Evolutionary psychologists, however, are faced with inferring psychological adaptations, not morphological ones. How should special design arguments about *psychological adaptation* be constructed?

Behaviors and psychological phenomena are often responses of the organism to aspects of the environment. They are effects of components of the nervous system interacting with each other or effects of the nervous system interacting with the muscular-skeletal system. Behaviors and psychological processes are like traits in that they produce effects of their own (e.g., the movement of a hand that shapes the environment to create a tool), and these effects are often functional. Psychological processes can qualify as adaptations—features of an organism shaped by selection because of their beneficial effects on the organism's fitness. To evaluate special design of psychological processes as adaptations, we need (a) a description of the psychological adaptation and (b) specifications of design of a psychological adaptation that would be good for producing a particular function.

Psychological adaptations are properties of nervous systems, and in theory it might one day be possible to describe them in terms of brain processes. (Indeed, some can probably be described at that level today.) But they can be described at a different level as well, at the information processing, cognitive, or decision rule level. Psychological processes act on information in the external or internal environment of an organism and produce behavioral responses (including ones that qualify as thinking, feeling, sensing, perceiving, preferring, and so on, as well as overt behaviors observable to others). One can describe an organism's responses or behavioral adjustments to information within its environment in terms of information processing algorithms or decision rules. Psychological theories generally try to describe psychological processes in this manner.

In the simplest of terms, a psychological adaptation might "look like" a rule of the following sort: "If environmental feature A is encountered, do X" (e.g., "If a snake is encountered, orient to it"). More complex rules add additional conditional statements (e.g., "If you are a child and you live with another child, be averse to sex with that person"). Some psychological adaptations lead to changes in behavioral responses over time (learning; e.g., "If behavior X is followed by reinforcer R in situation A, do X when in situation A again"). (See, for instance, Crawford [1998] for more discussion of the structure of psychological adaptations.) Psychological adaptations are not observed directly. They are typically what philosophers of science refer to as "dispositional" traits and must be inferred from repeated observations of individuals in relevant circumstances.

A special design argument about psychological adaptation is an argument about whether the decision rule or information processing algorithm of the alleged adaptation fits specifications of design of a psychological process that would perform a particular function well. Hence, an argument of design requires a specification of good design in addition to a description of adaptation.

Tooby and Cosmides (1992) proposed that researchers perform a "task analysis" to identify good design. This term is borrowed from Marr's (1980) usage of the term in perceptual psychology. In that context, a task analysis identifies what kinds of information available in the environment can solve a particular perceptual problem (e.g., object identification, depth perception, color constancy). Tooby and Cosmides generalized the term to refer to identification of what kinds of information would be needed to solve any adaptive problem. Hence, individuals must solve the problem of identifying siblings to avoid incest and know who to invest in relatively altruistically as one would a sibling. One possible cue is given by the Westermarck hypothesis: coresidence with another child during early life. Another possible cue is seeing one's mother (primary caretaker) breastfeed another child (though that cue would be available only to older siblings). On the basis of a task analysis, one can build hypotheses about what, specifically, kin discrimination adaptations might look like and then test those hypotheses. If one finds that individuals do indeed avoid sex with individuals with whom they co-reside during childhood, it seems reasonable to infer, based on a design argument, that this effect is due to psychological adaptations that evolved for the function of discriminating kin and avoiding incest (e.g., Lieberman, Tooby, & Cosmides, 2003).

By-products do not possess evidence for special design. Sometimes their evolutionary history can be understood, however, through special design analysis. If a trait with no apparent functional value itself appears to be linked to a feature that possesses design for a particular function, one can reasonably infer that the trait is a by-product. The belly button (and its linkage to the umbilical cord) illustrates this point.

The Problem of Exapted Learning Mechanisms

A special design argument claims that there is a sufficiently tight fit between a feature and specifications of design to solve an adaptive problem to rule out all explanations aside from natural selection for a purported function. In the case of psychological adaptation, a special problem of inference can arise, one due to learning. Learning is a process in which feedback from the environment modifies the neurological structures that give rise to behavior and cognition. Learning mechanisms are themselves adaptations that allow the organism to adaptively modulate behavior with changing environments (e.g., Crawford & Anderson, 1989). As adaptations they have functions (e.g., to learn a language, to fear a predator, etc.). By their very nature, however, learning mechanisms are somewhat flexible with respect to outcome. A learning mechanism can be so flexible that it can develop behavioral and cognitive traits that perform tasks that are not the function of the adaptation. For instance, being able to drive a car or play the stock market must in some sense represent the output of learning mechanisms that evolved for other purposes. Learning mechanisms have been exapted to new problems. Andrews et al. (2002) referred to these outcomes as outputs of an *exapted learning mechanism* (ELM).

ELMs pose a problem for special design arguments because they can lead an individual to behave in ways very consistent with specifications of good design for performing a task without any natural selection for the specific function of performing that task. Again, the ability to drive a car is a good example. So too is the ability to read. (In some sense, selection has been involved in shaping these task performances, but it is not natural selection on genes. As Skinner [1981] argued, both natural selection on gene frequencies over phylogenetic time scales and selection on behavior shaped ontogenetically through consequences involve selection. See Gould and Lewontin [1979] for a similar point.)

How can we discriminate cases in which proficiency for solving a task is due to selection, over phylogenetic time, for solving the task and cases in which proficiency is due to an ELM? In some instances, we can rule out selection over phylogenetic time, as we know that the task was not performing ancestrally. People cannot have adaptations for the functions of driving or reading because they haven't been doing those tasks long enough for selection to produce adaptations for them. This criterion is not enough, however. First, our knowledge of ancestral environments is very imperfect; we cannot always know whether a particular task was performed ancestrally. Second, we cannot rule out the possibility that our ancestors solved tasks relevant to them through ELMs. (After all, if humans still inhabit Earth 10,000 years from now and still read, they still, in all likelihood, will not have adaptations *for* the function of reading. Reading will still be achieved through an ELM.)

Andrews et al. (2002) proposed a provisional list of additional criteria that might be applied to demonstrate special design for a particular function, recognizing that not all criteria will be suitable for all adaptations:

1. *Developmental specificity and biased learning.* If a performance is achieved early, easily, and prior to other learned outcomes, special design for the performance is more likely. Children learn to speak words more readily and earlier than they learn to read (e.g., Pinker, 1994). Children learn "intuitive physics," expectations about the physical world, in ways that suggest they have not built up these expectations from repeated instances (e.g., Spelke, 1990). Individuals learn fears to specific stimuli (e.g., snakes, spiders) more readily than they learn fears to other and, currently, equally dangerous stimuli (e.g., electrical outlets; Öhman & Mineka, 2001).
2. *Mismatches with the current environment.* Some outcomes don't appear to be particularly useful in a current environment, though they may have been adaptive in an ancestral environment. Hence, people, and children in particular, exhibit cravings for foods high in sugar or fat content (e.g., Drewnowski, 1997). Eating these foods is now associated with poor health, and indeed, children are regularly exposed to models encouraging them to eat the "right" foods. Their adaptive utility in energy constrained ancestral populations, however, is understandable. These cravings may hence be likely to be outcomes of adaptations that evolved in ancestral environments.
3. *Empirical evidence difficult for an ELM to explain.* In general, any evidence that is difficult for an ELM to explain can bolster a special design argument. Developmental specificity is one kind of such evidence. But other kinds are possible. For instance, research has shown that women's mate preferences change across the ovulatory cycle. It is not at all obvious how these changes would be due to an ELM, and moreover, the constellation of preference shifts is broad (see, e.g., Gangestad & Thornhill, 2008).

As should be evident from this discussion, demonstrating that a feature exhibits special design for a particular effect, thereby ruling out all alternative to selection *for* that effect, requires evidence that goes beyond evidence for design per se. One must demonstrate design *plus* reasonably argue that the fit of the psychological process to the purported function did not arise through a broad-based learning mechanism (ELM).

Other Methodologies

Evolutionary functional analysis—the analysis of special design—is a methodology critical to the reconstruction of the events and processes that led to the evolution of particular traits. But it may be augmented by additional methodologies. Archaeological and paleontological records may leave traces that lead to hypotheses about selection pressures and exploration for design. Phylogenetic comparisons can allow understanding of when traits evolved and, through design analysis of traits in other species and phylogenetic inference, what functions those traits possessed ancestrally. Hence, for instance, this logic supports the inference that mammary glands, possessed by all mammals, first appeared as alterations of sweat glands (see Thornhill, 2007). Phylogenetic comparisons may be particularly useful to identify exaptations. A pattern of utility (even if not selection-shaped elegant design) in a modern species together with phylogenetic inference of adaptation for another function or by-product in an ancestral species can support an inference of exaptation (see Andrews et al., 2002, and commentaries; with respect to the evolution of culture, see Boesch & Tomasello, 1998).

THE ILLUSTRATION OF "SHARED INTENTIONALITY"

The foregoing discussion is general, and intentionally so. It applies generally to methodologies for assessing the evolutionary history of traits. But an illustration of how these methodologies have been and might further be applied to a trait that has been claimed to be one giving rise to human cultural phenomena is in order.

Dating to the early 1990s, Tomasello and colleagues have argued that part of what makes human culture possible is the capacity to understand others' intentional states—their perceptions, their knowledge states, their beliefs, their motives. Without the ability to understand others' intentional states, for instance, an individual cannot engage in meaningful use of human language, based in symbolic representation. Only if an individual understands that a partner can direct attention to outside objects or events can one engage in communication rooted in arbitrary symbolic representation with that other.

More recently, however, Tomasello et al. (2005) noted that intentional understanding is not sufficient to produce human culture. Other species possess forms of intentional understanding as well as culture (e.g., Laland & Hoppitt, 2003) but nonetheless do not engage in the kinds of cultural interactions that humans do. Tomasello et al. (2005) proposed that "shared intentionality"—interest in collaboration with others in pursuit of shared goals—results from human adaptations that permit cultural activities that nonhuman primates lack.

Tomasello et al. (2005) laid out a number of lines of evidence that humans possess adaptations for shared intentionality. For instance, even young infants express interest in sharing emotions with another person, engage in behavioral turn-taking, and participate in triadic engagement, in which two human partners interact with some outside object and coordination of their action requires one partner's understanding of the other's perspective (e.g., an adult and an infant rolling a ball back and forth between each other). Nonhuman apes, though possessing some remarkable abilities to infer intentional states (e.g., Hare et al., 2000), lack the abilities that even young children exhibit, even if they have had extensive histories of interactions with humans (see Tomasello et al., 2005).

So capacities for shared intentionality, Tomasello et al. (2005) argued, evolved in the hominin line since our lineage split from modern chimps, perhaps even in the past 150,000 years. But through what process? Tomasello et al. (2005) suggested that these capacities evolved as *adaptations* for the *function* of facilitating cooperative or collaborative interaction. That is, they suggested that, once originated (through developmental novelties), these capacities were maintained by selection because of benefits accrued through cooperative interaction.

How compelling is this adaptationist account? In large part, this question comes down to asking how compelling is the available evidence for design. As well, how compelling the account is depends partly on how solid the claims based on comparative data are. I'm not sufficiently expert in this area to address these issues, and this issue is beyond the scope of this chapter. The commentaries on Tomasello et al.'s (2005) target article, however, partly support as well as question the design evidence. In my mind, they ought to be read partly in that light.

CONCLUSION

Culture and genetic evolution are not alternative explanations for the same phenomena. One can pose different explanations for any particularly human practice, and explanations may entail different roles for culture and genetic evolution. Ultimately, however, cultural phenomena themselves have evolved underpinnings. A major task for cultural psychology is, in my mind, to explicate the precise developmental and psychological underpinnings of human cultural phenomena.

An appreciation of just what evolutionary outcomes—whether they be adaptations or by-products—give rise and shape to cultural processes will inform an understanding of nature of cultural phenomena. There's no denying that we learn important information through cultural practices. But in absence of a precise understanding of the processes through which we learn socially transmitted information (as well as the kinds of information we do learn through those processes), cultural explanations for behavior are bound to be inexact. The potential value of an evolutionary perspective on cultural phenomena is that it offers a principled framework that can guide inquiry into the nature of the processes that gave rise to cultural phenomena in the hominin line. I've tried here to describe the outlines of that framework and to the general kinds of questions that must be answered to explain the evolved bases of human culture.

REFERENCES

Andrews, P. W., Gangestad, S. W., & Matthews, D. (2002). Adaptationism: How to carry out an exaptationist program. *Behavioral and Brain Sciences*, *25*, 489–547.

Boesch, C., & Tomasello, M. (1998). Chimpanzee and human cultures. *Current Anthropology*, *39*, 591–614.

Boyd, R., & Richerson, P. (1985). *Culture and the evolutionary process*. Chicago: University of Chicago Press.

Brosnan, S. F., & de Waal, F. B. M. (2003). Capuchin monkeys reject unequal pay. *Nature, 425*, 297–299.

Crawford, C. B. (1998). The theory of evolution in the study of human behavior: An introduction and overview. In C. B. Crawford & D. L. Krebs (Eds.), *Handbook of evolutionary psychology* (pp. 3–41). Mahwah, NJ: Lawrence Erlbaum.

Crawford, C. B., & Anderson, J. L. (1989). Sociobiology: An environmentalist discipline? *American Psychologist, 12*, 1449–1459.

Dawkins, R. (1976). *The selfish gene.* Oxford, UK: Oxford University Press.

Dindo, M., Thierry, B., & Whiten, A. (2008). Social diffusion of novel foraging methods in brown capuchin monkeys (*Cebus apella*). *Proceedings of the Royal Society B, 275*, 187–193.

Drewnowski, A. (1997). Taste preferences and food intake. *Annual Review of Nutrition, 17*, 237–253.

Durham, W. H. (1991). *Coevolution: Genes, culture, and human diversity*. Stanford, CA: Stanford University Press.

Gangestad, S. W., & Thornhill, R. (2008). Human oestrus. *Proceedings of the Royal Society B, 275*, 991–1000.

Gintis, H. (2003). A hitchhiker's guide to altruism: Gene-culture coevolution and the internalization of norms. *Journal of Theoretical Biology, 220*, 408–418.

Gould, S. J., & Lewontin, R. C. (1979). The spandrels of San Marco: A critique of the adaptationist programme. *Proceedings of the Royal Society of London B, 205*, 581–598.

Gould, S. J., & Vrba, E. S. (1982). Exaptation: A missing term in the science of form. *Paleobiology, 8*, 4–15.

Hare, B., Call, J., Agnetta, B., & Tomasello, M. (2000). Chimpanzees know what conspecifics do and do not see. *Animal Behaviour, 59*, 771–785.

Hauser, M. D., Chomsky, N., & Fitch, W. T. (2002). The faculty of language: What is it, who has it, and how did it evolve? *Science, 298*, 1569–1579.

Hill, K. (2007). Evolutionary biology, cognitive adaptations, and human culture. In S. W. Gangestad & J. A. Simpson (Eds.), *The evolution of mind: Fundamental questions and controversies* (pp. 348–356). New York: Guilford.

Kroeber, A. L., & Kluckholn, C. (1952). Culture: A critical review of concepts and definitions. *Papers of the Peabody Museum of American Archaeology and Ethnology, 47*, 1–223.

Laland, K. N., & Brown, G. (2002). *Sense and nonsense: Evolutionary perspectives on human behaviour.* London: Oxford University Press.

Laland, K. N., & Hoppitt, W. (2003). Do animals have culture? *Evolutionary Anthropology, 12*, 150–159.

Laland, K. N., Odling-Smee, J., & Feldman, M. W. (2000). Niche construction, biological evolution, and cultural change. *Behavioral and Brain Sciences, 23*, 131–146.

Lewontin, R. C. (1983). Gene, organism and environment. In D. S. Bendall (Ed.), *Evolution from molecules to men* (pp. 273–285). Cambridge, UK: Cambridge University Press.

Lieberman, D., Tooby, J., & Cosmides, L. (2003). The architecture of human kind detection. *Nature, 445*, 727–731.

Lumsden, C. J., & Wilson, E. O. (1981). *Genes, mind, and culture: The coevolutionary process.* Cambridge, MA: Harvard University Press.

Lycett, S. J., Collard, M., & McGrew, W. C. (2007). Phylogenetic analyses of behavior support existence of culture among wild chimpanzees. *Proceedings of the National Academy of Sciences USA, 104*, 17588–17592.

Marr, D. (1980). *Vision*. New York: Freeman.

Norberg, U. M. L. (2002). Structure, form, and function of flight in engineering and the living world. *Journal of Morphology, 252*, 52–81.

Öhman, A., & Mineka, S. (2001). Fear, phobias and preparedness: Toward an evolved module of fear and fear learning. *Psychological Review*, *108*, 483–522.

Pinker, S. (1994). *The language instinct*. New York: HarperCollins.

Rendell, L., & Whitehead, H. (2001). Culture in whales and dolphins. *Behavioral and Brain Sciences*, *24*, 309–324.

Richerson, P. J., & Boyd, R. (2005). *Not by genes alone: How culture transformed human evolution*. Chicago: University of Chicago Press.

Sakura, O. (1989). Variability in contact calls between troops of Japanese macaques: A possible case of neutral evolution of animal culture. *Animal Behaviour*, *38*, 900–902.

Skinner, B. F. (1981). Selection by consequences. *Science*, *213*, 501–504.

Spelke, E. S. (1990). Principles of object perception. *Cognitive Science*, *14*, 25–56.

Sperber, D., & Hirschfeld, L. A. (2004). The cognitive foundations of cultural stability and diversity. *Trends in Cognitive Science*, *8*, 40–46.

Sterelny, K., & Griffiths, P. E. (1999). *Sex and death: An introduction to the philosophy of biology*. Chicago: University of Chicago Press.

Thornhill, R. (1990). The study of adaptation. In M. Bekoff & D. Jamieson (Eds.), *Interpretation and explanation in the study of behavior. Volume II: Explanation, evolution, and adaptation* (pp. 31–62). Boulder, CO: Westview.

Thornhill, R. (1997). The concept of an evolved adaptation. In Bock, G. R., & Cardew, G. (Eds.), *Characterizing human psychological adaptations* (pp. 4–13). New York: Wiley.

Thornhill, R. (2007). Comprehensive knowledge of human evolutionary history requires both adaptationism and phylogenetics. In S. W. Gangestad & J. A. Simpson (Eds.), *The evolution of mind: Fundamental questions and controversies* (pp. 31–37). New York: Guilford.

Tomasello, M. (2000). Culture and cognitive development. *Current Directions in Psychological Science*, *9*, 37–40.

Tomasello, M., Carpenter, M., Call, J., Behne, T., & Moll, H. (2005). Understanding and sharing intentions: The origins of cultural cognition. *Behavioral and Brain Sciences*, *28*, 675–691.

Tooby, J., & Cosmides, L. (1992). Psychological foundations of culture. In J. Barkow, L. Cosmides, & J. Tooby (Eds.), *The adapted mind* (pp. 19–136). New York: Oxford University Press.

Whiten, A., Spiteri, A., Horner, V., Bonnie, K. E., Lambeth, S. P., Schapiro, S. J., & de Waal, F. B. M. (2007). Transmission of multiple traditions within and between chimpanzee groups. *Current Biology*, *17*, 1038–1043.

Williams, G. C. (1966). *Adaptation and natural selection*. Princeton, NJ: Princeton University Press.

Williams, G. C. (1992). *Natural selection: Domains, levels and challenges*. New York: Oxford University Press.

Wrangham, R. W., McGrew, W., de Waal, F. B. M., & Heltne, P. G. (Eds.). (1996). *Chimpanzee cultures*. Cambridge, MA: Harvard University Press.

8

Teach These Souls to Fly
Supernatural as Human Adaptation

SHELDON SOLOMON, JEFF GREENBERG,
TOM PYSZCZYNSKI, FLORETTE COHEN,
and DANIEL M. OGILVIE

> A sublime ecstasy ... gave wings to the soul, and allowed it to soar from the obscure world to light and joy. (Mary Shelley, *Frankenstein*, 1818/1994, p. 120)

Fantasies of flight are a universal human longing across time and space (Ogilvie, 2004). From the first gods to the latest comic book superhero (e.g., the Flying Friar, which debuted in Britain and the United States in 2006), humans, young and old, around the world have yearned and still yearn to transcend their earthly confines—be it by levitating a few inches above ground, soaring like a bird, hurtling across the sky with planelike alacrity or more (Superman was after all "faster than a speeding bullet"), or beaming across galaxies in the blink of an eye. Why are such aerial forays so ubiquitous and psychologically uplifting? Do humans everywhere share a universal predilection for gravity-defying aerial transportation simply to avoid traffic and tolls? Or is there something more psychologically significant that underlies the lofty aspirations of the human animal?

In this chapter, we propose that flight fantasies are an archetypal example of humankind's imaginative construction of supernatural conceptions of reality in response to the awareness of, and unwillingness to accept, death. After a brief historical and geographical survey of fantasies of flight and review of literatures establishing that all human culture and religion consists of beliefs that include at least one significant violation of natural laws, we present research in support of the general proposition that reminders of fatality increase belief in the supernatural. We follow this with our own research demonstrating that flying fantasies are intensified by intimations of mortality and that imagining oneself in flight mitigates defensive reactions to mortality salience and may have therapeutic significance.

FANTASIES OF FLIGHT

Myths about flying deities are common to nearly all ancient cultures. The gods of ancient Egypt, Minoa, and Mesopotamia were often depicted with wings, and ancient Hebrews often placed wings on the seraphim and cherubim on the Ark of the Covenant. Neither the Jews nor the ancient Greeks and Romans, however, viewed wings as absolutely necessary for flight. Greek gods flew without any visible means, and biblical descriptions of angels, such as those who visited Abraham or the one who wrestled with Jacob, are wingless. According to Hindu mythology, the first elephants in the world had wings and cavorted and consorted with the clouds.

In China, there are many legends of emperors flying with wings or in chariots. As early as 2200 B.C., the emperor Shun is reported to have escaped a burning tower and flown over his territory with the aid of two large reed hats. In Northern Europe, Wayland the Smith was propelled into the sky by a shirt of feathers. In Africa, Kibaga the warrior flew invisibly over his enemies and dropped rocks on them. Flying ability plays an important role in many shamanistic traditions; often the "first" shamans appeared as shape-shifting bird messengers (Nauwald, 2004). For example, according to Cherokee folklore, First Boy and his brother Inage Utasuhi' ("The Boy Who Grew Up Wild") learned the secrets of hunting by spying on their father, Kanati ("The Lucky Hunter"), when Inage Utasuhi' "changed himself into a small bird and flew to a branch of the hollow tree. … Then the wild boy changed himself into a downy feather, floated through the wind, and landed unseen on Kanati's shoulder" (Bruchac, 1993, pp. 32–33). Analogous airborne figures are found in the mythology of nearly every civilization (even though not all flights ended well; e.g., Icarus's ignoring his father Daedalus's admonition to fly at a modest altitude, culminating in the first aerial "meltdown"). And mythic heroes with the power of flight are not just a thing of the past. Martial artists and superheroes who flout gravity continue to be enormously popular in comic books, toys, and movies. Whenever movies featuring Superman, Spiderman, Batman, or the X-Men come out, they are tremendously successful, and gravity-defying feats are prominently displayed by the superheroes; indeed, they are expected.

NATURAL AND SUPERNATURAL

In *Religion Explained: The Evolutionary Origins of Religious Thought*, Pascal Boyer (2001) observed that all cultural worldviews include a supernatural dimension in which at least one natural law is violated; for example, gods who fly or ghosts who walk through walls. But why do all cultural worldviews include at least one major counterintuitive violation of nature? Boyer and others (e.g., Bloom, 2005; Kirkpatrick, 2004; Mithen, 1996) have claimed that human beings' ubiquitous belief in supernatural phenomena has no adaptive significance whatsoever. Rather, they view supernatural beliefs as accidental: "an incidental by-product of cognitive functioning gone awry" (Bloom, 2005, p. 105)—the result of massive and enduring mental misunderstanding. For example, Mithen (1996, p. 177), in summary and support of Boyer's account, explained supernatural beliefs as:

> a mixing up of knowledge about different types of entities in the real world—knowledge which would have been "trapped" in separate cognitive domains with the Early Human mind. For example, Early Humans would have known that rocks are not born and do not die like living things. And Early Humans would also have known that people have intentions and desires, while inert nodules of stone do not. Because they had isolated cognitive domains, there was no risk of the Early Human mind getting these entities mixed up, and arriving at a concept of an inert object that is neither born nor dies, but which nevertheless has intentions and desires.

From this perspective, then, modern humans are, relative to their unbewildered ancestral forebears, a feeble and degenerate form of life who believe in immortal souls because they are confusing themselves with stones, who believe they can fly because birds and butterflies can, and who deify trees, plants, and other totem animals because they are mistakenly imbuing them with rationality and intentionality as if they were humans.

In contradistinction to this view of humankind's universal predilection for supernatural beliefs as nonadaptive by-products of dispassionate cognitive processes, terror management theory (TMT; Greenberg, Pyszczynski, & Solomon, 1986; Greenberg, Solomon, & Pyszczynski, 1997; Solomon, Greenberg, & Pyszczynski, 1991), following Ernest Becker (1962, 1971, 1973, 1975), posits that such beliefs serve the profoundly adaptive function of quelling the potentially overwhelming terror that arises from the uniquely human awareness of death. According to Becker and TMT, although humans share with all other forms of life a basic biological predisposition toward continued existence in the service of survival and reproduction, we are exceptional in our possession of cognitive capabilities that enable us to engage in abstract, symbolic thought (Deacon, 1997). Symbolization renders humans explicitly aware of their own existence (to consider themselves as objects of their own subjective inquiry) and enables them to ponder the past, imagine possible futures, and conceive of that which does not presently exist and then transform our imaginative conceptions into concrete reality—all obviously highly adaptive capacities.

These prodigious cognitive capabilities, however, also led to some problematic consequences, specifically the awareness (a) of the inevitability of death, (b) of the fact that death can occur at any time for unpredictable and uncontrollable reasons, and (c) that humans are, from a biological perspective, ephemeral creatures momentarily carrying the baton for a tiny stint in the relay race of life and thus no more durable or noteworthy than porcupines or pomegranates. These unsettling realizations conspired to render humans prone to debilitating terror that is not conducive to effective instrumental behavior: Quivering masses of biological protoplasm bathing in their own dread-induced urine do not make efficient hunters or mates.

Becker, following Rank (1941/1958) and Brown (1959), posited that humans solved the problem of death through the creation and maintenance of cultural worldviews: beliefs about the nature of reality shared by individuals in a group that provide an explanation of the origin of the universe, prescriptions for being of value, and assurances of invulnerability and immortality to those of value—either literally through various conceptions of afterlives common to almost all religions (Burkert, 1996) and/or symbolically through producing biological offspring, amassing great fortunes, producing great works of art or science, identifying with a culture that will persist after one's own demise, or reveling in identifying with nature (see Lifton, 1979, for an extended discussion of different forms of symbolic immortality).

Interestingly, the Rank-and-Becker-inspired central tenet of TMT that the uniquely human concern with death was (and is) the main psychological impetus for the formation and maintenance of culturally constructed conceptions of reality (Solomon, Greenberg, Schimel, Arndt, & Pyszczynski, 2004) has also been advanced by a number of modern evolutionary theorists (e.g., Atran, 2002; Burkert, 1996; Deacon, 1997; Langer, 1982; Premack, in Wilson, 1978).[1] Here are two of many examples:

> One of the essentially universal attributes of human culture is what might be called the mystical or religious inclination. There is no culture I know of that lacks a rich mythical, mystical, and religious tradition. And there is no culture that doesn't devote much of this intense interpretive enterprise to struggling with the very personal mystery of mortality. Knowledge of death, of the inconceivable possibility that the experiences of life will end, is a datum that only symbolic representation can impart. Other species may experience loss, and the pain of separation, and the difficulty of abandoning a dead companion; yet without the ability to represent this abstract counterfactual (at least for the moment) relationship, there can be no emotional connection to one's own future death. … What great efforts we exert trying to forget our future fate by submerging the constant angst with innumerable distractions, or trying to convince ourselves the end isn't really what it seems by weaving marvelous alternative interpretations of what will happen in "the undiscovered country" on the other side of death. (Deacon, 1997, pp. 436–437)

> Existential anxieties are by-products of evolved emotions, such as fear and the will to stay alive, and of evolved cognitive capacities, such as episodic memory and ability to track the self and others over time. For example, once you can track even the seasons—and anticipate that leaves will fall off the tree in autumn and that squirrels will bury nuts—you cannot avoid overwhelming inductive evidence favoring your own death and that of those you are emotionally bonded to. Emotions compel such inductions and make them salient, and terrifying. This is "the Tragedy of Cognition." Dying is by nature not a telic event because once the process of dying starts (from birth on) it cannot be stopped to avoid the inevitable end state. By introducing a supernatural agent, religion resolves the Tragedy of Cognition. Dying is converted into a telic event whose goal state is an extended afterlife. The result is, in part, an allaying of an otherwise recurring and interminable existential anxiety. (Atran, 2002, pp. 66–67)

From a TMT perspective, then, all cultural constructions must have a supernatural dimension to render immortality credible or, for that matter, conceivable. Consistent with this claim, biologist Robert Hinde, in *Why Gods Persist: A Scientific Approach to Religion* (1999, p. 96), argued that the primary function of symbols is to allow "counter-intuitive concepts to be taken in even though they cannot readily be assimilated to existing representations." Or as Becker put it in *Escape From Evil* (1975, p. 64),

> Culture *means* that which is super*natural*; all culture has the basic mandate to transcend the physical, to permanently transcend it. All human ideologies, then, are affairs that deal directly with *the sacredness of the individual or the group life*, whether it seems that way or not, whether they admit it or not, whether the person knows it himself or not.

EMPIRICAL ASSESSMENTS OF TMT

Over 300 experiments by independent researchers in at least 13 countries have produced findings in accord with hypotheses derived from TMT (for recent reviews, see Greenberg, Solomon, & Arndt, 2008; Pyszczynski, Solomon, & Greenberg, 2003; Solomon, Greenberg, & Pyszczynski, 2004). Most of this research is based on the *mortality salience hypothesis*, which states that if cultural worldviews and the personal significance they afford serve a death-denying function, then asking individuals to think about their own death (*mortality salience* [MS]) should increase their need for the protection normally afforded by the cultural worldview and the self-worth derived from it and, consequently, should provoke judgments and behaviors that uphold faith in that worldview and one's self-worth within the context of that worldview.

In a typical study, we tell participants entering the lab that we are studying personality traits and that they consequently would complete some standard personality assessments. Embedded in several standard personality inventories to obscure the true purpose of the study is what is described as a new projective measure consisting of two open-ended questions to render mortality momentarily salient: "Please briefly describe the emotions that the thought of your own death arouse in you" and "Jot down, as specifically as you can, what you think will happen to *you* as you physically die." Participants in control conditions complete parallel questions about other topics. Participants are then given an opportunity to evaluate others who either share their cultural worldviews or have differing cultural worldviews or to assert their self-worth.

For example, Greenberg et al. (1990, Study 1) had Christian participants rate Christian and Jewish targets (who were portrayed as quite similar except for religious background) after an MS or control induction. In the control condition there were no differences in participants' evaluations of the targets; a reminder of death in the experimental condition, however, produced increased affection for the fellow Christian target and exaggerated hostility for the Jewish target. Following an MS or control induction, Greenberg et al. (1990, Study 3) then exposed American college students to essays supposedly written by an American author who either praised or condemned the American way of life. Participants rated the author of the pro-U.S. essay more favorably than the author of the anti-U.S. essay in the control condition; in response to MS, however, this tendency was exaggerated in both directions (i.e., more positive and negative reactions to pro- and anti-U.S. authors, respectively).

Subsequent studies have demonstrated behavioral effects of MS. Ochsmann and Mathy (1994) found that German university students sat closer to a German confederate and further away from a Turkish confederate after MS, relative to a control condition in which there was no difference in physical distance as a function of the ethnicity of the confederate. McGregor et al. (1998) showed that MS produced greater physical aggression against someone who did not share one's political orientation, and Greenberg, Simon, Porteus, Pyszczynski, and Solomon (1995) demonstrated that participants were more uncomfortable sifting sand through an American flag or using a crucifix as a hammer following an MS induction. The general tenor of this and other related work is that MS increases conformity to and defense of the worldview to which the individual subscribes.

Goldenberg and colleagues (for reviews, see Goldenberg, 2005; Goldenberg, Pyszczynski, Greenberg, & Solomon, 2000) have also demonstrated that following MS, people attempt to avoid and distance themselves from activities and stimuli that remind them that they are animals. For example, participants found the physical aspects of sex less appealing after MS and an essay highlighting the similarities of humans to other animals. Similarly, Landau et al. (2006) found that MS reduced males' attraction to sexually attractive but not wholesomely attractive women. Furthermore, TMT-based MS hypotheses have been supported examining a variety of cognitive, attitudinal, and behavioral domains; for example, the desire for structure and meaning, political preferences, legal decision making, health promotion, romantic relationships, creativity, and consumerism (for a recent review, see Greenberg et al., 2008).

These and many other MS effects have been obtained using a wide variety of operationalizations of MS; for example, subliminal reminders of death (Arndt, Greenberg, Pyszczynski, & Solomon, 1997) or interviews in front of a funeral parlor (Pyszczynski et al., 1996). In addition, studies have found MS effects to be quite different from the effects of reminders of other aversive events, such as uncertainty, failure, intense pain, social exclusion, general anxieties and worries, public speaking, and paralysis.

Finally, we have developed and empirically corroborated a dual-process model of conscious and unconscious defenses to death (Pyszczynski, Greenberg, & Solomon, 1999) that provides an account of the cognitive processes that underlie MS effects (research has shown that subtle reminders of death do not engender negative affect or physiological arousal, and MS effects are not mediated by affect) and how they unfold over time, pinpointing heightened accessibility of death-related thought outside of consciousness as the cause of these effects (for a recent review, see Arndt, Cook, & Routledge, 2004). This work has also demonstrated that the function of terror management defenses is to avert the potential for anxiety engendered by heightened accessibility of death-related thought.

Consistent with the dual-process model, recent research has demonstrated that threats to a close relationship (Mikulincer, Florian, & Hirschberger, 2003), the belief that the world is just (Hirschberger, 2006; Landau et al., 2004), the righteousness of one's nation (Schimel, Hayes, Williams, & Jahrig, 2007), and the belief that humans are different from animals (Goldenberg et al., 2001) increase the accessibility of death-related thought and produce effects comparable to those obtained in response to making mortality salient.

In sum, the central claim of TMT that a substantial proportion of human behavior is motivated by concerns about death and efforts to obtain symbolic and literal immortality by confidently adhering to a cultural worldview and sustaining a sense of meaning and value from it has been corroborated by a considerable corpus of empirical research. Perhaps the primary value of this large body of work is that it explains why people are so prone to kill and die in defense of their worldviews and self-worth: These constructs provide fundamental psychological security to the animal that knows it is mortal.

BELIEF IN SUPERNATURAL AGENTS IN THE FACE OF DEATH

Recent research by Norenzayan and Hansen (2006) has established a direct causal connection between reminders of death and increased belief in the supernatural.[2] In Study 1, participants reminded of their mortality responded more positively to the questions "How religious are you?" and "How strongly do you believe in God?" than participants reminded of their favorite foods. In Study 2, following an MS or control induction (either neutral or religious prime), participants read a *New York Times* article about women in a Korean fertility clinic being twice as likely to become pregnant after being prayed for by Christians in the United States, Canada, and Australia. They then responded to questions assessing their belief that the God being prayed to by the Christians exists, that God or a higher power in general exists, that God or a higher power can answer prayers, and that the results of the experiment in the *New York Times* article offers evidence that God or a higher power can answer prayers. In accord with predictions, supernatural beliefs were higher in response to a reminder of death.

A third study then investigated if making mortality salient would increase belief in the efficacy of a supernatural agent not intrinsic to the Christian worldview, Buddha. After an MS or neutral control induction, participants read a slightly modified version of the article about the efficacy of prayer for increasing fertility in Korean women, but the article stated that instead of Christians praying from the United States, Canada, and Australia, Buddhists of different denominations in Thailand, Taiwan, and Japan performed the sacred entreaties. The same dependent measures from Study 2 were modified to allude to Buddha or a higher power rather than to God or a higher power. Results indicated greater belief in supernatural agency in response to the MS induction, suggesting that an enhanced belief in the potency of supernatural powers following a reminder of death above and beyond a Christian worldview. All but one item, however, referred to Buddha or a higher power; consequently participants may have felt their own deity had intervened rather than Buddha per se. Indeed, the one item specifically referring solely to belief in the Buddha, "The Buddha that prayer groups were praying to exists," showed no effect of MS.

Finally, Study 4 investigated whether reminders of death would increase belief in a completely alien and unfamiliar supernatural power, ancestral shamanistic spirits. After thinking about death or one of two control inductions (dental pain or neutral prime), participants read an article supposedly from the *South China Morning Post* describing the effective use of clairvoyant shamans by Russians during the cold war and then reported their faith in the power of paranormal clairvoyance, the existence of ancestral spirits, the reliability of the guidance such spirits might provide, whether the achievements of the Russian clairvoyant program constitute evidence of the efficacy of ancestral spirit guidance, and the existence of God and God's power to provide effective guidance. Although the results indicated no significant effect of MS in the overall analysis, when the religious and nonreligious samples were examined separately, there was some evidence of MS increasing belief in the supernatural in the religious sample, but there was no hint of this effect in the nonreligious sample.

Norenzayan and Hansen (2006) in their general discussion considered a number of possible explanations for the effects of MS on supernatural beliefs. They noted that their findings are inconsistent with the idea that MS leads to rejection of an alien worldview, and they proposed (pp. 184–185) that this may be because alien worldviews still validate aspects of any religious worldview—that a spirit world exists and that supernatural events do occur: "If Buddha and ancestral spirits exist after death, so may our spirits—whether we are Buddhists or Shaman or otherwise." This strikes us as a likely explanation for why MS increases belief in the supernatural: Any supernatural entities or occurrences offer the religious-worldview-validating possibility of a spirit surviving beyond death. As Becker (1962, 1971, 1973) argued, literal immortality is the most compelling basis for controlling the potential terror engendered by the awareness of mortality.

Norenzayan and Hansen also proposed that there might be some primitive first line of defense against the terror of death, psychologically distinct from symbolic terror management defenses, which involves overextending a tendency to view the world in agentic terms to believe in supernatural forces. This is an intriguing possibility, but the findings of their studies are at odds with this notion because MS did not induce increased belief specifically in Buddha in Study 3 and had no effects on supernatural belief in the nonreligious in Study 4. If this were a primitive cognitive inclination, rather than an aspect of validation of religious worldviews, why would it not be exhibited in the nonreligious?

FANTASIES OF FLIGHT

We (Cohen, Sullivan, Solomon, Greenberg, & Ogilvie, 2009) recently conducted a series of studies building on Norenzayan and Hansen's (2006) work to extend our understanding of the role of supernatural beliefs and experiences in the mitigation of existential terror. More specifically, combining Ogilvie's (2004) observation that flying fantasies are historically and cross-culturally universal with the TMT assumption that human beings are fundamentally motivated to deny death, we hypothesized that for human beings, flying is the ultimate defiance of nature, a denial of human limitation that provides a sense of invulnerability and immortality, and "a rebellion against the tyranny of time and space" (Frye, 1964, p. 30). As Freud (1919/1955, p. 242) trenchantly observed, "It is true that the statement 'All men are mortal' is paraded in textbooks of logic as an example of a general proposition; but no human being really grasps it, and our unconscious has as little use now as it ever had for the idea of its own mortality." Death is thus an unacceptable infringement on the uniquely human sense that we are entitled to live forever, resulting in what we call *existential reactance* (Brehm, 1966) and consequent efforts to reinstate our freedom by endowing ourselves with supernatural powers that confer immortality.

As Dissanayake (1992/1995, p. 176; see also in this regard Jaynes, 1976; Lakoff & Johnson, 1980) noted, there's good reason for people to equate *up* with positive outcomes (e.g., having one's spirits "lifted"; "the sky is the limit") and *down* with less favorable consequences (e.g., being "downtrodden," "crestfallen," or "deflated"):

> Being downward directed means physically "giving in" (to the forces of gravity), being inert, striving toward safety. Upward direction, in contrast, suggests … getting up, lifting, physically overcoming, making an effort, being proud and adventurous, and in general escaping the pull of gravity that decrees that states of rest must somehow involve movement down.

Consistent with this analysis, Meier and Robinson (2004) found that concepts high in the visual field are judged more quickly as good and those low in the visual field are more quickly judged as bad, and Schubert (2005) found that objects high in the visual field are viewed as more powerful than those low in the visual field. And of course heaven is generally viewed as up and hell as down. So flying is not only a denial of human limitation but a movement toward goodness, power, and eternal bliss.

Accordingly, in our first study, we hypothesized that reminders of death should increase participants' desire to fly. Following an MS or TV salience control induction, participants read the following statement:

> **Fantasies and Dreams of Flying**
>
> Some people wish that they could fly. In their daytime fantasies they sometimes think about how wonderful it would be to be able to float along with the clouds or rise higher into outer space. Some envy the ability of birds that fly alone or in graceful formation with other birds as they look down on the land beneath them. Some envision themselves flying over water, even great expanses of water, and landing on some remote island. Flight also sometimes enters into their dreams, and the sensations are so pleasant that, upon awakening, they are disappointed that they can fly only in their dreams.

Participants then responded to the following questions and statements: "To what extent did you ever have *fantasies* about being able to fly?" "To what extent have you *dreamed* about being able to fly?" "Right now, the thought of being able to fly is very attractive to me."

Consistent with our prediction, participants reminded of their mortality reported a greater desire to fly (see Figure 8.1 for a graphical depiction of these findings).[3] Thus, in addition to the Norenzayan and Hansen (2006) finding that reminders of death increase beliefs in the existence and efficacy of supernatural phenomena, the findings of this study demonstrate that MS also engenders a desire to actually engage in supernatural experiences.

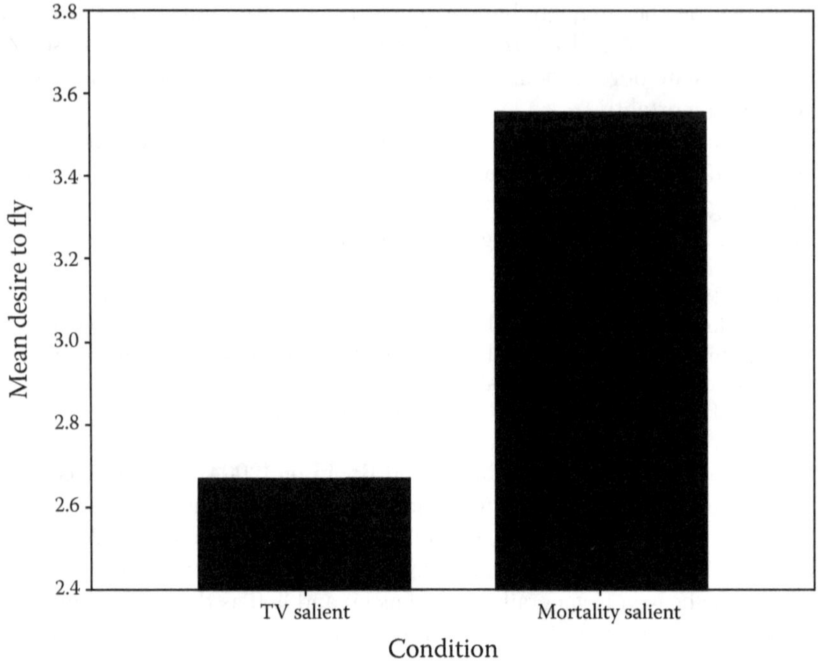

Figure 8.1 Mean desire to fly by condition.

We then reasoned that if engagement in supernatural experiences such as flying serves a death-denying function, then imagining oneself in flight should reduce or eliminate defensive reactions in response to MS. So in Study 2, which was conducted prior to the 2004 presidential election, we manipulated MS, asked participants to visualize flying or not, and then assessed the appeal of President George W. Bush, which we recently had shown was increased by MS (Landau et al., 2004). We expected to replicate the MS-induced increased liking for Bush but that imagining flying after MS would reduce or eliminate this effect.

To assess this hypothesis, we asked participants in the flight condition, following an MS or exam control induction, to "visualize yourself flying above a lush green mountain. Consider carefully each picture that comes before your mind's eye."

Trees ruffle as you soar above the mountaintop.
Snowcapped mountains spring up all around as you fly above.
There is a valley in view below as you hover in the air.
A river is running through the valley as you come back down to the ground.

We asked participants in the grounded condition to "visualize the rising sun. Consider carefully each picture that comes before your mind's eye."[4]

The sun is rising above the horizon into a hazy sky.
The sky clears and surrounds the sun with blueness.

Clouds. A storm blows up, with flashes of lightning.
A rainbow appears.

After each visualization, we asked participants to rate how vividly they could imagine the image and then asked them to look at an untitled print of T'ang Yin's (1470–1523) *Dreaming of Immortality in a Thatched Cottage*, which consists of a man meditating in a cottage on the side of a mountain with his mirror image floating in the air above the valley. We asked participants in the flight condition to "imagine yourself as the person floating in the valley over the mountain for about 10 seconds." We asked participants in the grounded condition to "imagine yourself as the person seated in the thatched hut on the mountain for about 10 seconds."

All participants then read a supposed "Opinion Survey" from Landau et al. (2004, Study 1), in which MS increased support for President Bush and his policies in Iraq:

> It is essential that our citizens band together and support the President of the United States in his efforts to secure our great Nation against the dangers of terrorism. Personally I endorse the actions of President Bush and the members of his administration who have taken bold action in Iraq. ... We need to stand behind our President and not be distracted by citizens who are less than patriotic. Ever since the attack on our country on September 11, 2001, Mr. Bush has been a source of strength and inspiration to us all. God bless him and God bless America.

Participants then rated their support for President Bush and his policies in Iraq. Analyses revealed main effects for MS and flight qualified by the predicted Mortality Salience × Flight interaction (results graphically depicted in Figure 8.2). In the grounded condition, MS engendered increased support for President Bush (replicating the Landau et al., 2004, finding; this effect, however, was completely eliminated in the flight condition). Looked at differently, although there was no difference in support for the president between flight and grounded participants in the exam salient condition, grounded participants had more favorable impressions of President Bush than those who imagined themselves flying in the MS condition.

Study 1 demonstrated that reminders of death increased participants' desire to fly. In Study 2, imagining oneself in flight eliminated a defensive reaction in response to an MS induction. These findings thus lend convergent support to the proposition that flying, as a specific supernatural phenomenon, serves a death-denying function. In a third study, we examined the possibility that if this is true, then, along with ameliorating an MS-induced worldview defense, flying fantasies should have a palliative effect on psychological trauma. Interestingly, this hypothesis is consistent with Atran's (2002, pp. 180–181) speculation about religion and post-traumatic stress disorder (PTSD):

> Emotionally eruptive existential anxieties motivate belief in the supernatural. ... Invocation of the supernatural not only cognitively validates these eruptive emotions, but is affectively validated by assuaging the very emotions that motivate belief in the supernatural. With this in mind, it is worth noting

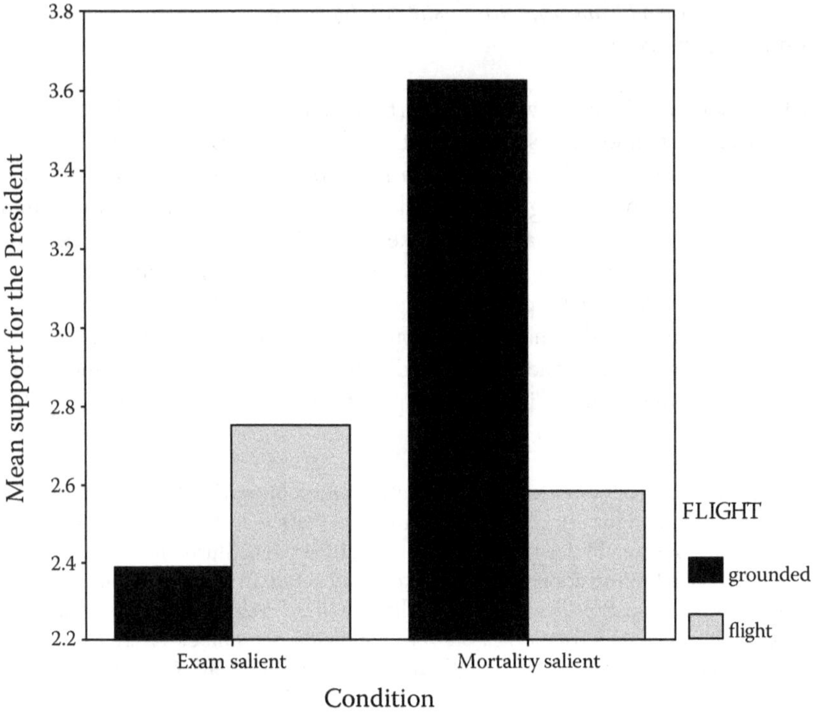

Figure 8.2 Mean support for the president by condition.

> that uncontrollable arousal mediated by adrenergic activation (as for subjects exposed to death scenes) may lead to PTSD if there is no lessening of terror and arousal within hours. … A possibility arises, then, that heightened expressions of religiosity following exposure to death scenes that provoke existential anxieties could also serve this blocking function.

Traumatic events constitute one set of particularly impactful experiences that make mortality salient; yet, in contrast to worldview defensive processes observed in typical terror management research, the trauma literature suggests that people may manage concerns with death aroused by traumatic experiences by dissociating from them (e.g., Gershuny & Thayer, 1999; Herman, 1997; Janoff-Bulman, 1992). Dissociation—separating the self from traumatic incidents and emotional reactions to them—is believed to serve a protective function by allowing individuals to avoid immediately experiencing the emotional impact of traumas. But unlike documented terror management defenses, dissociation has been found to contribute to vulnerability to anxiety-related problems (e.g., PTSD; Ozer, Best, Lipsey, & Weiss, 2003).

Accordingly, Kosloff et al. (2006) predicted and found in two studies that MS increases dissociation from a traumatic event (specifically, reflecting on the events of September 11, 2001) and that this increased dissociation mediates subsequent higher scores on a measure of anxiety sensitivity—the degree to which people are anxious about becoming anxious—a measure that is more predictive of the development of anxiety disorders than traditional self-report measures of state anxiety (Reiss, Peterson, Gursky, & McNally, 1986). To assess a possible palliative effect of flying fantasies on reactions to trauma, we therefore hypothesized that having participants imagine themselves flying would eliminate the MS-induced dissociation in response to reflecting on the events of September 11, 2001, along with subsequent heightened anxiety sensitivity.

The design and procedure for this study were identical to those in Study 2 until the administration of the dependent measures; specifically, participants were momentarily reminded of death or an upcoming exam and imagined themselves either in flight or on the ground. Participants were then asked to reflect on their feelings and experience at the time of and immediately after the September 11, 2001, attack on the World Trade Center and to complete the *Peritraumatic Dissociative Experiences Questionnaire* (PDEQ; Marmar, Weiss, & Metzler, 1997), as modified by Marshall, Orlando, Jaycox, Foy, and Belzberg (2002), assessing the extent of dissociation during that event (e.g., "I 'blanked out' or 'spaced out' or in some way felt that I was not part of what was going on"). Following the PDEQ, participants completed the *Anxiety Sensitivity Index* (Reiss et al., 1986; e.g., "When I notice that my heart is beating rapidly, I worry that I might have a heart attack").

Analyses of the PDEQ responses revealed main effects for MS and flight qualified by the predicted Mortality Salience × Flight interaction (results graphically depicted in Figure 8.3). In the grounded condition, MS increased dissociation, replicating the Kosloff et al. (2006, Study 1) finding; this effect,

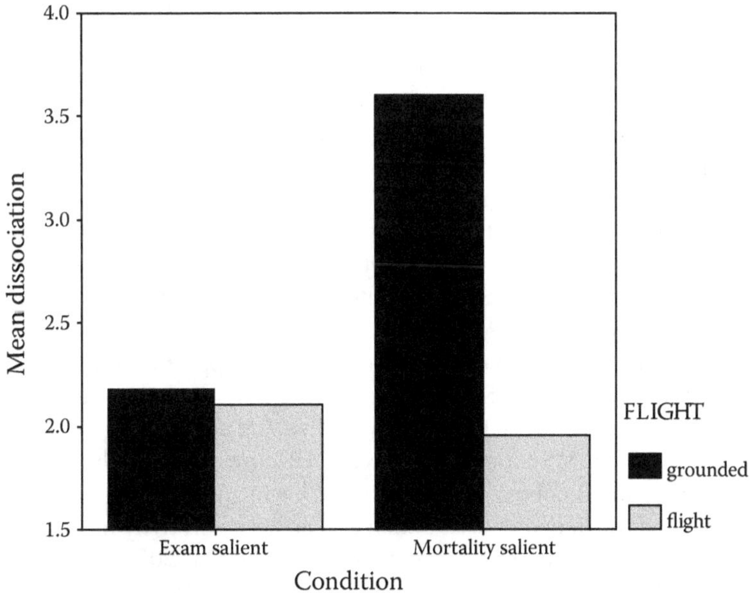

Figure 8.3 Mean dissociation by condition.

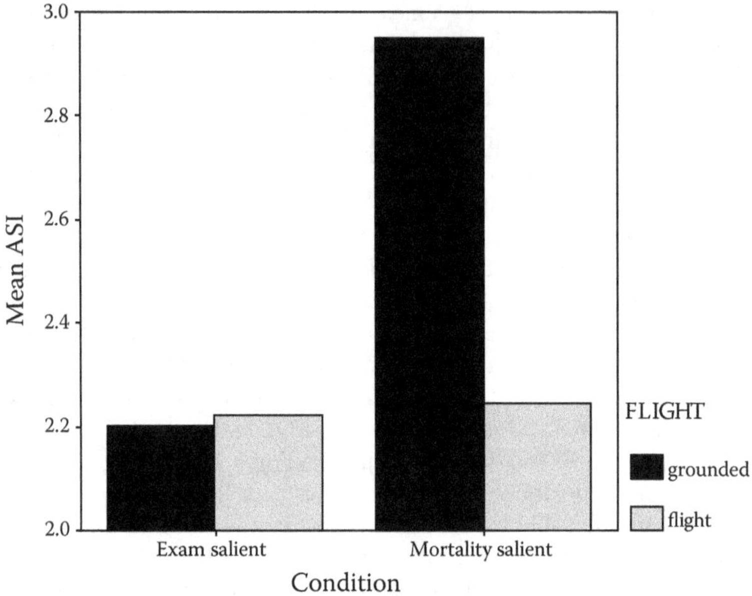

Figure 8.4 Mean anxiety sensitivity by condition. *Note.* ASI = *Anxiety Sensitivity Index.*

however, was completely eliminated in the flight condition. Looked at differently, although in the exam salient control condition there was no difference in dissociation between flight and grounded participants, in the MS condition, grounded participants reported higher dissociation than those who imagined themselves flying.

Analyses of the responses to the *Anxiety Sensitivity Index* revealed main effects for MS and flight, qualified by the predicted Mortality Salience × Flight interaction (results graphically depicted in Figure 8.4). Parallel with the dissociation findings, in the grounded condition, MS increased anxiety sensitivity, replicating the Kosloff et al. (2006, Study 1) finding; this effect, however, was completely eliminated in the flight condition. Looked at differently, although in the exam condition there was no difference in anxiety sensitivity between flight and grounded participants, after MS, grounded participants reported higher anxiety sensitivity than those who imagined themselves flying.

Taken together, these studies provide compelling support for the role of flying fantasies in the amelioration of defensive reactions to reminders of death. Making mortality salient increased the desire to fly, and having people imagine themselves flying eradicated an MS-induced worldview defense (specifically, support for President Bush and his policies in Iraq). Finally, having people imagine themselves flying eradicated MS-induced dissociation and anxiety sensitivity in response to a retrospective consideration of the events of September 11, 2001, suggesting that there may be therapeutic applications of these findings. It is ironic that having traumatized individuals "take flight" may be one way to help them get their psychological feet back on the ground!

SUMMARY

> Nor must we overlook the probability of the constant inculcation in a belief in God on the minds of children producing so strong and perhaps an inherited effect on their brains not yet fully developed, that it would be as difficult for them to throw off their belief in God, as for a monkey to throw off its instinctive fear and hatred of a snake. (Darwin, 1958, p. 93)

> It is traditional to say that the supernatural is what must be resorted to when we cannot explain things by natural processes. That is probably true, but the supernatural … is much more. The supernatural … is a powerful psychic force in maintaining emotional health and physical well-being and is something … unparalleled in the rest of human behavior. Behavior centering on the supernatural has life consequences that are too important for natural selection to have ignored. (Guthrie, 2005, p. 438)

In this chapter, we argued that the universal human propensity to take flight is a specific manifestation of a basic yearning for immortality that is the psychological impetus for the creative development of a supernatural dimension common to all cultural constructions. Clearly, additional research is in order to determine if flying is a uniquely effective means of death transcendence by comparing "fantasy flights" with other supernatural experiences, for example, walking through a wall or burrowing to the center of the earth, and if flying or other supernatural beliefs serve a terror management function for atheists and agnostics to address the important question of whether these effects are due to defense of religious cultural worldviews or presymbolic defenses as proposed by Norenzayan and Hansen (2006). But meanwhile, we believe the research presented here supports our contention (also advanced in Solomon, Greenberg, Schimel, et al., 2004) that supernatural beliefs are not accidental incidental by-products of other cognitive adaptations as claimed by some contemporary evolutionary psychologists. Rather, we view supernatural elements of culture as uniquely human cultural contrivances that serve an important psychological function and thus essential for sustaining consciousness as a viable form of mental organization in the wake of the explicit knowledge of death. As Isaacs (1948, p. 94) observed, it is precisely because of our fondness for fantasy that we can obtain a state of psychological equanimity that makes rational thought possible:

> In their developed forms, phantasy thinking and reality thinking are distinct mental processes, different modes of obtaining satisfaction. The fact that they have a distinct character when fully developed, however, does not necessarily imply that reality thinking operates quite independently of unconscious phantasy. It is not merely that they "blend and interweave"; their relationship is something less adventitious than this. On our view, reality thinking cannot operate without concurrent and supporting unconscious phantasies.

We are not proposing here that supernatural beliefs spontaneously arose to solve the adaptive problems engendered by the uniquely human awareness of death (although they might have); rather, we argue "that many new adaptations begin as by-products or modifications of characters that were originally selected for very different functions" (Jablonka & Lamb, 2005, p. 320). So in this sense we have no quarrel with theorists such as Atran, Bloom, Kirkpatrick, and Mithen, who view supernatural beliefs as by-products of agent detection, causal reasoning, and/or theory of mind. Where we part company is our insistence that once in existence, supernatural beliefs were especially well suited to quell existential anxieties and were extensively elaborated on to do so most compellingly (e.g., the Bible, the Koran, the Egyptian pyramids, gothic churches, rituals of human sacrifice). Therefore, such supernatural beliefs were particularly likely to be central components of successful (in the sense of being perpetuated over time) cultural worldviews, and the propensity to have faith in them most likely became highly adaptive (Baldwin, 1896):

> Learning and behavioral flexibility enables organisms to move into niches that differ from those their ancestors occupied, with the consequence that succeeding generations will face a new set of selection pressures. … By temporarily adjusting behaviors or physiological responses during its lifespan in response to novel conditions, an animal could produce irreversible changes in the adaptive context of future generations. Though no new genetic change is immediately produced in the process, the change in conditions will alter which among the existing or subsequently modified genetic predispositions will be favored in the future. (Deacon, 1997, pp. 322–323)

Consistent with this claim, Guthrie (2005) argued that supernatural beliefs meet the standard criteria for evolutionary adaptations: (a) ubiquitous, improbably complex, and tenaciously persistent; (b) demonstrable heritable variation; (c) viable selection pressure(s) that bestows reproductive advantages favoring genes underlying the expression of the putative adaptation; and (d) conferrable fitness benefits despite substantial biological costs. Specifically, supernatural beliefs are universal (although, like languages, they are manifested differentially across cultures) and inordinately sophisticated, and they involve extraordinary commanding emotional states rarely found in other human activities; for example, trance states, visions, speaking in tongues, and mystic experiences. In addition, studies of identical twins reared apart find moderately high heritability for religiosity (e.g., .49 by Bouchard, Lykken, McGue, Segal, & Tellegen, 1990), and recent research (e.g., Powell, Schahabi, & Thoresen, 2003) reveals an inverse relationship between spirituality and mortality (even after controlling for gender, unhealthy behaviors, social support, and prior health problems) and demonstrates (Jonas & Fischer, 2006) that intrinsic religiosity eliminates a worldview defense in response to MS. Finally, supernatural beliefs are sustained despite considerable physical (time- and resource-consuming rituals) and psychological ("major distortions of objective reality, which can produce debilitating stresses of hexes, curses, and guilt"; Guthrie, 2005, pp. 441–442) costs.

CONCLUSION

There is compelling evidence that people are prone to perceive agency and that supernatural beliefs serve important psychological functions. Thus, further theory and research should focus on how human cognitive predilections and motivational and affective concerns operate, independently or in concert, to shape cultural worldviews and influence human behavior. Our hope is that the ideas presented in this chapter will foster fruitful inquiry along these lines, especially because this is much more than an academic concern. In a world of clashing worldviews and anxiety-riddled citizens, we urgently need to develop worldviews that suit our cognitive proclivities and serve our existential concerns in ways that mitigate the violence and disquiet of our times—before they spiral completely out of control.

NOTES

1. Navarrete and colleagues (Kirkpatrick & Navarrete, 2006; Navarrete & Fessler, 2005) have recently claimed that TMT must be wrong because it is inconsistent with contemporary evolutionary thinking. They argue that concerns about mortality have nothing to do with the formation and maintenance of cultural worldviews; instead, they posit an evolved proclivity to respond to adaptive threats by seeking coalitions or demonstrating one's value as a coalitional partner. Perhaps it is not surprising that we disagree. We believe that TMT is entirely consistent with modern evolutionary principles and that the coalitional psychology advanced as an alternative to TMT is theoretically challenged, empirically unsubstantiated, and unable to account for a host of anthropological facts or empirical data in support of hypotheses derived from TMT (for detailed reactions to coalitional psychology, see Landau, Solomon, Pyszczynski, & Greenberg, 2007; Pyszczynski, Greenberg, Solomon, & Maxfield, 2006).
2. The first three studies used samples from the United States, and the fourth study used a sample from Canada. Although religious affiliation was not assessed in Study 1, in the other three studies, the majority of those who prescribed to a religion were Christian.
3. There were no gender effects in any of the studies described in this chapter.
4. We tried to make the grounded images involve movement to render them parallel to the flight condition.

REFERENCES

Arndt, J., Cook, A., & Routledge, C. (2004). The blueprint of terror management: Understanding the cognitive architecture of psychological defense against the awareness of death. In J. Greenberg, S. L., Koole, & T. Pyszczynski (Eds.), *Handbook of experimental existential psychology* (pp. 35–53). New York: Guilford.

Arndt, J., Greenberg, J., Pyszczynski, T., & Solomon, S. (1997). Subliminal presentation of death reminders leads to increased defense of the cultural worldview. *Psychological Science, 8*, 379–385.

Atran, S. (2002). *In gods we trust: The evolutionary landscape of religion*. New York: Oxford University Press.

Baldwin, J. M. (1896). A new factor in evolution. *American Naturalist, 30*, 441–451.

Becker, E. (1962). *The birth and death of meaning*. New York: Free Press.

Becker, E. (1971). *The birth and death of meaning: An interdisciplinary perspective on the problem of man* (2nd ed.). New York: Free Press.

Becker, E. (1973). *The denial of death*. New York: Free Press.

Becker, E. (1975). *Escape from evil*. New York: Free Press.

Bloom, P. (2005). Is god an accident? *Atlantic Monthly, 296*, 105–112.

Bouchard, T. J., Lykken, D. T., McGue, M., Segal, N. L., & Tellegen, A. (1990). Sources of human psychological differences: The Minnesota study of twins reared apart. *Science, 250*, 223–227.

Boyer, P. (2001). *Religion explained: The evolutionary origins of religious thought*. New York: Basic Books.

Brehm, J. W. (1966). *A theory of psychological reactance*. New York: Academic Press.

Brown, N. O. (1959). *Life against death: The psychoanalytical meaning of history*. Middletown, CT: Wesleyan University Press.

Bruchac, J. (1993). *Flying with the eagle, racing the great bear: Stories from Native America*. Mahwah, NJ: BridgeWater Books.

Burkert, W. (1996). *Creation of the sacred: Tracks of biology in early religions*. Cambridge, MA: Harvard University Press.

Cohen, F., Sullivan, D., Solomon, S., Greenberg, J., & Ogilvie, D. M. (2009). Finding everland: Flight fantasies and the desire to transcend mortality. *Under editorial review*.

Darwin, C. (1958). *The autobiography of Charles Darwin* (N. Barlow, Ed.). London: Collins.

Deacon, T. (1997). *The symbolic species: The co-evolution of language and the brain*. New York: Norton.

Dissanayake, E. (1995). *Homo aestheticus: Where art comes from and why*. Seattle: University of Washington Press. (Original work published 1992)

Freud, S. (1955). The uncanny. In *The complete psychological works of Sigmund Freud* (Vol. 17, pp. 219–256). London: Hogarth. (Original work published 1919)

Frye, N. (1964). *The educated imagination*. Bloomington: Indiana University Press.

Gershuny, B., & Thayer, J. (1999). Relations among psychological trauma, dissociative phenomena, and trauma-related distress: A review and integration. *Clinical Psychology Review*, *19*, 631–657.

Goldenberg, J. L. (2005). The body stripped down: An existential account of ambivalence toward the physical body. *Current Directions in Psychological Science*, *14*, 224–228.

Goldenberg, J. L., Pyszczynski, T., Greenberg, J., & Solomon, S. (2000). Fleeing the body: A terror management perspective on the problem of human corporeality. *Personality and Social Psychology Review*, *4*, 200–218.

Goldenberg, J. L., Pyszczynski, T., Greenberg, J., Solomon, S., Kluck, B., & Cornwell, R. (2001). I am NOT an animal: Mortality salience, disgust, and the denial of human creatureliness. *Journal of Experimental Psychology: General*, *130*, 427–435.

Greenberg, J., Pyszczynski, T., & Solomon, S. (1986). The causes and consequences of a need for self-esteem: A terror management theory. In R. F. Baumeister (Ed.), *Public self and private self* (pp. 189–212). New York: Springer-Verlag.

Greenberg, J., Pyszczynski, T., Solomon, S., Rosenblatt, A., Veeder, M., Kirkland, S., & Lyon, D. (1990). Evidence for terror management theory II: The effects of mortality salience on reactions to those who threaten or bolster the cultural worldview. *Journal of Personality and Social Psychology*, *58*, 308–318.

Greenberg, J., Simon, L., Porteus, J., Pyszczynski, T., & Solomon, S. (1995). Evidence of a terror management function of cultural icons: The effects of mortality salience on the inappropriate use of cherished cultural symbols. *Personality and Social Psychology Bulletin*, *21*, 1221–1228.

Greenberg., J., Solomon, S., & Arndt, J. (2008). A basic but uniquely human motivation: Terror management. In J. Y. Shah & W. L. Gardner (Eds.), *Handbook of motivational science* (pp. 114–134). New York: Guilford Press.

Greenberg, J., Solomon, S., & Pyszczynski, T. (1997). Terror management theory of self-esteem and cultural worldviews: Empirical assessments and conceptual refinements. In M. Zanna (Ed.), *Advances in experimental social psychology* (Vol. 29, pp. 61–139). Orlando, FL: Academic Press.

Guthrie, R. D. (2005). *The nature of Paleolithic art*. Chicago: University of Chicago Press.

Herman, J. (1997). *Trauma and recovery*. New York: Basic Books.

Hinde, R. A. (1999). *Why gods persist: A scientific approach to religion*. New York: Routledge.

Hirschberger, G. (2006). Terror management and attributions of blame to innocent victims: Reconciling compassionate and defensive responses. *Journal of Personality and Social Psychology*, *91*, 832–844.

Isaacs, S. (1948). The nature and function of phantasy. *International Journal of Psycho-Analysis*, *29*, 73–97.

Jablonka, E., & Lamb, M. J. (2005). *Evolution in four dimensions: Genetic, epigenetic, behavioral, and symbolic variation in the history of life*. Cambridge, MA: MIT Press.

Janoff-Bulman, R. (1992). *Shattered assumptions: Towards a new psychology of trauma*. New York: Free Press.

Jaynes, J. (1976). *The origin of consciousness in the breakdown of the bicameral mind*. Boston: Houghton Mifflin.

Jonas, E., & Fischer, P. (2006). Terror management and religion: Evidence that intrinsic religiousness mitigates worldview defense following mortality salience. *Journal of Personality and Social Psychology*, *91*, 553–567.

Kirkpatrick, L. A. (2004). *Attachment, evolution, and the psychology of religion*. New York: Guilford.

Kirkpatrick, L. A., & Navarrete, C. (2006). Reports of my death anxiety have been greatly exaggerated: A critique of terror management theory from an evolutionary perspective. *Psychological Inquiry*, *17*, 288–298.

Kosloff, S., Solomon, S., Greenberg, J., Cohen, F., Gershuny, B., Routledge, C., & Pyszczynski, T. (2006). Fatal distraction: The impact of mortality salience on dissociative responses to 9/11 and subsequent anxiety sensitivity. *Basic and Applied Social Psychology*, *28*, 349–356.

Lakoff, G., & Johnson, M. (1980). *Metaphors we live by*. Chicago: University of Chicago Press.

Landau, M. J., Goldenberg, J. L., Greenberg, J., Gillath, O., Solomon, S., Cox, C., Martens, A., & Pyszczynski, T. (2006). The siren's call: Terror management and the threat of sexual attraction. *Journal of Personality and Social Psychology*, *90*, 129–146.

Landau, M. J., Solomon, S., Greenberg, J., Cohen, F., Pyszczynski, T., Arndt, J., Miller, C. H., Ogilvie, D. M., & Cook, A. (2004). Deliver us from evil: The effects of mortality salience and reminders of 9/11 on support for President George W. Bush. *Personality and Social Psychology Bulletin*, *30*, 1136–1150.

Landau, M. Solomon, S., Pyszczynski, T., & Greenberg, J. (2007). On the compatibility of terror management theory and perspectives on human evolution. *Evolutionary Psychology*, *5*, 476–519.

Langer, S. K. (1982). *Mind: An essay on human feeling* (Vol. III). Baltimore: Johns Hopkins University Press.

Lifton, R. J. (1979). *The broken connection: On death and the continuity of life*. New York: Simon and Schuster.

Marmar, C. R., Weiss, D. S., & Metzler, T. J. (1997). The *Peritraumatic Dissociative Experiences Questionnaire*. In J. P. Wilson & T. M. Keane (Eds.), *Assessing psychological trauma and PTSD* (pp. 412–428). New York: Guilford.

Marshall, G. N., Orlando, M., Jaycox, L. H., Foy, D. W., & Belzberg, H. (2002). Development and validation of a modified version of the *Peritraumatic Dissociative Experiences Questionnaire*. *Psychological Assessment*, *14*, 123–134.

McGregor, H. A., Lieberman, J. D., Greenberg, J., Solomon, S., Simon, L., Arndt, J., & Pyszczynski, T. (1998). Terror management and aggression: Evidence that mortality salience motivates aggression against worldview-threatening others. *Journal of Personality and Social Psychology*, *74*, 590–605.

Meier, B. P., & Robinson, M. D. (2004). Why the sunny side is up. *Psychological Science*, *15*, 243–247.

Mikulincer, M., Florian, V., & Hirschberger, G. (2003). The existential function of close relationships: Introducing death into the science of love. *Personality and Social Psychology Review*, *7*, 20–40.

Mithen, S. (1996). *The prehistory of the mind: The cognitive origins of art, religion and science*. London: Thames and Hudson.
Nauwald, N. (2004). *Flying with shamans in fairy tales and myths*. Havelte, the Netherlands: Binkey Kok.
Navarrete, C. D., & Fessler, D. M. T. (2005). Normative bias and adaptive challenges: A relational approach to coalitional psychology and a critique of terror management theory. *Evolutionary Psychology*, *3*, 297–325.
Norenzayan, A., & Hansen, I. G. (2006). Belief in supernatural agents in the face of death. *Personality and Social Psychology Bulletin*, *32*, 174–187.
Ochsmann, R., & Mathy, M. (1994). *Depreciating of and distancing from foreigners: Effects of mortality salience*. Unpublished manuscript, Universität Mainz, Germany.
Ogilvie, D. M. (2004). *Fantasies of flight*. New York: Oxford University Press.
Ozer, E. J., Best, S. R., Lipsey, T. L., & Weiss, D. S. (2003). Predictors of posttraumatic stress disorder and symptoms in adults: A meta-analysis. *Psychological Bulletin*, *129*, 52–73.
Powell, L. H., Schahabi, L., & Thoresen, C. E. (2003). Religion and spirituality: Linkages to physical health. *American Psychologist*, *58*, 36–52.
Pyszczynski, T., Greenberg, J., & Solomon, S. (1999). A dual process model of defense against conscious and unconscious death-related thoughts: An extension of terror management theory. *Psychological Review*, *106*, 835–845.
Pyszczynski, T., Greenberg, J., Solomon, S., & Maxfield, M. (2006). On the unique psychological import of the human awareness of mortality: Theme and variations. *Psychological Inquiry*, *17*, 328–356.
Pyszczynski, T., Solomon, S., & Greenberg, J. (2003). *In the wake of 9/11: The psychology of terror.* Washington, DC: American Psychological Association.
Pyszczynski, T., Wicklund, R., Floresku, S., Gauch, G., Koch, H., Solomon, S., & Greenberg, J. (1996). Whistling in the dark: Exaggerated consensus estimates in response to incidental reminders of mortality. *Psychological Science*, *7*, 332–336.
Rank, O. (1958). *Beyond psychology*. New York: Dover Books. (Original work published 1941)
Reiss, S., Peterson, R. A., Gursky, D. M., & McNally, R. J. (1986). Anxiety sensitivity, anxiety frequency, and predictions of fearfulness. *Behaviour Research and Therapy*, *24*, 1–8.
Schimel, J., Hayes, J., Williams, T., & Jahrig, J. (2007). Is death really the worm at the core? Converging evidence that worldview threat increases death-thought accessibility. *Journal of Personality and Social Psychology*, *92*, 789–803.
Schubert, T. W. (2005). Your highness: Vertical positions as perceptual symbols of power. *Journal of Personality and Social Psychology*, *89*, 1–21.
Shelley, M. (1994). *Frankenstein*. New York: Penguin Books. (Originally work published 1818)
Solomon, S., Greenberg, J., & Pyszczynski, T. (1991). A terror management theory of social behavior: The psychological functions of self-esteem and cultural worldviews. In M. Zanna (Ed.), *Advances in experimental social psychology* (Vol. 24, pp. 91–159). Orlando, FL: Academic Press.
Solomon, S., Greenberg, J., & Pyszczynski, T. (2004). The cultural animal: Twenty years of terror management theory and research. In J. Greenberg, S. Koole, & T. Pyszczynski (Eds.), *Handbook of existential experimental social psychology* (pp. 13–34). New York: Guilford.
Solomon, S., Greenberg, J., Schimel, J., Arndt, J., & Pyszczynski, T. (2004). Human awareness of death and the evolution of culture. In M. Schaller & C. Crandal (Eds.), *The psychological foundations of culture* (pp. 15–40). Mahwah, NJ: Lawrence Erlbaum.
Wilson, E. O. (1978). *On human nature*. Cambridge, MA: Harvard University Press.

9

The Birth of High Gods
How the Cultural Evolution of Supernatural Policing Influenced the Emergence of Complex, Cooperative Human Societies, Paving the Way for Civilization

AZIM F. SHARIFF, ARA NORENZAYAN,
and JOSEPH HENRICH

Voltaire's well-tread quote, "If God did not exist, it would be necessary to invent him," was written with direct reference to the effectiveness of God as a supernatural policing agent (Voltaire & Redman, 1977). We argue that indeed supernatural policing was a driving force for the invention of God, but this invention—like so many other cultural products—was not the product of a brilliant religious mind or a committee of Machiavellian priests. Instead, omniscient, moralizing supernatural agents derived from a suite of religious beliefs that were culturally selected for their ability to galvanize cooperation in larger groups, promote in-group cohesion, and foster competition with other social groups. The emergence of religions, and modern world religions in particular, has been a cumulative process involving myriad interacting individuals that stretched over hundreds of generations of interacting individuals within the context of intergroup competition.

Humans are not just social, group-living animals but also highly cultural animals (Henrich, in press; Norenzayan, Schaller, & Heine, 2006). The cognitive and behavioral capacities that make human culture possible—complex communication skills, social learning mechanisms, and biased information processing that favors common traits and prestigious individuals—evolved because they allow individuals to readily adapt their behavior to the novel and changing environments at rates much faster than genetic evolution (Boyd & Richerson, 1998; Henrich & Boyd, 1998; Henrich & Gil-White, 2001; Richerson & Boyd, 2005; Tomasello, 1999; Tomasello, Kruger, & Ratner, 1993).

Natural selection has equipped many species with both individual and social learning capacities. As individuals of these species confront the challenges of survival and reproduction, they use their naturally evolved learning capacities to locally adapt. When encountering an evolutionarily novel food, crows and chimpanzees (just to name two) can individually figure out how to use tools for extracting the food (Hunt, 1996; McGrew, 1974). Chimps and dolphins can learn about these tools from conspecifics, who have already figured out the problems individually (Boesch & Tomasello, 1998; Rendell & Whiten, 2001). This means that evolutionary problems are often tackled first, in many species, by learning. Cultural evolution in humans has solved a vast range of evolutionary challenges, as the insights and accidents of generations accumulate and populations become increasingly better adapted (Boyd & Richerson, 1995). Clothing is a cultural adaptation to cold weather. Fire is an energy-saving and nutrient-releasing cultural adaptation to acquiring high-quality food that was shaped the subsequent evolution of our digestive system (Wrangham, Conkin-Brittain, 2003). The use of different spices across human societies shows that spicing, including tastes and recipes, is a cultural adaptation to meat-borne pathogens that are particularly dangerous in hot climates (Sherman & Billing, 1999). Inuit kayaks are culturally evolved engineering marvels that adapt this tropical primate to arctic hunting. These are true adaptations in the evolutionary psychological sense, because they are complex, functionally integrated solutions to recurrent ecological problems. But they are not directly the product of natural selection acting on genes (Richerson & Boyd, 2005) or evoked from domain-specific modules.

On the one hand, genetically evolved aspects of our minds and bodies can constrain cultural developments. And certainly genetic evolution laid the groundwork for the emergence of cultural learning and cultural evolution. On the other hand, however, cultural traits can arise and spread to address environment social problems, which in other species could be dealt with only by genetic evolution. For example, the omnivore's dilemma (Rozin, 1987) suggests that the human capacity to eat a wide range of plant and animal products dramatically increased calorie intake and hence survival but also gave rise to selective pressures to avoid harmful substances (such as rotten meat, poisonous plants) that could have been lethal. Along with evolved psychological adaptations (e.g., the emotion of disgust), an interlocking set of culturally evolved beliefs, practices, and institutions (food taboos, hygiene rules, eating rituals) has shaped human diets in adaptive ways. Careful mathematical modeling of the interaction between cultural and genetic evolutionary processes shows that culture need not be on a tight "genetic leash." Sometimes the cultural tail wags the genetic dog (Rogers, 1988), meaning that cultural evolution can drive genetic evolution by altering the selective environment faced by genes.

In this chapter we explore the idea that some of the central features of religion, and in particular those features that have spread so successfully since the origins of agriculture, have emerged via competition among different cultural groups, bearing different religious beliefs and practices. An integrated suite of religious beliefs, rituals, practices, and institutional forms thus evolved to address the evolutionary challenge of sustaining large-scale cooperation and exchange among nonrelatives. We further consider the possibility that these cultural evolutionary processes, if they have occurred over a sufficiently long time span, may have influenced the course of human genetic evolution in a process known as culture–gene coevolution. Culture and genes may have interacted to make certain aspects of religion—such as big gods—more "thinkable."

To begin, we lay a foundation for this effort by summarizing an account of the cognitive capacities that underlie supernatural agent beliefs (gods, ghosts, ancestor spirits) as evolutionary by-products—natural selection did not favor these capacities because they gave rise to supernatural beliefs. Then we argue that the human capacity for deep commitment to such beliefs was exploited through the mechanisms of cultural evolution to serve as supernatural policing agents to solve the evolutionary problems associated with cooperative behavior in large, genetically unrelated groups. Although we are not the first to advance the idea that religion galvanizes cooperation within groups (for early discussions of religion and social cohesion, see Durkheim, 1912/1995; for recent treatments, see Irons, 1991; Johnson & Krueger, 2004; Sosis & Alcorta, 2003; Wilson, 2002), our aim is to argue for the central role of belief in supernatural agents (in addition to religious ritual) and a culturally evolved (rather than genetically evolved) explanation for these innovations. To do so, we must first visit the current discussion of religion's place within the story of human evolution.

THE COGNITIVE ARCHITECTURE OF GOD CONCEPTS

Several theorists of religion (e.g., Johnson & Bering, 2006; Landau, Greenberg, & Solomon, 2005) have argued that religion is a naturally selected genetic adaptation—a trait complex, in the same way that the vertebrate eye, or echolocation in bats, is an adaptation that has conferred a reproductive advantage to ancestral organisms. Such arguments need to fulfill the strict criteria of adaptive design that are the standard in evolutionary biology: compelling adaptive function in ancestral environments, unitary and complex design, efficiency, precision, specificity, economy, and reliability (cf. Williams, 1966). Such a model also needs to rule out both the possibility that religion is a cultural by-product of adaptive design (Atran & Norenzayan, 2004) and the possibility that it is not a product of adaptive cultural learning processes (Henrich & McElreath, 2006; Richerson & Boyd, 2005), of the kind that produced adaptations such as kayaks and spicing in food preparation recipes. As we argue in this chapter, religion fulfills none of these criteria (for similar views, see Atran, 2002; Bloom, 2005; Boyer, 2001; Kirkpatrick, 1999).

Instead, we argue that religion is not an evolutionary adaptation per se. In fact religion is not a unitary thing; it simply points to a family resemblance category of converging sets of cultural by-products, rooted in innate psychological tendencies that constrain and channel the transmission and survival of religious beliefs and practices. These four converging paths are counterintuition (supernatural agents), commitment (motivation belief in counterintuitive agents, displays in costly sacrifices), compassion (relieving existential anxieties), and communion (ritual) (Atran & Norenzayan, 2004). These psychological criteria—the four Cs of religion—are themselves cultural manipulations of psychological adaptations (agency detection, costly commitment) or panhuman existential concerns (fear of death, of social deception), and many belief systems in many places do not even have all four (Johnson, 2003). Religions evolve along culturally distinct though partially convergent paths that are constrained by a complex evolutionary landscape reflecting cognitive, emotional, and material conditions for ordinary social life. Given the mental and social realities of this landscape, certain religious elements are more likely to proliferate. For example, in terms of what supernatural agents come to be believed, there is an optimal balance of how much these beings conform to and how much they violate our intuitive assumptions about physical, biological, and psychological phenomena. The proliferation sweet spot is a *minimally* counterintuitive supernatural being—super enough to capture attention, and natural enough to still make sense.

The combination of an intuitive conceptual grounding and an interesting nonintuitiveness makes beliefs more likely to be transmitted and retained in a population than random departures from common sense. On the one hand, category violations that shake basic notions of ontology are attention arresting and hence resistant to memory degradation. Only if the resultant impossible worlds remain bridged to the everyday world, however, can information be stored, evoked, and transmitted (Atran & Norenzayan, 2004; Atran & Sperber, 1991; Boyer, 1996). Several lines of experiments support these assertions, indicating that minimally counterintuitive concepts (Barrett & Nyhof, 2001; Boyer & Ramble, 2001) as well as minimally counterintuitive narrative structures such as folktales (Norenzayan, Atran, Faulkner, & Schaller, 2006) have a cognitive advantage over other cognitive templates, be they entirely intuitive or maximally counterintuitive. Once these beliefs are cognitive selected, they are available to undergo cultural selection and stabilization. In what follows, we explore how cultural evolutionary processes may have selected among the potential pool of readily transmittable beliefs to expand and galvanize cooperative behavior in large social groups.

COOPERATION IN LARGE GROUPS

The social environment of religion's infancy was one likely characterized by relatively small groups. These groups were held together by a few behavioral mechanisms that have genetically evolved in nonhuman species to permit limited amounts of cooperation. Social organisms confront a tension between the stability and cooperativeness of the social group, on the one hand, and the selfishness of the individual, on the other. Although group living conveys many advantages to individual members (e.g., avoidance of and protection from predators), there are many potentially cooperative circumstances in which it is more advantageous for individuals to evade contributing to the collective and free riding on the contributions of others. This strategy will, unchecked, prove so successful that it will overrun an entire population, making group living an impossibility.

As a result, the evolutionary mechanisms of kin selection and reciprocal altruism have favored the emergence of altruism toward relatives and in reciprocal dyads or very small groups. Among humans, indirect reciprocity, wherein reputations can be ascertained by third parties rather than only through personal interactions, has increased the number of potential dyadic partners. Indirect reciprocity, however, does not increase the size of the cooperative group and operates effectively only so far as these reputations can be very reliably transmitted and recalled for most potential partners (Henrich & Henrich, 2007). None of these mechanisms permits large-scale cooperation.

Thus, though humans have evolved to use each of these strategies, the extent of human social interaction was still, for much of human history, limited to cooperation in very small groups. There are two ways in which human sociality was limited. First, kin selection and reciprocity are limited to small cooperative units of two or three individuals and cannot explain interactions in which large of numbers cooperate in the same unit, such as in warfare, group hunting and food sharing, recycling, blood donation, voting, or community house construction. Second, because groups were likely regulated by reputational information and personal relationship, this caps the size at which individuals can maintain a generalized sense of trust toward fellow group members. Extrapolating from neocortex size, Dunbar (2003) estimated that human brains were designed to manage ancestral groups of about 150 members. Beyond this number, unfamiliarity abounds, trust disintegrates, reciprocity is compromised, and groups divide or collapse. Although this specific number can be disputed (e.g., Smith, 1996), it is apparent today from the size of modern human settlements that solutions have been found to the limitations that used to make such settlements unstable. This effect is demonstrated in ethnographic work in part of New Guinea, where villages routinely split once they exceed about 300 people (i.e., 150 adults). Tuzin (1976, 2001) detailed the historical emergence of an anomalous village of 1,500 people and showed how culturally evolved beliefs about social organization, marriage, norms, rituals, and supernatural agents converged to maintain harmony and galvanize cooperation in a locale where this scale was previously unknown.

Archaeological evidence makes clear that human societies had begun to "scale up" group size and the scale of cooperation between 14,000 and 12,000 years ago, as the Pleistocene gave way to the Holocene and the preagricultural villages of the Natufians gave way to towns such as Jericho (Cauvin, 1999). A number of innovations—all necessary, none sufficient—emerged around this time that allowed larger populations to live relatively harmoniously in cohesive groups. Revolutions in agriculture, hierarchical political organization, and, we argue, religious beliefs and associated costly rituals made such settlements sustainable.[1]

THE ROLE OF GODS IN PROMOTING COOPERATION

Emerging religious belief systems, we suggest, increased trust among unrelated individuals, allowing cooperation to expand beyond the small groups to which it had been previously limited. There is empirical evidence that religion, today, facilitates trust and cooperation among genetically unrelated individuals. Recently, Tan and Vogel (2005) examined religiosity in the context of a trust game. The results were clear: Religious trustees were trusted more, particularly by religious trusters, and religious trustees were indeed more cooperative in turn toward the trusters. Importantly, these findings were not reducible to ingroup–outgroup processes. Consistent with these results, findings by Gervais, Shariff, and Norenzayan (2007) reveal that prejudice toward atheists is mainly driven by moral distrust rather than by visceral antipathy, as is the case for ethnic prejudice (Allport, 1954). Sosis and Ruffle (2003) examined the link between religion and cooperative behavior in Israeli kibbutzim. They found that religious kibbutz members were more cooperative than secular members, and religious attendance predicted cooperative decision making, controlling for a number of variables. In a different analysis, Sosis and colleagues compared the longevity of religious and secular communes in 19th-century America (Sosis & Bressler, 2003). For any given year, religious communes were found to outlast those driven by secular ideologies, such as socialism, by a factor of four. The remarkable survival value of religion could be explained by the cooperative advantages that it confers to groups. But what accounts for these seemingly religiously derived cooperation and trust benefits?[2]

We hypothesize that cultural evolution favored the emergence of an interrelated suite of beliefs about the traits of supernatural agents. As background, the religions of small-scale societies including foragers often do not have one or two powerful gods who are markedly associated with moral behavior (Roes & Raymond, 2003). Many gods are ambivalent or whimsical, even creator gods. Gods, in most small-scale societies, are not omniscient or omnipotent. Notions of a pleasant afterlife appear to be a relatively recent innovation (McNeill, 1991). We suggest that moralizing high gods gradually moved to the forefront of religious systems as cultural evolution—driven by processes favoring larger, more cooperative, more harmonious groups—favored rituals and practices that instill greater degrees of committed belief in people about gods who (a) cared about cooperative- and harmony-enhancing behavior (the group's moral norms), (b) could and would reward and punish appropriately, and (c) had the power to monitor all behavior all the time. These religious beliefs helped expand the sphere of human cooperation. In particular, we suggest that the fear of imagined supernatural policing agents helped overcome the constraints imposed on the scale of human social interaction and cooperation by our kin and reciprocity-based psychologies.

The omniscience of these agents extends one's vulnerability of "being caught" to all times and all places. Some gods can even read people's thoughts. Moreover, there are no restrictions on how many transgressions these supernatural agents can keep track of. The consequence is that "hidden defection," which was still a viable individual strategy in groups with indirect reciprocity, is markedly reduced.

Partially outsourcing not only the monitoring but also the punishing aspects of cheater detection to supernatural agents also contributes to addressing the problem of costly punishment (Johnson & Bering, 2006). The costliness of punishing cheaters (through both the act of punishing and the potential retribution for this act) itself creates a second order of cheaters—those who free ride on their punishing duties. This is a problem that can extend, at least theoretically, ad infinitum (Henrich & Boyd, 2001). Because supernatural agents are not generally thought to be privy to the same concerns as men, they can be seen to punish without cost or fear of retribution. Finally, the belief that the punishments of moralizing high gods are accurate and complete is favored by cultural evolution. The idea that no one escapes the omniscient judge may help satisfy human intuitions about fairness and justice (Haidt & Joseph, in press). The belief in a supernatural watcher can extend the otherwise limited scope of human cooperation, effectively infinitely, provided that the fear of these supernatural beings reaches a near-ubiquitous distribution in the group.

A growing body of empirical support bolsters these claims (see Norenzayan & Shariff, 2008). Snarey (1996) examined the features of god concepts across cultures as a function of life-threatening water scarcity. Societies with high water scarcity were more likely to have morally concerned deities who encouraged the prosocial use of natural resources. This finding held even when controlling for cultural diffusion of high gods via missionary activities. Thus, high gods were culturally selected when freeloading was particularly detrimental to the cohesiveness of the social group.

In a similar cross-cultural analysis, Roes and Raymond (2003) predicted, and found, that across cultures, large societies are associated with moralizing high gods—group size was correlated with the existence of supernatural watchers who are concerned about the morality of human interactions. This finding held controlling for the cultural diffusion of high gods via missionary activity, as well as for societal inequality.

In societies with moralizing gods, a fear of supernatural agents among individuals can be evoked simply to enforce moral norms. In one study, children were explicitly told not to look inside a box and then left alone in the room with it (Bering, 2003). Those who were previously told that a fictional supernatural agent, Princess Alice, is watching were significantly less likely to peek inside the forbidden box. A later study (Bering, 2006) found a similar effect in university students. Those who were casually told that the ghost of a dead student had been spotted in the experimental room were less willing to cheat on a rigged computer task.

If reminders of a supernatural agent can reduce cheating, reminders of a moralizing high god may reduce selfish behavior and increase generosity, even toward strangers. Shariff and Norenzayan (2007) tested this possibility. Participants who were implicitly primed with god concepts behaved more altruistically in an economic game measuring fair behavior than those receiving either a neutral prime or no prime at all. In an anonymous, non-iterated version of the "dictator game," participants were randomly assigned to be either the giver or the receiver. Those assigned to the role of the giver were allotted $10, which they were given the opportunity to share—in any amount they saw fit—with the receiver, who would otherwise receive nothing. Assured anonymity from the other player and confidentiality in their decision, 38% of givers in the control conditions kept all the money for themselves. This figure fell to 14% for participants implicitly primed with god concepts. At the same time, the proportion offering $5 to the receiver—an even half of the money—rose from 20% in the control conditions to 48% in the religiously primed condition. Among non-student atheists, however, the god primes had no effect. Subsequent studies showed that this effect is not explainable in terms of changes in positive or negative mood or in terms of increases in feelings of empathy.

Although other interpretations are possible, these results suggest that the imagined presence of supernatural watchers can reduce selfishness and increase the adherence to fairness norms, even among anonymous strangers. Throughout history, this combination of cheating reduction and generosity fostering would have proved even more effective at stabilizing large societies than cheating reduction on its own. But is this suite of beliefs surrounding moralizing high gods a product of long-term cultural evolution or a reliably developing product of genetic evolution and thus a piece of human nature? Like most of human thought and behavior, there will undoubtedly be influences from both genetic evolution and cultural evolution on these beliefs. Certainly, as discussed earlier, the mental capacities that make such beliefs plausible, even thinkable, are the product of the genetic evolution. Equally certainly, the specific content of religious beliefs, such as the belief in Old Man Coyote, Vishnu, or the Abrahamic God, is transmitted culturally. A better question, then, is *to what extent* and *which* specific details of religious beliefs in supernatural watchers are culturally rather than biologically evolved. This is where the debate begins.

SUPERNATURAL PUNISHING AGENTS: CULTURAL OR GENETIC ADAPTION?

A number of theorists (e.g., Harris & McNamara, 2008; Johnson & Bering, 2006) have proposed that religious beliefs, such as those associated with supernatural watchers, are genuine genetically evolved adaptations for enhancing human cooperation. That is, they suggested that there are modules for religious beliefs that originated in genetic mutations and have been favored by natural selection because of their cooperation-enhancing abilities. Johnson and Bering (2006), specifically, suggested that the belief in supernatural agents served the adaptive purpose of the wholesale suppression of selfish behavior.

Although we are in agreement with much of Johnson and Bering's (2006) argument regarding the effects of moralizing supernatural agents on cooperation, we disagree with their suggestion that these beliefs emerged as genetic adaptations. The position that we endorse places many of the important details of religious beliefs in general, and the beliefs about the characteristics of supernatural agents more specifically, in greater debt to cultural evolution (see also Atran & Norenzayan, 2004; Henrich, 2007). We argue that the fear of punishing supernatural policing agents, instead of being a specific genetic adaptation, developed as evolutionary by-products honed over generations by cultural evolution. The evolved structure of the brain resulted in a mind that was very receptive to ideas about supernatural agents, a receptivity that was capitalized on by competing cultural variants of supernatural agents.

There are a number of factors that favor our approach over that of the "god beliefs as genetic adaptation." First, theoretically, the reputational models of cooperation verbally described by these authors (Bering, 2006) are actually unlikely to favor or explain larger scale cooperation in purely genetic evolutionary models, although they can work well for cultural evolution. Second, it not clear how beliefs in supernatural agents could be encoded in DNA, and even if they can be, it's not clear why natural selection would resort to programming supernatural beliefs into the human genome, as opposed to pursuing a variety of other, seemingly less costly routes to addressing the adaptive problem created by reputation management. Third, the genetic adaptation approach seems to flounder with the empirical evidence indicating that many small-scale societies lack moralizing high gods that act as omniscient supernatural punishers. We briefly discuss each of these issues next.

Evolutionary Modeling: The Selection Between Multiple Stable Strategies

Formal genetic evolutionary models based on purely within-group natural selection do not provide a solution to larger scale cooperative dilemmas (Henrich, 2006; Henrich & Henrich, 2007; Panchanathan & Boyd, 2004). These models—whether they involve costly punishment or reputation-based withdrawal of help—show that the same process can stabilize *any* costly behavior (including costly maladaptive behaviors that hurt the group and the individual), not merely cooperative behaviors. This means that these approaches suffer from an "equilibrium selection problem," and we have no theoretical reason to expect within-group genetic selection to favor larger scale cooperation. Within-group transmission processes, therefore, cannot provide a complete solution to the dilemma of larger scale cooperation.

If we consider cultural evolution, however, and allow these alternative stable equilibria to compete in a process called cultural group selection, cultural evolution can favor norms and beliefs that lead to larger scale cooperation. This process, described next, is well modeled and does not suffer from the problems often associated with arguments for the genetic group selection of cooperation (Henrich & Henrich, 2007).

The previous description of reputation and cooperation may be surprising, as some psychologists have repeatedly claimed that "individual-level selection" based on reputation can favor larger scale cooperation (Bering, 2006). There are three issues that seem to need clarifying. First, we emphasize that we are referring to the analysis of mathematical models, not verbal models. Whenever theorists, deploying the mathematical tools that have long formed the bedrock of the study of evolutionary processes (Nowak, 2006), have sought to model reputation-based processes for solving larger scale cooperative dilemma, the previously mentioned issue of equilibrium selection emerges (Panchanathan & Boyd, 2004). That is, there is simply no mathematical model that supports the purely verbal models that some evolutionary psychologists have so frequently asserted: All such models generate multiple stable equilibria that include cooperative outcomes along with numerous noncooperative ones. Viewed as a genetic evolutionary process, these models require some mechanism, such as genetic group selection, to shift among these equilibria.

Second, part of this confusion may result from a failure to distinguish cooperation in dyads from larger scale cooperation in big groups. Reputation can favor cooperation in dyads (Leimar & Hammerstein, 2001; Panchanathan & Boyd, 2003), but this is not the kind of cooperation at issue. The models typically cited by psychologists, if any are cited at all, are limited to dyadic cooperation and do not extend to larger cooperative groups. Reputation-based reciprocity can provide a foundation for human concerns about reputation (Nowak & Sigmund, 2005), but either cultural evolution or culture–gene coevolution is needed to explain why reputation extends to cover all manners of social norms, including those that stabilize larger scale cooperation.

Third, the kind of cultural group selection we are discussing involves groups stable in equilibria, some of which are cooperative and some of which are not. This is not the kind of between-group influence on individual fitness that most non-specialists are accustomed to reading about, and it is not susceptible to the usual concerns that target the genetic group selection of altruism. In an ecology of different groups, defectors entering cooperative groups are suppressed by within-group selective processes (via punishment or reputational damage). This is unlike the usual case of genetic group selection in which defectors reap a fitness bonanza when they enter cooperative groups (lots of people to free ride on). The effect of this suppression of free riding is to maximize the importance of the variation between groups and to magnify the importance of competition between groups (Henrich & Boyd, 2001).

Evolutionary Fit: Wholesale Versus Selective Suppression of Selfishness

Our second concern is the suggestion that the fear of supernatural policing agents was a genetic adaptation rests heavily on the assumption that such a belief could be genetically encoded, an assumption that can by no means be casually overlooked. Despite rampant speculation, there is no evidence to support the idea that modules evolve at the level of particular beliefs. Moreover, many have criticized the extension of biological evolutionary explanations to this level of specificity on theoretical and empirical grounds (e.g., Fodor, 1987; Panksepp & Panksepp, 2000).

Granting, however, that beliefs could develop as mutations and ignoring the empirical record of religion in small-scale societies, is it plausible that such a mutation would proliferate? According to Johnson and Bering (2006), the fear of supernatural watchers emerged in response to the ability that humans developed to communicate information about reputations. In this new environment where one's slights and transgressions could be broadcast beyond the offended party, the selfish strategies of yore became much more costly. As a result, those possessing the mutation of a fear of omniscient watchers would have acted less selfishly in general, as they were in constant fear of being judged by the watcher and thereby would be less likely to attract the negative repercussions of being caught and exposed as a selfish operator. The wholesale suppression of selfish behavior, they concluded, would be ultimately adaptive at the individual level.

Theoretically, the introduction of the omniscient, punishing supernatural agents is both a roundabout and a suboptimal strategy to be genetically selected for. True, evolution does not always take the shortest distance between two points, but there is considerable evidence to indicate that more direct and effective strategies did develop to overcome this new threat of public exposure, not the least of which is keeping one's selfish freeloading hidden. Why would beliefs emerge that cause one to improperly calibrate to the threat of reputational damage? Why not simply do what natural selection has so often done in nonhumans and select for domain-specific forms of risk aversion?

Instead of the undiscerning strategy associated with the fear of supernatural watchers, it appears that humans have evolved a discriminate strategy wherein selfish, freeloading behavior was suppressed in those situations where one's reputation was vulnerable. These types of clandestine strategies seem to be present in rudimentary forms in chimpanzees and are significantly more elaborate in humans (Byrne & Whiten, 1988). The obvious advantage of this adaptation is that even if a very conservative, hypersensitive approach to protecting one's reputation is taken, it avoids more false positive errors where one could have gotten away with acting selfishly while still managing to keep false negatives to a tolerable minimum.

Recent empirical evidence demonstrates this hypersensitivity with which people hide their selfishness. Two studies, in particular, show how people in what are rationally understood as anonymous situations act less selfishly when they are in the mere presence of images of eyes, or eyespots. Haley and Fessler (2005) found that people are more likely to act prosocially on a computer-based economic game when stylized eyespots were subtly embedded on the computer's desktop. Bateson, Nettle, and Roberts (2006) showed that people were less likely to cheat on paying at a self-serve coffee station based on the honor system when a pair of eyes was conspicuously featured on the price list poster. This sensitivity to eyes is an evolutionary ancient adaptation down to the level of birds (Stevens, 2005) and fish (Neudecker, 1989), which has, in humans, been exploited for reputation protection.

What is also notable about these studies is that in the control conditions, where anonymity is more securely simulated, selfish behavior and cheating behavior are rampant. In the Shariff and Norenzayan (2007) study, student participants in the control conditions generally acted exceedingly selfishly in the dictator game when the purported anonymity protected their reputations, an effect found for both religious and nonreligious players. And this is not unusual behavior for students. Hoffman et al. (1994) showed that as students feel more and more secure in their anonymity, prosocial behavior drops steeply. We can only speculate about the past, but it is clear today that selfishness has been not wholly suppressed but ardently, adeptly, and adaptively hidden.

In addition, the fear of supernatural agents can carry with it substantial costs, which, again, would make alternative, simpler methods of suppressing selfish behavior much more compelling genetic bargains. Examples, such as voodoo culture (e.g., Rigaud, 1985) where the spread of the paranoia over supernatural agents becomes nearly paralyzing, demonstrate supernatural devotion so apparently costly that it tests the resolve of the most fervent panadaptationist. Similarly, widespread witchcraft beliefs are notorious for sowing hatred and disharmony in communities and often inhibit the adoption of health-enhancing medical practices. These examples demonstrate, again, that the invention of supernatural agents is both an overly roundabout and an overly inefficient means of attaining the ends for which it has been suggested.

In sum, adaptations that honed people's ability to cheat, defect, or act selfishly without getting caught would have proved not only more adaptive but also more likely to have been genetically evolved. This casts the development of a fear of supernatural agents in a new light. We suggest that supernatural agents, instead of emerging in an environment where the existing strategies were openly selfish and liable to get one socially exiled, emerged in an environment where selfish behavior was carefully hidden. In this environment, belief in supernatural agents would have actually proved *maladaptive.* Those fearing their gods or desiring their rewards would have curtailed their tendencies to lie, cheat, and steal even when they could get away with it. Natural selection operating within groups in a noncultural world should have, if anything, led humanity away from these beliefs, instead of toward them. And yet, here we are.

Empirical Problems: Many Religions Do Not Have Gods Like That

Finally, the genetic adaptation approach to supernatural beliefs faces some straightforward empirical problems. The big problem is that not only do millions of atheists not believe in supernatural agents but people of many small-scale societies don't believe in the types of moralizing high gods that the purely genetic approach predicts. Fans of the genetic adaptation approach typically presume that atheists are not really unbelievers, in a deep emotional sense. This has yet to garner empirical support, though emerging evidence shows that if this deep belief does actually exist, it does not translate into meaningful behavior. Self-described atheists are, for example, differently affected by unconscious religious primes (Shariff & Norenzayan, 2007, Study 2). Moreover, this approach provides no ready explanation for the lack of moralizing high gods in many small societies or the historical association of moralizing high gods and complex, cooperative societies (Roes & Raymond, 2003).[3]

Highlighting the fact that high god concepts are the product of cultural, not genetic, innovation, we think it is worth noting that ancestor gods found in many village societies often cannot observe people beyond village boundaries, are sometimes limited to observing people's behavior (they cannot read minds and intent), may lack the power to act, cannot give a heavenly reward, and may care only about specific acts and not general principles. Because they are not omniscient mind readers, they can misinterpret human intentions and goof, resulting in unjust punishment.

The logic, then, of supernatural policing agents is better found in cultural evolution. The persistence of the fear of supernatural watchers in the face of immediate individual costs can be compellingly explained through cultural group selection, without direct reference to a specialized genetic adaptation. We explore this explanation in detail in the next section.

CULTURAL EVOLUTION CAN ALTER THE SELECTIVE ENVIRONMENT FACED BY GENES IN ANCESTRAL HUMAN ENVIRONMENTS

In the scenario we propose, cultural group selection favored those culturally transmitted social norms that best promoted cooperation within the group and success in competition with other groups. The evolution of such norms, which has been extensively modeled, can stabilize costly behaviors through the effects of reputation on the withdrawal of help and through direct costly punishment (as well as some other mechanisms). Cultural group selection merely favors the combinations of particular norms that are most beneficial to the group.

As this process continues, however, it favors larger and larger cooperative groups (Roes & Raymond, 2003). As group size increases, it begins to stress the limits of reputational information and diffuse punishment's capacity for stabilizing cooperation and maintaining within-group harmony. We argue that widespread beliefs in certain kinds of supernatural agents can help extend the potency of social norms by covering the expanding opportunities for cheating and free riding that emerge as the group expands and coverage of reputational information begin to crack. Eventually, these groups, with widespread commitment to powerful, omniscient moralizing gods, would become larger and generally more competitive than groups whose belief structures did not increase cooperation. Henrich (2007) discussed several case examples, with supporting empirical data, of the cultural group selection of religion in action.

Within these groups, it is generally not optimal to free ride because combinations of reputation and punishment have stabilized cooperation and other group-beneficial norms. Beliefs in supernatural agents could be disadvantageous because they could prevent an individual from free riding in situations when he or she might in fact get away with it. These same religious systems, however, tend to punish belief in culturally foreign gods or lack of belief. As a result, to get the benefits of free riding in the occasional opportunistic (unmonitored) situations, nonbelievers would need to pay most of the costs associated with believing (e.g., helping, participating in rituals) in all monitored situations (to avoid punishment) to access those probably rare situations for free riding. And if feigning belief (i.e., nonbelief) increases one's likelihood of botching the divinely required practices, words, and actions of believers (and getting caught in nonbelief), then the relative advantage of nonbelief could be outweighed by the extra cost of being more likely to get fingered as an apostate or heretic—and as a result being punished or socially excluded.

At their most extreme, examples of such punishment in *this* world include Muslim theocracies such as those in Iran that place the penalty of death on apostasy. In the next world, those punishments become even more severe; both doctrinal Islam and Christianity promise eternal hellfire for those whose doubt exceeds their belief. But these aren't the only costs that atheists face for their lack of belief. Polling data on social attitudes continue to show atheists to be the least accepted of various major minority groups, including the typically marginalized groups of African Americans, Muslims, and homosexuals (Edgell, Gerteis, & Hartmann, 2006). When polled individuals were asked if they would disapprove of their child marrying an atheist—a standard measure of prejudice—over 47% admitted that they would (the rates if the child was marrying an African American or Muslim were 27% and 34%, respectively). Research exploring the psychology of antiatheist prejudice finds that this aversion is driven primarily by moral distrust (Gervais et al., 2007), a finding consistent with the theory that nonbelievers pose a perceived threat to a moral system policed by supernatural agents.

Because it is at least plausible that cultural groups with different forms of social organization and different religious beliefs have been competing for tens of thousands of years (Richerson & Boyd, 1998), our approach opens the possibility that cultural evolution could have altered the selection pressures faced by genes and favor the evolution of a psychology that is more susceptible to believing in and committing to god beliefs. Thus, a psychological predisposition to believing in moralizing gods could then be favored by natural selection within groups (and between groups) as a consequence of the ways that cultural evolution (via social norms) shaped social environments. Cultural evolution may have favored genes that make these gods easier to believe in and commit to. Of course, there may not have been time for much genetic evolution in this regard, but it is nonetheless important to note that natural selection need not oppose such beliefs once cultural group selection has shaped the selective social environment.

CONCLUSION

A combination of findings from cognitive science and an understanding of cultural evolutionary processes give us the best chance to understand the phenomenon of religion in the world today. Modern religious beliefs are deeply rooted in our evolutionary history, yet they are not the necessary and ineradicable consequences of our genetic makeup but part of a much more fluid and responsive cultural system. In short, religions are a coevolutionary phenomenon. The case we have made suggests that religions are both a cognitive by-product of reliably developing aspects of our cognition and a consequence of long-term cultural evolutionary forces, including those very forces that shaped the complex, large-scale, cooperative institutions that dominate the modern world. Our evolved cognition strongly constrains the forms of religious representations. Not all of the possible representations, however, have the same consequence of outcomes in the lives of individuals and societies. Cultural evolution is influenced by outcomes and thus can create a force that favors particular kinds of representations of others. Although the few great monotheisms in the world are a historically recent phenomenon, it is no coincidence that the world is now dominated by them and that much human behavior is influenced by the belief in a few high gods. To achieve a civilization of this scale, it was necessary to invent them.

NOTES

1. Although the scaling up of human societal size and degree of cooperation was clearly moving up at the beginning of the Holocene, it is perfectly plausible that cultural evolution, driven by competition among cultural groups, has been occurring for tens

of thousands of years prior to the agricultural revolution (Richerson & Boyd, 2005). From around 45,000 to 20,000 years ago in southern Europe, cultural complexity was clearly flourishing, with cave art, figurines, sophisticated tools, and ceremonial burials. Populations may have been dense and semisedentary, and there is evidence of societal complexity greater than that typically associated with foragers throughout the Upper Paleolithic (Kelly, 1995; Price & Brown, 1988).

2. Sosis and colleagues applied "costly signaling" to argue such ritual signals can curb the free-rider problem and thus facilitate cooperation and trust. We don't dispute the importance of ritual, but here we would like to focus on explaining the nature of the gods involved. Costly signaling has nothing to contribute toward explaining the nature religious beliefs.
3. Johnson (2008) has since argued that even atheists, and members of cultures with high gods, still maintain some mechanisms of supernatural—though not deistic—punishment. Karma, or Just-World Beliefs, for instance, may fill the same policing functions as high gods. This may be true; it succeeds, however, only at pushing back the issue of what exactly is being genetically selected for. The data suggesting the relative success of cultures that have the same human genes but then culturally select paradigmatic high gods still demonstrate that there are fitness benefits to certain cultural variants, that is, cultural selection.

REFERENCES

Allport, G. (1954). *The nature of prejudice.* Cambridge, MA: Addison-Wesley.

Atran, S. (2002). *In gods we trust: The evolutionary landscape of religion.* New York: Oxford University Press.

Atran, S., & Norenzayan, A. (2004). Religion's evolutionary landscape: Counterintuition, commitment, compassion, communion. *Behavioral and Brain Sciences, 27*, 713–770.

Atran, S., & Sperber, D. (1991). Learning without teaching. In L. Tolchinsky-Landsmann (Ed.), *Culture, schooling and psychological development* (pp. 39–55). Norwood, NJ: Ablex.

Barrett, J. L., & Nyhof, M. A. (2001). Spreading non-natural concepts: The role of intuitive conceptual structures in memory and transmission of cultural materials. *Journal of Cognition and Culture, 1*, 69–100.

Bateson, M., Nettle, D., & Roberts, G. (2006). Cues of being watched enhance cooperation in a real-world setting. *Biology Letters, 2*, 412–414.

Bering, J. M. (2003, August). *On reading symbolic random events: Children's causal reasoning about unexpected occurrences.* Paper presented at the Psychological and Cognitive Foundations of Religiosity Conference, Atlanta, GA.

Bering, J. M. (2006). The folk psychology of souls. *Brain and Behavioral Sciences, 29*, 453–462.

Bloom, P. (2005). Is God an accident? *Atlantic Monthly, 296*(5), 105–112.

Boesch, C., & Tomasello, M. (1998). Chimpanzee and human cultures. *Current Anthropology, 39*, 591–614.

Boyd, R., & Richerson, P. J. (1995). Why does culture increase adaptability? *Ethology and Sociobiology, 16*(2), 125–143.

Boyer, P. (1996). What makes anthropomorphism natural: Intuitive ontology and cultural representations. *Journal of the Royal Anthropological Institute (N.S.), 2*, 83–97.

Boyer, P. (2001). *Religion explained: The evolutionary origins of religious thought.* New York: Basic Books.

Boyer, P., & Ramble, C. (2001). Cognitive templates for religious concepts: Cross-cultural evidence for recall of counter-intuitive representations. *Cognitive Science, 25*, 535–564.

Byrne, R. W., & Whiten, A. (Eds.). (1988). *Machiavellian intelligence: Social expertise and the evolution of intellect in monkeys, apes, and humans.* New York: Oxford University Press.

Cauvin, J. (1999). *The birth of the gods and the origins of agriculture* (T. Watkins, Trans.). Cambridge, UK: Cambridge University Press.

Dunbar, R. I. M. (2003). The social brain: Mind, language, and society in evolutionary perspective. *Annual Review of Anthropology*, 32, 163–181.

Durkheim, E. (1912/1995). *The elementary forms of religious life*. New York: Free Press.

Edgell, P., Gerteis, J., & Hartmann, D. (2006). Atheists as "other": Moral boundaries and cultural membership in American society. *American Sociological Review, 71*(2), 211–239.

Fodor, J.A. (1987). Modules, frames, fridgeons, sleeping dogs and the music of the spheres. In J. Garfield (Ed.), *Modularity in knowledge representation and natural language understanding* (pp. 25-36). Cambridge, MA: MIT Press.

Gervais, W., Shariff, A. F., & Norenzayan, A. (2007). *Beware the infidel: The implicit mistrust of atheists*. Unpublished manuscript.

Haidt, J., & Joseph, C. (in press). The moral mind: How 5 sets of innate moral intuitions guide the development of many culture-specific virtues, and perhaps even modules. In P. Carruthers, S. Laurence, & S. Stich (Eds.), *The innate mind* (Vol. 3).

Haley, K. J., & Fessler, D. M. T. (2005). Nobody's watching? Subtle cues affect generosity in an anonymous economic game. *Evolution and Human Behavior, 26*, 245–256.

Harris, E., & McNamara, P. (2008). Is religiousness a biocultural adaptation? In J. Bulbulia, R. Sosis, E. Harris, R. Genet, C. Genet, & K. Wyman, (Eds.), *The evolution of religion: Studies, theories, and critiques* (pp. 79–85). Santa Margarita, CA: Collins Foundation Press.

Henrich, J. (2006). Cooperation, punishment, and the evolution of human institutions. *Science, 312*, 60–61.

Henrich, J. (2007). *The evolutionary links between cooperation, religion, and ritual.* Unpublished manuscript.

Henrich, J. (in press). A cultural species. In M. Brown (Ed.), *Explaining culture scientifically.* Seattle: University of Washington Press.

Henrich, J. & Boyd, R. (1998). The evolution of conformist transmission and the emergence of between-group differences. *Evolution and Human Behavior, 19*, 215–242.

Henrich, J., & Boyd, R. (2001). Why people punish defectors: Weak conformist transmission can stabilize costly enforcement of norms in cooperative dilemmas. *Journal of Theoretical Biology, 208*, 79–89.

Henrich, J., & Gil-White, F. (2001). The evolution of prestige: Freely conferred status as a mechanism for enhancing the benefits of cultural transmission. *Evolution and Human Behavior, 22*, 1–32.

Henrich, J., & McElreath, R. (2006). Dual inheritance theory: The evolution of human cultural capacities and cultural evolution. In R. Dunbar & L. Barrett (Eds.), *Oxford handbook of evolutionary psychology* (pp. 555–570). Oxford, UK: Oxford University Press.

Henrich, N. S., & Henrich, J. (2007). *Why humans cooperate: A cultural and evolutionary explanation.* Oxford, UK: Oxford University Press.

Hoffman, E., McCabe, K., Shachat, K., Smith, V. (1994). Preferences, property rights and anonymity in bargaining games. *Games and Economic Behavior, 7*, 346–380.

Hunt, G. R. (1996). Manufacture and use of hook-tools by New Caledonian crows. *Nature, 379*, 249–251.

Irons, W. (1991). How did morality evolve? *Zygon: Journal of Religion and Science, 26*, 49–89.
Johnson, A. (2003). *Families of the forest: Matsigenka Indians of the Peruvian Amazon*. Berkeley: University of California Press.
Johnson, D.P. (2008). The supernatural punishment hypothesis. *Oral presentation at The Society for Personality and Social Psychology Annual Meeting*. Albuquerque, New Mexico. (February, 2008).
Johnson, D. D. P., & Bering, J. M. (2006). Hand of God, mind of man: Punishment and cognition in the evolution of cooperation. *Evolutionary Psychology*, *4*, 219–233.
Johnson, D. D. P., & Krueger, O. (2004). Supernatural punishment and the evolution of cooperation. *Political Theology*, *5*(2), 159–176.
Kelly, R. L. (1995). *The foraging spectrum: Diversity in hunter-gatherer lifeways*. Washington, DC: Smithsonian Press.
Kirkpatrick, L. (1999). Toward an evolutionary psychology of religion. *Journal of Personality*, *67*, 921–952.
Landau, M. J., Greenberg, J., & Solomon, S. (2005). The motivational underpinnings of religion. *Behavioral and Brain Sciences*, *27*, 743–744.
Leimar, O., & Hammerstein, P. (2001). Evolution of cooperation through indirect reciprocity. *Proceedings: Biological Sciences*, *268*(1468), 745–753.
McGrew, W. C. (1974). Tool use by wild chimpanzees feeding on driver ants. *Journal of Human Evolution*, *3*, 501–508.
McNeill, W. H. (1991). *The rise of the West: A history of the human community*. Chicago: University of Chicago Press.
Neudecker, S. (1989). Eye comouflage and false eyespots: Chaetodontid responses to predators. *Environmental Biology of Fishes*, *25*, 143–157.
Norenzayan, A., Atran, S., Faulkner, J., & Schaller, M. (2006). Memory and mystery: The cultural selection of minimally counterintuitive narratives. *Cognitive Science, 30*, 531–553.
Norenzayan, A., Schaller, M., & Heine, S. (2006). Evolution and culture. In M. Schaller, J. Simpson, & D. Kenrick (Eds.), *Evolution and social psychology* (pp. 343–366). New York: Psychology Press.
Norenzayan, A., & Shariff, A. F. (2008). *The origin and evolution of religious prosociality*. Unpublished manuscript, University of British Columbia.
Nowak, M. A. (2006). *Evolutionary dynamics: Exploring the equations of life*. Cambridge, MA: Belknap Press of Harvard University Press.
Nowak, M. A., & Sigmund, K. (2005). Indirect reciprocity. *Nature*, *437*, 1291–1298.
Panchanathan, K., & Boyd, R. (2003). A tale of two defectors: The importance of standing for the evolution of indirect reciprocity. *Journal of Theoretical Biology*, *224*, 115–126.
Panchanathan, K., & Boyd, R. (2004). Indirect reciprocity can stabilize cooperation without the second-order free rider problem. *Nature*, *432*, 499–502.
Panksepp, J., Panksepp, J.B. (2000). The seven sins of evolutionary psychology. *Evolution and Cognition, 6*, 108–131.
Price, D. T., & Brown, J. A. (Eds.). (1988). *Prehistoric hunter gathers: The emergence of cultural complexity*. New York: Academic Press.
Rendell, L. E., & Whiten, H. (2001). Culture in whales and dolphins. *Behavioral and Brain Sciences*, *24*, 309–382.
Richerson, P., & Boyd, R. (1998). The evolution of ultrasociality. In I. Eibl-Eibesfeldt & F. K. Salter (Eds.), *Indoctrinability, ideology and warfare* (pp. 71–96). New York: Berghahn Books.
Richerson, P. J., & Boyd, R. (2005). *Not by genes alone: How culture transformed human evolution*. Chicago: University of Chicago Press.
Rigaud, M. (1985). *Secrets of voodoo* (R. B. Cross, Trans.). San Francisco: City Light Books.

Roes, F., & Raymond, M. (2003). Belief in moralizing gods. *Evolution and Human Behavior*, *24*, 126–135.

Rogers, A. A. (1988). Does biology constrain culture? *American Anthropologist*, *90*, 819–831.

Rozin, P. (1987). Psychobiological perspectives on food preferences and avoidances. In M. Harris & E. B. Ross (Eds.), *Food and evolution: Toward a theory of human food habits* (pp. 181–205). Philadelphia: Temple University Press.

Shariff, A. F., & Norenzayan, A. (2007). God is watching you: Priming God concepts increases prosocial behavior in an anonymous economic game. *Psychological Science*, *18*(9), 803–809.

Sherman, P. W., & Billing, J. (1999). Darwinian gastronomy: Why we use spices. *BioScience*, *49*, 453–463.

Smith, R. J. (1996). Biology and body size in human evolution: Statistical inference misapplied. *Current Anthropology*, *37*(3), 451–481.

Snarey, J. (1996). The natural environment's impact upon religious ethics: A cross-cultural study. *Journal for the Scientific Study of Religion*, *80*, 85–96.

Sosis, R., & Alcorta, C. (2003). Signaling, solidarity, and the sacred: The evolution of religious behavior. *Evolutionary Anthropology*, *12*, 264–274.

Sosis, R., & Bressler, E. R. (2003). Cooperation and commune longevity: A test of the costly signaling theory of religion. *Cross-Cultural Research*, *37*(2), 211–239.

Sosis, R., & Ruffle, B. J. (2003). Religious ritual and cooperation: Testing for a relationship on Israeli religious and secular kibbutzim. *Current Anthropology*, *44*, 713–722.

Stevens, M. (2005). The role of eyespots as anti-predator mechanisms, principally demonstrated in the Lepidoptera. *Biological Reviews*, *80*, 573–588.

Tan, J. H. W., & Vogel, C. (2005, September). *Religion and trust: An experimental study*. Paper presented to the annual congress of the International Association for Research in Economic Psychology, Prague.

Tomasello, M. (1999). *The cultural origins of human cognition*. Cambridge, MA: Harvard University Press.

Tomasello, M., Kruger, A., & Ratner, H. (1993). Cultural learning. *Behavioral and Brain Sciences*, *16*, 495–511.

Tuzin, D. (1976). *The Ilahita Arapesh*. Berkeley: University of California Press.

Tuzin, D. (2001). *Social complexity in the making: A case study among the Arapesh of New Guinea*. London: Routledge.

Voltaire, F. M. A. D., & Redman, B. R. (1977). *The portable Voltaire*. New York: Penguin Books.

Williams, G. (1966). *Adaptation and natural selection*. Princeton, NJ: Princeton University Press.

Wilson, D. S. (2002). *Darwin's cathedral*. Chicago: University of Chicago Press.

Wrangham, R., & Conklin-Brittain, N. (2003). Cooking as a biological trait. *Comparative Biochemistry and Physiology Part A: Molecular and Integrative Physiology*, *136*, 35–46.

10

Social Selection and the Origins of Culture

RANDOLPH M. NESSE

Something extraordinary happened on the evolutionary path that gave rise to creatures capable of culture. The changes are so profound it is as if we humans were somehow domesticated. Levels of violence are drastically lower than for the other great apes. We are born helpless, we require extended care, and we actively teach each other. We pay exquisite attention to each other's wishes and emotional states. We not only cooperate in ways other great apes cannot, we also form deep attachments to nonrelatives and groups that result in altruistic behaviors obviously harmful to fitness. Even our bones are different from our ancestors in ways typical of a domesticated species (Leach, 2003).

Domestication does not require planning. Self-interested behaviors are sufficient. Chasing away aggressive wolves allows friendly ones to gain an advantage by scavenging scraps. After only a thousand generations, this has transformed wolves into the prosocial, loyal, and helpful dogs we now love. Of course, humans were not domesticated by choices made by some other species. Nonetheless, many human social characteristics would be easy to understand if we had somehow been domesticated. Aspects of culture now select for prosociality and capacities for complex social cognition. But what happened before there was culture? What got the process going?

We are understandably curious about what happened on our evolutionary path that made us capable of culture. The sequence likely involved so many interacting factors and recursive causal cycles that any description that satisfies our evolved minds will inevitably oversimplify the actual process. Nonetheless, as illustrated by the chapters in this book, an enormous amount of thought and research has advanced our understanding of how selection shaped capacities for culture. Old arguments pitting evolution and culture as alternatives have been replaced by formulations that recognize both as essential to any full explanation of human behavior (Alexander, 1979; Axelrod, 1984; Boyd & Richerson, 2005; Hammerstein, 2003; Lehman, Chiu, & Schaller, 2004; Levinson & Jaisson, 2006; Norenzayan, Schaller, & Heine, 2006; Ridley, 1997). Furthermore, it is increasingly recognized that the causal pathways are multidirectional; selection forces shape capacities for culture, and culture gives rise to new selection forces that further shaped our ancestors in a cascading process that runs away to vast complexity (Henrich & Henrich, 2006; Lumsden & Wilson, 1981; Richerson & Boyd, 2005).

This general framework is uncontroversial, but the details are perplexing at best. To begin with, there are so many of them! Selection has shaped many traits closely related to culture, including bipedality, hunting, language, tool making, manual dexterity, agriculture, cooperation, emotions, facial expressions, foresight, inhibition, guilt, conformity, imitation, social learning, norm enforcement, morality, empathy, and theory of mind, among others (Levinson & Jaisson, 2006; Norenzayan et al., 2006). Many selection forces have been proposed to shape these traits, including kin selection, group selection, sexual selection, cultural group selection, and the benefits of exchange relationships (Hammerstein, 2003; Lehmann & Keller, 2006). This chapter builds on this knowledge that human capacities for culture result from many selection forces shaping many traits, but it proposes that these factors, even taken together, are insufficient to account for some aspects of human social cognition and cooperation.

Some of these aspects may be explained by social selection. Social selection is the subtype of natural selection in which fitness is influenced by the behaviors of others. The benefits of being preferred as a sexual partner are well recognized in sexual selection, as are the resulting extraordinary traits. The fitness benefits of being preferred as a social partner are also well recognized in basic evolutionary biology, but the effects of social partner selection in shaping human social cognition is just now being explored systematically (Alexander, 2005; Nesse, 2007). This chapter reviews the basic principles of social selection and applies them to the origins of human capacities for social cognition, several of which are essential prerequisites for culture.

COOPERATION, MOSTLY EXPLAINED

Social selection shapes many human capacities in addition to prosociality, but its special contribution is explaining some uniquely human phenomena that are otherwise difficult to account for. So, we begin with a brief overview of evolutionary explanations of cooperation, the limits of those explanations, and what social selection offers that is new.

That human capacities for cooperation are central to culture, and explained in part by culture, is not controversial. No other species demonstrates complex relationships and exchange networks on the scale of humans (Stevens, Cushman, & Hauser, 2005). Explaining such capacities for cooperation has become a major focus at the intersection of biology and social science. The problem and solutions will be familiar to many. Prior to 1966, capacities for cooperation were explained as obviously beneficial for groups, but everything changed when Williams (1966) pointed out and Dawkins (1976) emphasized that selection would tend to eliminate genes that resulted in greater fitness benefits to others than the self. The problem this insight created has now been mostly solved by kin selection, mutual benefits, and various versions of reciprocity theory.

Kin selection explains much individual sacrifice that benefits others (Hamilton, 1964). Genes that harm individual fitness can nonetheless be selected for if they give a sufficient benefit to relatives who share genes in common. How much benefit is sufficient depends on the genetic closeness of the relationship. For siblings (who share 50% of genes in common), a behavior will increase fitness if its costs to the actor are half or less than the benefits to the sib. This principle is the key to explaining much costly cooperation among kin of all species, not just humans.

Cooperation between nonrelatives has a different evolutionary explanation (Trivers, 1971). Most of this cooperation arises from mutually beneficial exchanges that are easy to explain because no enforcement is needed. For instance, birds benefit directly from picking parasites from the hide of a rhinoceros. Trading favors or exchanging goods yields more than the investment for both parties; when exchanges are delayed, however, the system changes because there is a danger that the other will defect, leaving the cooperator at a disadvantage. This has been extensively modeled using an artificial game called the prisoner's dilemma, in which players on each move choose to cooperate or defect. Mutual cooperation yields maximum net benefits, but on any given move, a player who defects gets a large reward while the cooperating player gets none (Axelrod, 1984). Hundreds of studies based on this model have yielded remarkable insights into how cooperation works (Axelrod, 1997).

Mutual exchange is ubiquitous in human interactions. It was originally thought that such reciprocal exchange was also common in animals, but new attempts to document the same process in animals have come up remarkably empty-handed (Stevens et al., 2005). This is very important for the present thesis. Human patterns of exchange, and the associated fitness benefits, are not found in other animals. They may be crucial steps on the road to culture.

Another conclusion from recent research is that most people do not play behavioral economic games according to the strategy that maximizes their payoffs. Instead, they cooperate too much at the start, and if others defect, they tend to impose spiteful punishment (Fehr & Fischbacher, 2003). Yet another important conclusion is that cultures differ dramatically in the extent of cooperation individuals demonstrate on related behavioral tasks (Henrich et al., 2005).

These basic models for the origins of cooperation have been extended and investigated to create a spectacularly fast-developing area of human knowledge (Hammerstein, 2003; Henrich, 2006). There is no way to review this huge area of work here, but it is essential to describe both the success of some main directions in this enterprise and its inability to provide fully satisfactory explanations for some human traits.

Many mathematical models outline how selection can shape traits that give advantages to groups without necessarily framing the process in terms of group selection (Lehmann & Keller, 2006; Queller & Strassmann, 2006; West, Griffin, & Gardner, 2007). Another useful line of work has modeled how cooperators can identify and selectively associate with each other (Noë & Hammerstein, 1994).

Closely related is work on how reputation effects can maintain tendencies to sacrifice, and much recent work has been done on how social systems (and natural selection) can maintain tendencies to punish defectors (Axelrod, Hammond, & Grafen, 2004; Henrich & Boyd, 2001). Many of the most profitable directions have investigated how social norms that benefit groups are preserved and transmitted (Fehr & Fischbacher, 2004). Cultural group selection models also describe how complex exchange networks within groups create new selection forces that shape human capacities for culture (Boyd & Richerson, 1985, 2005). Extensive behavioral economic investigations have explored human behavior in laboratory settings generally characterized by anonymity and small cash payments, and they are allowing initial investigations about brain mechanisms for social capacities (Fehr & Fischbacher, 2003).

Despite all this progress, there is a consensus that something is missing. Simple economic and biological views predict that individuals will maximize their personal and genetic interests, but observations from the laboratory and everyday life reveal much difficult-to-explain human generosity, moral behavior, and spitefulness. Investigations about the possible role of strong reciprocity (Gintis, 2000) and commitment strategies (Nesse, 2001) address only a tiny fraction of the problem. Two leaders in the field summarized by saying, "At the ultimate level, the evolution and role of altruistic rewarding for cooperation in larger groups remains in the dark" (Fehr & Fischbacher, 2003, p. 790). Gintis, Bowles, Boyd, and Fehr went further:

> We suspect, on the basis of the many studies completed over the past several years, that the new knowledge obtained will give us a picture of prosociality (and its obverse, antisociality) that is fundamentally incompatible with the economist's model of the self-interested actor and the biologists' model of the self-regarding reciprocal altruist. (Gintis et al., 2003, p. 169)

This chapter proposes that social selection offers an important part of the answer.

SOCIAL SELECTION

Social selection is the subtype of natural selection in which choices made by other individuals influence fitness and change gene frequencies. The next few pages outline the history of the concept and where it stands now.

To prevent confusion, note that social selection has a different meaning in sociology and epidemiology, where it refers to factors that result in overrepresentation of certain kinds of individuals in a group or neighborhood. For instance, the prevalence of schizophrenia tends to be higher in low-income neighborhoods because many people with this disease cannot afford to live elsewhere. Note also that recent use of social selection as an alternative to sexual selection is confused (Roughgarden, Oishi, & Akcay, 2006); social selection is not an alternative to sexual selection but the superordinate category of which sexual selection is a subtype. The question of whether social selection for social partners can explain traits previously attributed to social selection for sexual partners remains interesting.

The full range of social selection's applications is increasingly recognized and modeled (Frank, 2006; Simon, 1990; Tanaka, 1996; Wolf, Brodie, & Moore, 1999), but it was first explored in depth in a pair of papers by West-Eberhard (West-Eberhard, 1979, 1983). She defined social selection in relation to its exemplar and subtype, sexual selection:

> *Sexual selection* refers to the subset of social competition in which the resource at stake is mates. And *social selection* is differential reproductive success (ultimately, differential gene replication) due to differential success in social competition, whatever the resource at stake. (West-Eberhard, 1983, p. 158)

In sexual selection, females choose males with extreme displays, giving them a fitness payoff of increased matings, which shapes yet more extreme male displays and stronger female preferences for mates who can muster extreme displays. As already noted, social selection is the same, except both sexes choose, and the payoffs are social resources other than matings.

Geoffrey Miller argued that sexual selection shapes capacities for morality and culture along with extreme displays of prowess and fitness (Miller, 2000, 2007). This seems plausible, and it is surprising that the idea is not more widely discussed. Miller emphasized, however, the dominance of sexual selection over other kinds of social selection, saying, "Other forms of social selection are important, but mostly because they change the social scenery behind sexual selection" (2000, p. 13). In contrast, the thesis here is that nonsexual forms of social selection are independently powerful whether or not they influence mating success. Being preferred as a social partner can give manifold benefits that increase fitness by routes other than directly getting more or better mates. The relative importance of sexual and nonsexual social selection is an empirical question.

Darwin recognized sexual selection as a kind of social selection resulting from mate choices, and he attributed moral and cognitive capacities to this process (Darwin, 1871), but it took a long time for this lead to be further developed (Cronin, 1991). Wynne-Edwards offered many examples of social selection, especially those related to hierarchy, although he misidentified many as results of group selection:

> The hierarchy is a purely internal phenomenon arising among the members of a society, but it can nevertheless enormously affect their individual expectations of life and reproduction. Its establishment places in their own hands, therefore, a powerful selective force which can conveniently be described as social selection. It is similar in character to the process Darwin believed to apply in the more restricted field of sexual selection. (Wynne-Edwards, 1962, p. 139)

The idea was further developed by Crook (1972), who mentioned it only briefly:

> Social selection, then, is that process leading to the evolutionary enhancement of morphological allesthetic and behavioral characteristics that function within a social system to provide biological advantages to the individual in relation to survival prior to reproduction, the formation of zygotes, and the birth and rearing to maturity of young or the progeny of close kin. Direct

> competition, often by means of ritualized display, is usually involved. Social selection results from (a) the effects of competition between the subject and others of either sex with respect of commodities essential to survival in a situation that will allow an attempt at reproduction, (b) competition for access to preferred members of the opposite sex for mating, and (c) effects of competition between subjects for access to commodities of the environment and social for the reading of the young in the troop below reduction age. Of these b is the process most commonly referred to as sexual selection. Social selection is undoubtedly one of the main evolutionary processes responsible for the emergence of both individual and group behavioral characteristics. (Crook, 1972, p. 264)

West-Eberhard was the first to explore the power of social selection to explain many traits in plants and nonhuman animals. For instance, she noted that the extraordinary diversity and colorful profusion of floral extravagances result from competition for pollinators. She also attended closely to the runaway characteristics of social selection:

> Social characters—the weapons used in fighting, and the signal used in ritualized combat and competitive display—should evolve rapidly, for they are under a special strong selection (slight variations in these characters are associated with great variance and reproductive success). … There is a potential for mutually accelerating selection for attractiveness and choice whenever one class of individuals is in a position to choose the winners among those competing. (West-Eberhard, 1983, pp. 158, 160)

In the earlier paper, she suggested that this runaway process could explain primate intelligence:

> It is tempting to speculate that the explosive evolutionary increase in the proto hominid brain size, which had the appearance of a "runaway" process, was associated with the advantage of intelligence in the maneuvering and plasticity associated with social competition in primates. (West-Eberhard, 1979, p. 228)
>
> Thus, the evolution of higher levels of social integration can be an "emergent" and result of selection on competing individuals. … As Wilson (1975) has foreseen, a major synthesis regarding social behavior and natural selection is in the making. I believe, and the intuitions of other authors confirm, that it will develop along the lines crudely sketched in this essay. The main theme is competition within groups, and its special consequences: competitive rituals and displays, "runaway" specialization in traits contributing to social success, intraspecific character divergence (the evolution of alternative strategies), mutually exclusive specializations, divisions of labor, mutual dependence, and social integration. (West-Eberhard, 1979, p. 233)

Competition within groups, rituals and displays, divisions of labor, and social integration—this list could hardly be more germane to the origins of human capacities for culture. Almost 30 years later, a full application of social selection to the evolution of human behavioral characteristics is a project still getting underway.

SOCIAL SELECTION FOR PARTNERS

There seems to have been some tipping point beyond which human social cognitive evolution took off fast. Once beyond this point, some process entered a positive feedback cycle that increased the strength of selection. Although many traits evolved together to make advanced sociality possible, one candidate for a tipping point is the moment when the number and quality of relationships began to strongly influence fitness. In chimpanzees, alliances influence mating success (de Waal, 1982; Smuts, 1987), and bonobos seem to spend much of their time negotiating less hierarchical but more complex relationships (de Waal & Lanting, 1997). Laboratory studies find that chimpanzees consistently prefer a cooperative partner in a task that requires mutual action (Melis, Hare, & Tomasello, 2006). Only humans, however, get major fitness benefits from exchanging multiple different resources with many other individuals with time delays that make defection a potential problem. Humans are, consequently, exquisitely sensitive to social rejection (Maner, DeWall, Baumeister, & Schaller, 2007). Once fitness is significantly influenced by one's desirability as a social partner, individuals with the best displays of resources and generosity will get increasing advantages. At this point, social selection for displays of partner value and generosity could enter a runaway cycle that shapes extreme human social traits not found in other species.

What kinds of traits should we expect social selection to shape? It should shape traits that make an individual preferred as a social partner, including (a) high levels of resources (health, vigor, personal skills, powerful allies, status, territory, and other resources), (b) tendencies to share those resources reliably and selectively with relationship partners, (c) accurate intuitions about what others are seeking in a partner, and (d) strong motivations to please partners and other in-group members.

These characteristics are not culture itself, of course, but they are traits that make humans capable of culture. Note that the goal here is not to explain culture itself. Culture emerges from human cognitive capacities and takes on a life of its own. It could emerge, however, only after humans gained the requisite cognitive and motivational capacities. As already noted, social structures, from dyads to groups to cultures, create new selection forces that further shape capacities for social cognition. Our goal here is to assess the role of social selection at the initiation of that process, and its subsequent role after culture itself influences the fitness of variations in cognitive capacities.

Social selection offers an explanation for the central role of display of resources and generosity in every culture. From potlatch ceremonies to conspicuous consumption (Veblen, 1899), they are often referred to as "status displays," but they may increase fitness by attracting and keeping the best possible social partners. The importance of reliable resource sharing is reflected in the ubiquity of social norms (Fehr & Fischbacher, 2004), gossip about violators (Dunbar, 1996), and preoccupation with group boundaries, conformity, and the rights and responsibilities of group members.

The motivation to please others is obvious and ubiquitous, but selfishness and norm violations get more attention. We are attuned to detect such violations for the very good reason that deception is ubiquitous (Cosmides & Tooby, 1992). However, social selection and prisoner's dilemma models, highlight different risks. In models based on the prisoner's dilemma model, every interaction involves a risk that the other will defect. Constant vigilance is warranted, as is constant assessment about whether one's interests are served best by cooperating or defecting. In contrast, social selection shapes extreme vigilance about how others are judging you. This is not only to preserve reputation to maximize personal gain but because selection has directly shaped intense wishes to please others by acting in whatever ways will make one a preferred partner. This should not be pushed too far. Cultures, and individuals within cultures, vary greatly in strategies for maintaining cooperation (Henrich et al., 2006) and how much they care about what others think about them. There is no one essential human nature to be explained. Nonetheless, social selection may have played important roles in shaping human capacities that make us capable of culture.

SOCIAL SELECTION FOR CAPACITIES FOR CULTURE

I have emphasized the importance of social selection as an explanation for human prosociality. This is congruent with tendencies to emphasize the role of cooperation in aspects crucial to culture, including getting mates, hunting, agriculture, warfare, or tool making. In contrast to approaches that emphasize one or another product of selection, however, social selection calls attention to one process that shapes a whole suite of traits. It pulls the focus away from the products of selection and their utility and toward the process that influences fitness for individuals who vary on certain behavioral traits. Although this process depends on an individual's behavior, fitness is directly influenced by what happens in the brains of other individuals. Fitness changes depending on whether others choose to accept or reject, help or walk away.

This suggests that we should expect extraordinarily pervasive, subtle, and expensive attempts to influence others to prefer one as a partner. I use the word *influence* instead of *manipulate* because others are on the lookout for self-serving manipulators. The perspective from social selection is, in some respects, the opposite of that from the prisoner's dilemma. An appearance of being Machiavellian toward relationship partners dramatically decreases one's value as a partner. Conversely, the appearance of genuine caring about the welfare of partners increases one's value and should be selected for. Although false displays and deceptions are inevitable, when partner choice has high payoffs, deception strategies will be especially risky. For every sociopath in modern Western societies, there are 10 overly conscientious, socially anxious individuals who are constantly preoccupied with pleasing others and maintaining relationships (Grant et al., 2005).

In short, social selection can shape tendencies to genuine altruism, by which I mean helping motives shaped directly by selection, with calculation of costs and benefits as a secondary, not a primary, consideration. Such genuine altruism can give a net long-term payoff, offering a possible solution to the "mystery of altruism." If a relationship gets badly out of balance, altruistic motivations tend to fade, as expected. The perspective from social selection, however, emphasizes that inexpensive decisions made by others have major influences on an individual's fitness.

The result may be motives to try to discern what will please others and to provide it if that is possible at reasonable cost—the very model of prosociality.

As noted already, prosociality is but one trait expected from social selection for social partners. Deep enduring relationships are something beyond cooperation, and having powerful motives for creating and maintaining enduring close relationships is distinctly human. Individual relationships in baboons and chimpanzees can be important (Smuts, 1985), but they are nowhere near the intensity or importance seen in humans (Mills & Clark, 1994). Likewise, chimpanzees and baboons chose alliance partners based on kinship, size, strength, and social status (Smuts, 1987). Only humans, however, make discerning choices among potential partners based also on their ability to help with special skills or knowledge, their wealth, and their moral character. Although studies of human relationships tend to neglect the role of simple alliances, many relationships involve much more than mere alliance (Mills & Clark, 1994). Displays of power and resources are ubiquitous among animals, and, as noted by Wynne-Edwards (1962), they are important for competing in hierarchies that exist because of patterns of social choices. But only humans invest heavily in costly displays of generosity and conformity to social norms.

A variety of other mental traits give advantages when social selection influences fitness. Ability to recognize and remember other individuals is essential, thus helping to explain specialized mechanisms for face recognition and for remembering cheaters (Mealey, Daood, & Krage, 1996). The benefits of knowing others' expectations and needs should shape close attention to others' preferences and a theory of mind to anticipate what will please or annoy them (Povinelli & Giambrone, 2001; Wellman & Cross, 2001). The benefits of being able to predict how complex social relationships will play out have been proposed as an early engine of social evolution (Humphrey, 1976). Because the costs of being rejected or excluded are high, strong motivation to inhibit selfish and impulsive behavior will give advantages in most situations. Social anxiety inhibits and punishes actions disliked by others (Nesse, 1990). A tendency to conformity may arise by social selection if individuals prefer to associate with those like themselves and if they discriminate against outsiders. Social selection magnifies the benefits of language skills for communicating and understanding others, especially by making promises and threats effective means of social influence (Nesse, 2001).

Social selection shapes much more than prosocial tendencies. Flares of temper, for instance, give a selective benefit by influencing the behaviors of others. Others can be strongly influenced if they can be convinced that an emotional individual will act in ways contrary to his or her self-interest. This shapes emotions that are extremely emotional and unpredictable. Likewise, spiteful behavior gives advantages by influencing the behavior of others who quickly recognize that defecting from such a partner will be costly. These examples are at the intersection of social selection and commitment theory (Nesse, 2001).

SOCIAL SELECTION AND HUMAN DOMESTICATION

I began with the observation that humans are so exquisitely attuned to the wishes of others that we seem almost to have been domesticated. Of course, no other species domesticated us. It seems useful, however, to consider the possibility that we domesticated ourselves (Leach, 2003). Self-interested social choices gave selective advantages to those of our ancestors who were less aggressive, more generous, and better able to intuit what others wanted. Preferred partners got advantages, those who preferred the best partners got advantages, and the preference and displays ran away to shape extreme prosociality that is otherwise difficult to explain.

Adam Smith's (Smith, 1759/1976) "invisible hand" offers an interesting parallel. It refers to how self-interested individual choices give rise to mutually beneficial economic systems. Sellers who charge too much lose customers. Employers who pay too little lose employees. Goods in short supply command increased prices that induce increased production of those goods so supply meets demand. The result is mutually beneficial cooperation beyond what any planning process can create. Such capitalistic systems are, of course, notable for their unfairness and inefficiencies. The large picture, however, is one of remarkably efficient and smoothly operating markets that emerge from the actions of individuals pursing their own self-interest.

Markets work because the self-interested choices of agents create incentives that shape markets to the mutual benefit of most participants. The role of market models in social biology and the crucial role of partner choice in fostering cooperation are just now being recognized (Noë, 1990, 2001), but antecedents go back a long way. Adam Smith used the phrase "invisible hand" only once in *The Theory of Moral Sentiments* (Smith, 1759/1976), but his idea offers a remarkable parallel to how social selection can shape human capacities for cooperation and culture.

Individuals advertise what they have to offer in relationships, and others make choices among potential partners. Those who offer others the most, while asking the least in return, are chosen more often by better partners and gain correspondingly increased benefits. Deceptive display and defection are always possible, but reputation monitoring and the ability to switch partners limit their utility. Trustworthy individuals are preferred not only because defection is less likely but because they reduce monitoring costs. If individuals can convince each other that their mutual commitment goes beyond mere exchange to genuine friendship, they get considerable additional benefits by having help available when it is most needed (Nesse, 2001).

On the individual level, this process also unfolds during development. The dog owner rewards desired behavior and punishes disobedience. By simple reinforcement, the dog's behavior maximizes not only its own interests but those of its owner as well. Children are socialized not only by such rewards but by their evolved tendencies to conform and to take in cultural norms as their own. The cascading effects of culture have been though so many iterations that it is hard to even describe how selection shaped the mechanisms that facilitate enculturation. It is easy, however, to see that it did, and it is remarkable to observe the reliability of these mechanisms.

Viewing human social evolution as domestication by social selection offers a perspective quite different from that of economic models that must confront costly cooperation as a mystery. Individuals pursuing self-interest make thousands of social choices. These choices are the invisible hand that domesticated our species. Each lifetime recapitulates the process; tiny social cues act on mechanisms shaped by selection to detect and respond to them, steadily but firmly rewarding increasingly prosocial behavior.

Of course, it does not always work. Some people have few relationships; others get trapped in unsatisfactory relationships. Some people spend so much time and effort trying to please others that their life is a burden. Others mistrust others so completely that their relationships can make use of only crude exchanges. Understanding the origins of such individual differences is a task different from trying to understand the selection forces that shaped human social capacities. They are related, and the perspective of social selection should illuminate phenomena such as sociopathy and excess scrupulosity, but these are topics for another occasion. For now, it is important mainly to note the wide variation of human social phenotypes, the good evolutionary reasons for this, and the implication that attempts to explain essentialized human nature are inconsistent with biology.

CONCLUSION

Social selection is an important subtype of natural selection that can explain some human traits that are otherwise difficult to understand. Its focus on the role of partner choice calls attention to the fitness effects of decisions made by other individuals and thus to the fitness benefits of trying to understand what others want and how to get them to prefer one as a partner. The benefits of such displays and choices create escalating positive feedback cycles that result in extreme traits with high costs. These traits, such as strong motives to please others, give a net long-term benefit on the average, but like the peacock's tail, they can also have substantial negative effects. Sexual selection increases the magnitude of a display until its fitness advantages from increased matings are balanced by other costs such as energy expenditures and increased vulnerability to predation. Social selection increases the magnitude of prosocial traits until the benefits of getting more and better partners are balanced by personal costs incurred by creating displays, following norms, fulfilling commitments, and helping others. The positive feedback in this process offers an explanation for how selection could have shaped such extraordinarily costly social traits.

Culture is by no means explained by this process alone. It seems likely, however, that social selection was of particular importance at the transition point where culture became possible. Once complex culture emerges, it becomes, as many have noted (Barkow, 1989; Boyd & Richerson, 1985; Dunbar, Knight, & Power, 1999; Durham, 1991; Henrich & Henrich, 2006; Lancaster, 1975; Lumsden & Wilson, 1981; Norenzayan et al., 2006), a force of selection in its own right, shaping yet more complex social capacities that result in yet more complex cultures in which social selection becomes even more important in shaping "endless forms most beautiful and most wonderful" (Darwin, 1859).

REFERENCES

Alexander, R. D. (1979). *Darwinism and human affairs*. Seattle: University of Washington Press.

Alexander, R. D. (2005). Evolutionary selection and the nature of humanity. In V. Hösle & C. Illies (Eds.), *Darwinism and philosophy* (pp. 301–348). Notre Dame, IN: University of Notre Dame Press.

Axelrod, R. M. (1984). *The evolution of cooperation*. New York: Basic Books.

Axelrod, R. M. (1997). *The complexity of cooperation: Agent-based models of competition and collaboration*. Princeton, NJ: Princeton University Press.

Axelrod, R., Hammond, R. A., & Grafen, A. (2004). Altruism via kin-selection strategies that rely on arbitrary tags with which they coevolve. *Evolution*, *58*(8), 1833–1838.

Barkow, J. H. (1989). *Darwin, sex, and status: Biological approaches to mind and culture*. Toronto, ON: University of Toronto Press.

Boyd, R., & Richerson, P. J. (1985). *Culture and the evolutionary process*. Chicago: University of Chicago Press.

Boyd, R., & Richerson, P. J. (2005). *The origin and evolution of cultures*. New York: Oxford University Press.

Cosmides, L., & Tooby, J. (1992). Cognitive adaptations for social exchange. In J. H. Barkow, L. Cosmides, & J. Tooby (Eds.), *The adapted mind: Evolutionary psychology and the generation of culture* (pp. 163–229). New York: Oxford University Press.

Cronin, H. (1991). *The ant and the peacock: Altruism and sexual selection from Darwin to today*. New York: Cambridge University Press.

Crook, J. H. (1972). Sexual selection, dimorphism, and social organization in the primates. In B. Campbell (Ed.), *Sexual selection and the descent of man* (pp. 231–281). Chicago: Aldine.

Darwin, C. (1859). *The origin of species*. London: John Murray.

Darwin, C. (1871). *The descent of man and selection in relation to sex*. London: John Murray.

Dawkins, R. (1976). *The selfish gene*. Oxford, UK: Oxford University Press.

de Waal, F. B. M. (1982). *Chimpanzee politics: Power and sex among apes*. New York: Harper and Row.

de Waal, F. B. M., & Lanting, F. (1997). *Bonobo: The forgotten ape*. Berkeley: University of California Press.

Dunbar, R. I. M. (1996). *Grooming, gossip, and the evolution of language*. Cambridge, MA: Harvard University Press.

Dunbar, R. I. M., Knight, C., & Power, C. (1999). *The evolution of culture: An interdisciplinary view*. New Brunswick, NJ: Rutgers University Press.

Durham, W. H. (1991). *Coevolution: Genes, culture, and human diversity*. Stanford, CA: Stanford University Press.

Fehr, E., & Fischbacher, U. (2003). The nature of human altruism. *Nature*, *425*(6960), 785–791.

Fehr, E., & Fischbacher, U. (2004). Social norms and human cooperation. *Trends in Cognitive Science*, *8*(4), 185–190.

Frank, S. A. (2006). Social selection. In C. W. Fox & J. B. Wolf (Eds.), *Evolutionary genetics: Concepts and case studies* (pp. 350–363). New York: Oxford University Press.

Gintis, H. (2000). Strong reciprocity and human sociality. *Journal of Theoretical Biology*, *206*, 169–179.

Gintis, H., Bowles, S., Boyd, R., & Fehr, E. (2003). Explaining altruistic behavior in humans. *Evolution and Human Behavior*, *24*(3), 153–172.

Grant, B. F., Hasin, D. S., Blanco, C., Stinson, F. S., Chou, S. P., Goldstein, R. B., et al. (2005). The epidemiology of social anxiety disorder in the United States: Results from the National Epidemiologic Survey on alcohol and related conditions. *Journal of Clinical Psychiatry*, *66*(11), 1351–1361.

Hamilton, W. D. (1964). The genetical evolution of social behavior I, and II. *Journal of Theoretical Biology*, *7*, 1–52.

Hammerstein, P. (2003). *Genetic and cultural evolution of cooperation*. Cambridge, MA: MIT Press in cooperation with Dahlem University Press.

Henrich, J. (2006). Social science: Cooperation, punishment, and the evolution of human institutions. *Science*, *312*(5770), 60–61.

Henrich, J., & Boyd, R. (2001). Why people punish defectors: Weak conformist transmission can stabilize costly enforcement of norms in cooperative dilemmas. *Journal of Theoretical Biology*, *208*(1), 79–89.

Henrich, J., Boyd, R., Bowles, S., Camerer, C., Fehr, E., Gintis, H., et al. (2005). "Economic man" in cross-cultural perspective: Behavioral experiments in 15 small-scale societies. *Behavioral and Brain Sciences*, *28*(6), 795–815.

Henrich, J., & Henrich, N. (2006). Culture, evolution and the puzzle of human cooperation. *Cognitive Systems Research*, *7*, 220–245.

Henrich, J., McElreath, R., Barr, A., Ensminger, J., Barrett, C., Bolyanatz, A., et al. (2006). Costly punishment across human societies. *Science*, *312*(5781), 1767–1770.

Humphrey, N. K. (1976). The social function of intellect. In P. G. Bateson & R. A. Hinde (Eds.), *Growing points in ethology* (pp. 303–318). London: Cambridge University Press.

Lancaster, J. B. (1975). *Primate behavior and the emergence of human culture*. New York: Holt, Rinehart and Winston.

Leach, H. M. (2003). Human domestication reconsidered. *Current Anthropology*, *44*(3), 349–368.

Lehman, D. R., Chiu, C.-Y., & Schaller, M. (2004). Psychology and culture. *Annual Review of Psychology*, *55*, 689–714.

Lehmann, L., & Keller, L. (2006). The evolution of cooperation and altruism: A general framework and a classification of models. *Journal of Evolutionary Biology*, *19*(5), 1365–1376.

Levinson, S. C., & Jaisson, P. (2006). *Evolution and culture*. Cambridge, MA: MIT Press.

Lumsden, C. J., & Wilson, E. O. (1981). *Genes, mind, and culture*. Cambridge, MA: Harvard University Press.

Maner, J. K., DeWall, C. N., Baumeister, R. F., & Schaller, M. (2007). Does social exclusion motivate interpersonal reconnection? Resolving the "porcupine problem." *Journal of Personality and Social Psychology*, *92*(1), 42–55.

Mealey, L., Daood, C., & Krage, M. (1996). Enhanced memory for faces of cheaters. *Ethology and Sociobiology*, *17*(2), 119–128.

Melis, A. P., Hare, B., & Tomasello, M. (2006). Chimpanzees recruit the best collaborators. *Science*, *311*(5765), 1297–1300.

Miller, G. F. (2000). *The mating mind: How sexual choice shaped the evolution of human nature*. New York: Doubleday.

Miller, G. (2007). Sexual selection for moral virtues. *Quarterly Review of Biology*, *82*(2), 97–126.

Mills, J., & Clark, M. S. (1994). Communal and exchange relationships: Controversies and research. In R. Erber & R. Gilmour (Eds.), *Theoretical frameworks for personal relationships* (pp. 29–42). Hillsdale, NJ: Lawrence Erlbaum.

Nesse, R. M. (1990). Evolutionary explanations of emotions. *Human Nature*, *1*(3), 261–289.

Nesse, R. M. (2001). *Evolution and the capacity for commitment*. New York: Russell Sage Foundation.

Nesse, R. M. (2007). Runaway social selection for displays of partner value and altruism. *Biological Theory*, *2*(2), 1–13.

Noë, R. (1990). A veto game played by baboons: A challenge to the use of the prisoner's dilemma as a paradigm for reciprocity and cooperation. *Animal Behavior*, *39*, 78–90.

Noë, R. (2001). Biological markets: Partner choice as the driving force behind the evolution of mutualisms. In R. Noë, J. A. R. A. M. van Hooff, & P. Hammerstein (Eds.), *Economics in nature: Social dilemmas, mate choice and biological markets* (pp. 93–118). Cambridge, UK, and New York: Cambridge University Press.

Noë, R., & Hammerstein, P. (1994). Biological markets: Supply and demand determine the effect of partner choice in cooperation, mutualism and mating. *Behavioral Ecology and Sociobiology*, *35*, 1–11.

Norenzayan, A., Schaller, M., & Heine, S. (2006). Evolution and culture. In J. S. M. Schaller & D. Kenrick (Ed.), *Evolution and social psychology* (pp. 343–366). New York: Psychology Press.

Povinelli, D. J., & Giambrone, S. (2001). Reasoning about beliefs: A human specialization? *Child Development*, *72*(3), 691–695.

Queller, D. C., & Strassmann, J. E. (2006). Models of cooperation. *Journal of Evolutionary Biology*, *19*(5), 1410–1412.

Richerson, P. J., & Boyd, R. (2005). *Not by genes alone: How culture transformed human evolution*. Chicago: University of Chicago Press.

Ridley, M. (1997). *The origins of virtue: Human instincts and the evolution of cooperation.* New York: Viking.

Roughgarden, J., Oishi, M., & Akcay, E. (2006). Reproductive social behavior: Cooperative games to replace sexual selection. *Science*, *311*(5763), 965–969.

Simon, H. A. (1990). A mechanism for social selection and successful altruism. *Science*, *250*, 1665–1668.

Smith, A. (1976). *The theory of moral sentiments*. Oxford, UK: Clarendon. (Original work published 1759)

Smuts, B. B. (1985). *Sex and friendship in baboons*. New York: Aldine.

Smuts, B. B. (1987). *Primate societies*. Chicago: University of Chicago Press.

Stevens, J. R., Cushman, F. A., & Hauser, M. D. (2005). Evolving the psychological mechanisms for cooperation. *Annual Review of Ecology, Evolution, and Systematics*, *36*(1), 499–518.

Tanaka, Y. (1996). Social selection and the evolution of animal signals. *Evolution*, *50*, 512–523.

Trivers, R. L. (1971). The evolution of reciprocal altruism. *Quarterly Review of Biology*, *46*, 35–57.

Veblen, T. (1899). *The theory of the leisure class: An economic study in the evolution of institutions*. New York: Macmillan.

Wellman, H. M., & Cross, D. (2001). Theory of mind and conceptual change. *Child Development*, *72*(3), 702–707.

West, S. A., Griffin, A. S., & Gardner, A. (2007). Social semantics: Altruism, cooperation, mutualism, strong reciprocity and group selection. *Journal of Evolutionary Biology*, *20*(2), 415–432.

West-Eberhard, M. J. (1979). Sexual selection, social competition, and evolution. *Proceedings of the American Philosophical Society*, *123*(4), 222–234.

West-Eberhard, M. J. (1983). Sexual selection, social competition, and speciation. *Quarterly Review of Biology*, *58*(2), 155–183.

Williams, G. C. (1966). *Adaptation and natural selection: A critique of some current evolutionary thought*. Princeton, NJ: Princeton University Press.

Wolf, J. B., Brodie, E. D., III, & Moore, A. J. (1999). Interacting phenotypes and the evolutionary process. II: Selection resulting from social interactions. *American Naturalist*, *153*, 254–266.

Wynne-Edwards, V. C. (1962). *Animal dispersion in relation to social behavior*. Edinburgh, UK: Oliver and Boyd.

11

Are Our Minds Fundamentally Egalitarian? Adaptive Bases of Different Sociocultural Models About Distributive Justice

TATSUYA KAMEDA, MASANORI TAKEZAWA, YOHSUKE OHTSUBO, and REID HASTIE

Egalitarianism is a morally and ideologically charged concept with many facets. It has been relied on for solutions to moral dilemmas in almost every aspect of human life, including economic, political, legal, racial, gender, and religious domains. As far as material egalitarianism is concerned, however, the ideology that dictates equal sharing of key resources regardless of individual contributions seems to be out of place in today's world, when meritocracy and related capitalist ideologies appear to be becoming more and more dominant. In line with such observations, Francis Fukuyama, an influential American political economist and social philosopher, has argued that the progression of human history as a struggle between ideologies has ended, with the world settling on a capitalist liberal democracy since the end of the cold war and the collapse of the Berlin Wall in 1989 (Fukuyama, 1992). Yet, more than 15 years after the publication of *The End of History and the Last Man*, we (including Fukuyama himself, 2006) are not so sure if the prophecy has been (or will be) fulfilled in a world that is divided by so many religious, cultural, and political barriers.

This chapter takes an adaptationist approach to the ubiquity of egalitarian sharing in human societies. We argue that, although income inequality sharply divides industrialized societies all over the world, material egalitarianism may still operate as a fundamental principle affecting social sharing and exchange under uncertainty in many domains of human activity. We present some preliminary evidence for this claim from an original social survey, ethnographies in anthropology, evolutionary computer simulations, and behavioral experiments. We then discuss the broader implications of these data sets, concluding with a conjecture that human minds may be structured fundamentally as egalitarian sharers and that merit-based ideologies which pervade our modern lives may be seen as an adjustment on top of such a psychological foundation.

IS EGALITARIAN SHARING WORKING IN COMPETITIVE SOCIETIES?

Where can we find egalitarianism alive in the modern, competitive societies? One obvious example, morally motivated by this notion, is found in our tax systems. Progressive taxation whereby people with more income pay a higher percentage of it in taxes is common across many liberal democratic societies including the United States, Japan, and the United Kingdom. In principle, income redistribution through such tax systems in combination with various social welfare policies reduces gaps between the rich and the poor, trending toward ultimate equality. Other less obvious examples may be found in medical policies concerning organ transplantation and in workers' attitudes about job layoffs (Elster, 1992). In most countries, organ recipients are selected not only on the basis of need and compatibility but also on a somewhat egalitarian scheme that considers how long the patients have been on the waiting list. Likewise, there is some evidence that people prefer work-sharing (i.e., working fewer hours) to layoffs as a solution to a shrinking labor market. In either case, a purely market-driven solution (e.g., buying and selling of organs) is avoided.

DIFFERENTIAL ENDORSEMENT OF EGALITARIAN IDEOLOGY

As illustrated by these examples, material egalitarianism seems to be still alive and sometimes institutionalized in some key institutions of our societies. Do we actually endorse egalitarian sharing as a general social policy? What factors underlie people's (possibly different) preferences for distribution rules? We conducted a preliminary survey to address this question with students from seven Japanese universities. These students answered a series of questions regarding their preferences for various resource allocation schemes. One example question is shown in Figure 11.1.

Division A is an allocation proportional to individual production levels (referred to as an equity rule in social psychology; Adams, 1965), whereas Division B follows an egalitarian rule. Notice that the recipient's (Y's) share is the same in the two divisions. Note this question asked about preferences for resource allocation beyond direct self-interest—preferences for a desirable *social* state rather than a desirable *personal* state. Among a total of 1,322 Japanese students who answered the question, 73% chose Division B, which dictated an egalitarian allocation. More important, the proportion of egalitarian-ideology endorsers differed substantively across the seven schools, ranging from 63% to 83%. How can we explain these

> Imagine that you participated in a TV quiz show as a member of a 3-person team. You were teamed with two other persons, *X* and *Z*, whom you met for the first time at the show. In the quiz, *X* gave 2 answers correctly, you (=*Y*) gave 3 answers correctly, and *Z* gave 4 answers correctly. Thus, in total, your team answered 9 questions correctly, winning the first prize, 180,000 yen.
>
> As to how to share the award, which division would you prefer between the following two?

	X	*Y* (=*You*)	*Z*
Division A:	40,000	60,000	80,000
Division B:	60,000	60,000	60,000

Figure 11.1 An example question about distributive ideologies used in the survey.

differences? Although these universities differed along many dimensions, including urban versus rural, size of the student body, private versus public, and so on, only one factor was correlated with the differences in the proportion of egalitarian endorsers—the social rank of the university. Students in the less prestigious schools, who tended to be from working-class families, endorsed the egalitarian ideology at higher rates. The correlation between the social rank of the university and the proportion of egalitarian-ideology endorsers was substantial ($r = -.85$). A hierarchical linear model analysis further confirmed this observation; after controlling for the macrolevel (university) factor, socioeconomic status (e.g., income, education) of each student's parents was the strongest predictor of the student's attitude toward the egalitarian ideology.

WHERE DO THESE SOCIAL-CLASS DIFFERENCES COME FROM?

These results suggest that distributive ideologies may be different in middle-class versus working-class populations. Then, why do blue-collar, working-class citizens endorse the egalitarian ideology more strongly than white-collar, middle-class citizens? Although the two classes could vary on many dimensions, we conjecture that the differential degree of uncertainty concerning the supply of vital resources in life (e.g., jobs, housing, health) may be a key. It is true that in modern societies various buffers operate to manage uncertainty about resources, including pension funds, health insurance, and so on. Yet, the availability of such buffers may differ across individuals within the same society, along with the availability of other *personal* defenses against uncertain fate (e.g., personal wealth, education). Compared to white-collar citizens, blue-collar citizens have less access to such buffers and are more susceptible to injury from various life uncertainties. An egalitarian system, based on a distributive ideology dictating equal allocation of resources regardless of members' different production levels, would buffer some of the damage that unexpected life events inflict on individual welfare.

Thus, we argue that egalitarian ideology, as endorsed by blue-collar, working-class citizens, may be seen as a *collective* solution to cope with life uncertainties, when *personal* solutions (e.g., wealth, education) are unavailable. We also think that this type of collective solution for life uncertainties may have operated in many human societies until quite recently, possibly supporting the evolution of psychological algorithms designed to deal with resource uncertainty and sharing. In the following, we develop this argument with several lines of evidence, including ethnographies in anthropology, evolutionary computer simulations, and behavioral experiments.

ETHNOGRAPHIES IN ANTHROPOLOGY

Sharing important resources, such as food, with nonkin associates is a general practice in human societies. Although a primitive form of food sharing is known in some primates (de Waal, 1996), no primate other than humans has broad social-sharing habits. Anthropologists have studied social exchange and sharing in various hunter-gatherer societies to explore their origins and early forms. Kaplan and Hill (1985) observed that food transfers among the Ache foragers, who live in subtropical eastern Paraguay, show markedly different patterns for hunted meat (e.g., peccary) and collected resources (e.g., cassava). Although some collected resources are shared with nonfamily members, hunted meat is much more likely to be the target of communal sharing. Because sharing with kin is a universal practice across many species (Hamilton, 1963), the central question here is why hunted meat is shared communally beyond the acquirer's direct kin and why different sharing norms apply to different resources within the same culture.

Risk-Reduction Hypothesis

Kaplan and Hill (1985) explained these differences in terms of the degree of uncertainty involved in resource acquisition. Although provision of vegetables and fruits is relatively stable and dependable, acquisition of meat is a highly variable, uncertain prospect. On average, there is a 40% chance that an Ache hunter will come back from a hunt empty-handed. It is essential for the Ache to manage the variance associated with meat acquisition, securing a stable supply of the precious resource. Storage by freezing or other preservation methods is not efficient in a hunter-gatherer situation. Kaplan and Hill (1985) argued that, instead, the sharing system functions as a collective risk-reduction device. By including many individuals in the risk-pooling group, the variance in meat supply decreases exponentially. Once established and maintained, the communal-sharing system will buffer the variance in the meat supply.

Recently, Gurven (2004) provided a comprehensive review of worldwide ethnographic examples about food-transfer patterns. According to his review,

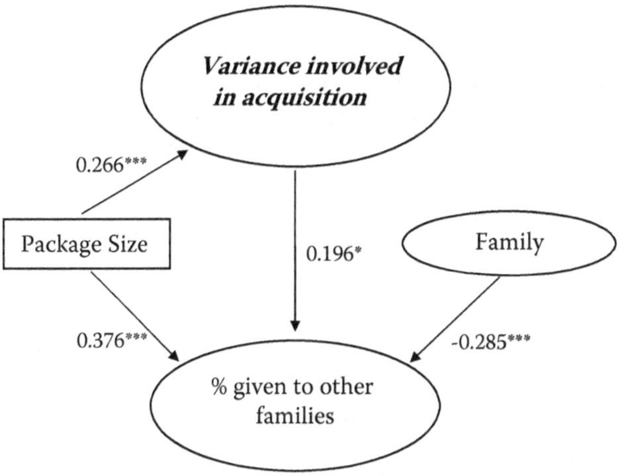

Figure 11.2 What determines how much a Hiwi acquirer gives to other families? *Note.* "Variance involved in acquisition" refers to a composite index involving the degree of asynchrony in acquisition of the resource among individuals, variation of encounter rates in search, and variation in energy per pursuit. Only significant paths are shown in the figure (based on Gurven, 2004, with minor modifications). * $p \leq .05$, *** $p < .001$.

communal sharing of hunted meat is robust across many primordial societies, including forager-agriculturalists and hunter-gatherers. Figure 11.2 depicts results of a path analysis about food transfers among the Hiwi foragers who used to live in southwestern Venezuela (based on Gurven, 2004, with minor modifications by the present authors). As can be seen in the figure, variance involved in acquisition is a significant predictor of the percentage of the food resource given to other families, when controlling for package size (e.g., big game versus small game) of the resource as well as family size of the acquirer.

Although communal sharing of hunted meat in hunter-gatherer societies does not necessarily mean that the food items are shared exactly equally (Gurven, 2004), the notion of risk reduction fits closely with our observation that the egalitarian ideology was endorsed more strongly by working-class citizens than by middle-class citizens. Yet, is the communal-sharing system robust and sustainable as a collective risk-reduction device under uncertainty?

EVOLUTIONARY COMPUTER SIMULATIONS

Problem of Egoism

The risk-reduction hypothesis is a functional explanation; the communal-sharing system serves the survival of the whole group. If the system exists, everybody is better off under uncertainty. Yet, from the adaptationist perspective that focuses on fitness outcomes to each individual, this explanation leaves one critical question unanswered—the problem of egoism in social dilemmas (Dawes, 1980). Hunted meat, especially when a large portion is acquired, is regarded as a common property in most hunter-gatherer societies; the process of meat distribution is treated as appropriation from the public domain. Then, what if some individuals behave as egoists who share in other people's acquisitions but are declining to share their own acquisitions with others? Such egoists might be better off in terms of individual fitness than are those who are loyal to the communal-sharing norm. If so, Darwinian logic implies that such egoists would proliferate, eventually coming to dominate the group. The risk-reduction explanation is incomplete, because it is silent about how the proliferation of such egoists is suppressed.

Evolutionary Games

The social dynamics as illustrated previously are analogous to biological competition for an ecological niche, in that a behavioral trait producing more beneficial outcomes spreads and eventually dominates in a population. Biologists and economists have developed a mathematical tool, evolutionary game theory, for modeling such adaptive dynamics (Gintis, 2000; Maynard Smith, 1982). Evolutionary game theory is different from classical game theory in that it does not assume that players are superhuman information processors. Instead, it represents various behavioral tendencies as strategies in a game and examines how each strategy performs against other strategies in terms of net profit. Even though a given strategy may be limited by players' information-processing capacity, it will proliferate gradually in the population if it can outperform other strategies.

We (Kameda, Takezawa, & Hastie, 2003) developed an evolutionary game theory model for the emergence of a communal-sharing norm when foraging under conditions of uncertainty. Our model assumed that, because of the highly uncertain nature of meat acquisition, an individual hunter constantly faces two kinds of decision problems: how to behave when successful, and how to behave when unsuccessful. This analysis yields four mutually exclusive and exhaustive behavioral strategies, depicted in Table 11.1; each individual in a group is assumed to behave according to one of these strategies. The model also posits that, because of the highly uncertain nature of hunting, acquisition of meat by some members yields a large asymmetry in resource level between haves and have-nots (cf. the "twists of fate" situation as conceived of by Kelley et al., 2003). A hunter's attempt to monopolize the meat under such situations can lead to fights with other community members who demand communal sharing, imposing a cost on each loser. In other words, those communal sharers, the purest supporters of an egalitarian ideology (see Table 11.1), try to punish the violator and enforce the sharing norm by engaging in costly fights. The theoretical question then becomes whether the communal sharers outperform members pursuing alternative strategies (the egoists, in particular). If the communal sharers perform well, the evolutionary logic implies that they will proliferate and dominate in the group, resulting in the establishment of a communal-sharing system.

Computer Simulations

A series of evolutionary simulations in which model parameters (group size, resource value, fighting cost) were varied systematically revealed the following results. First,

TABLE 11.1 Four Behavioral Strategies in the Evolutionary Game Model When Resource Acquisition (Hunted Meat) Is Uncertain

	When one is an unsuccessful hunter	
When one is a successful hunter	**Demands share of meat as a common property**	**Grants successful hunter's private ownership**
Provisions as a common property	*Communal sharer*	*Saint*
Claims private ownership of meat	*Egoist*	*Bourgeois*

even when communal sharers were introduced as a rare "mutant" strategy into an egoist-dominant group, they overcame the initial handicap in frequency and dominated the group rather quickly, within a few hundred iterations ("generations"). Second, once dominant, the communal sharers continued to outperform any other mutant strategies (egoist, saint, bourgeois; see Table 11.1) in fitness, thus blocking their intrusion into the group. In all simulation runs, the dominance of communal sharers continued over thousands of generations. Figure 11.3 displays a representative result of such simulation runs (see Kameda et al., 2003, for details).

In terms of evolutionary game theory, these results imply that communal sharing is an *evolutionarily stable strategy*. The egoist strategy does not qualify as such, because egoists' attempts to defend their own acquisitions against many have-nots (including other egoists who were unsuccessful) tax them heavily in fighting costs. But how sensitive is this result to model parameters such as group size, resource value, and fighting cost? A sensitivity analysis, whereby we varied the parameters systematically, revealed that the communal-sharing strategy was indeed robust.

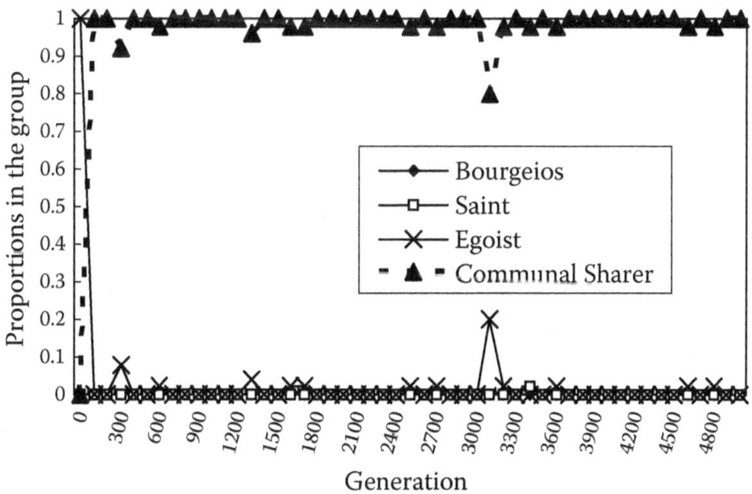

Figure 11.3 Representative results of a simulation run: Communal sharers can invade an egoist-dominant group and resist invasion by other "mutant" strategies (cited from Kameda, Takezawa, & Hastie, 2003).

For instance, except in rather unrealistic conditions in which the cost of potential injury accruing from a fight was essentially nonexistent (i.e., smaller than 0.3% of the resource value), the communal-sharing strategy always qualified as an evolutionarily stable strategy. In other words, in terms of individual fitness maximization, the communal-sharing norm consistently emerged and persisted under a broad parameter conditions, overcoming the problems of egoism and free riding in norm enforcement (Axelrod, 1986; Yamagishi, 1986).

RESOURCE-SPECIFIC ALTRUISM

Windfall as a Common Property?

Thus far, the argument has been strictly ecological: An adaptive strategy (e.g., communal sharing) should emerge in response to local ecology (e.g., a hunter-gatherer environment). Given that uncertainty in resource supply was a recurrent, adaptive problem faced by archaic hominids in the ancestral environment (Cosmides & Tooby, 1992), however, it is likely that human psychology is equipped with evolved algorithms designed specifically to deal with resource uncertainty and sharing. People's reactions to "windfall profits" may provide a case in point: People use windfall money, more often than money acquired by labor, for altruistic purposes such as treating friends or donating to charities. Although the fungible resource under consideration is the same (money), different habits seem to be triggered depending on how the resource is acquired. A common explanation for this phenomenon is provided by the labor theory of value ("money earned without effort is valued less"). Our evolutionary game theory analysis suggests, however, that the key factor triggering sharing may be the uncertainty associated with the acquisition of the resource rather than the absence of effort. As Cosmides and Tooby (1992) noted, "Information about variance in foraging success should activate different modes of operation of these algorithms, with high variance due to chance triggering a psychology of sharing" (p. 213).

Behavioral Experiments

We (Kameda, Takezawa, Tindale, & Smith, 2002) tested this possibility by conducting vignette experiments in which the uncertainty factor was manipulated independently of the effort factor. We provided Japanese and American participants with a series of hypothetical scenarios in which they (or a friend) obtained some money, either (a) contingent on investing substantial effort, (b) unexpectedly but after investing substantial effort (i.e., low contingency between effort and outcome; chance was another key factor for success, yielding high outcome variance in the situation), or (c) unexpectedly with almost no effort. Table 11.2 presents one of the scenarios used in the experiment.

Participants were then asked to rate their willingness to share the money with a friend (or the extent to which they would demand some share from a friend). The results showed that the Japanese sample tended to be more generous and demanding than the American sample, essentially replicating the previous

TABLE 11.2 Example Scenarios ("Prize-Giveaway") Used in Kameda, Takezawa, Tindale, and Smith (2002)

Condition	Scenario
Certain/High-Effort Condition	An acquaintance requested you to fill out application forms for a prize-giveaway. It was a tedious job to fill out the forms. You completed 50 forms in total. Your acquaintance paid you $100 for this service.
Uncertain/High-Effort Condition	You decided to apply for a prize-giveaway. Although it was a tedious job to fill out the application forms, you completed 50 of them to increase the chance to win. Later, you found that you won a prize of $100.
Uncertain/Low-Effort Condition	You decided to apply for a prize-giveaway and submitted one application form. Later, you found that you won a prize of $100.

finding that Japanese prefer equal distributions more than Americans (cf. Bond, Leung, & Wan, 1982). But more important, both Japanese and American participants were more willing to share (and demand more sharing for) the unexpected money, even when the amount of effort invested was identical for expected and unexpected gains. These differences were significant, even when personal ideologies about desirable distribution were controlled for; endorsers of merit-based ideology and of egalitarian ideology were both affected by the uncertainty factor.

This was also confirmed by a laboratory experiment (Study 4 in Kameda et al., 2002). After participants were paid for their work during the experiment, they were solicited to donate some money to help participants in another unrelated experiment. Even though they had received the identical amount of money for the identical amount of work, participants whose final rewards were determined in a random manner by using a roulette wheel of fortune made a greater donation than those who were rewarded in a deterministic manner. Notice that the modern notion of property rights makes no distinction between the legitimacy of entitled ownership in these two conditions.

Is Social Sharing Under Uncertainty Always Conducted Willingly?

The "just" view of egalitarianism as proposed by several moral philosophers (e.g., Rawls, 1971) implies that this rule is internalized as a basic moral value that binds us, unconditionally, under uncertainty. The "windfall psychology" could be interpreted as a manifestation of such a basic moral principle. Yet, we feel that this is probably an overstatement. Instead, we believe that an acquirer of a resource under uncertainty shows some egalitarian tendency behaviorally but not always willingly. In other words, an image of a "reluctant or cautious sharer" may be a more accurate description of the behavior than the image of a moralistic sharer.

Several lines of research provide support for this view. Eckblad and von der Lippe (1995) investigated 261 lottery winners of prizes of 1 million Norwegian kroner ($150,000). Those winners were asked about various psychological reactions after winning the prizes. The results revealed that a wish for anonymity, together with fear of envy from others, was one of the most frequent reactions among the respondents. Social sharing there, if any, could thus be characterized as a "vigilant sharing" (Erdal & Whiten, 1994), whereby distribution of the prize resulted from the vigilance of envious nonwinners who would immediately benefit from sharing.

Research on the ultimatum bargaining game by experimental economists is also illuminative of this point. Ultimatum bargaining is a two-person game in which Player 1 (proposer) divides a resource, and Player 2 (responder) then decides to either reject or accept the division. If Player 2 rejects the proposed division, both players receive nothing (Güth, Schmittberger, & Schwarze, 1982). Reviewing those studies of "one-shot" ultimatum games (played with a partner only once without switching roles), Camerer (2003) summarized the findings as follows:

> The results … are very regular. Modal and median ultimatum offers are usually 40–50 percent and means are 30–40 percent. There are hardly any offers in the outlying categories of 0, 1–10, and the hyper-fair category 51–100. Offers of 40–50 percent are rarely rejected. Offers below 20 percent or so are rejected about half the time. (p. 49)

Although exact frequencies of offers vary across societies, depending on the local cultural norms and ecological condition (see Henrich et al., 2004), no study has ever demonstrated that extremely small offers (1% to 10%) were a modal response. Notice, given that small offers are frequently rejected in the ultimatum game, it is vital for proposers to anticipate responders' expectation for fairness and to offer more equitable divisions of the resource. As we conceive it, this situation is parallel to meat sharing under uncertainty—to the extent that some (many) have-nots in the group expect a share of the meat, it is to one's personal advantage to share the meat with them.

To summarize, acquirers (lottery winners, proposers in the ultimatum game, successful hunters) should be highly sensitive to the expectations of nonacquirers. Put differently, it may be the case that nonacquirers play an active, initiative role in social sharing, either implicitly or explicitly (cf. Bliege Bird & Bird, 1997; Blurtorn Jones, 1987; Hawkes, O'Connell, & Blurton Jones, 2001). This reasoning implies that there may be some asymmetry between acquirers and nonacquirers, with the psychology of windfalls being more easily and more vigorously evoked among nonacquirers. Given that the modern notions of property rights operate in exactly the opposite way (i.e., sharing is at the acquirer's discretion), this poses an intriguing possibility awaiting future investigations.

CONCLUSION

In this chapter, we reviewed four sets of empirical and theoretical results with implications for the adaptive bases of material egalitarianism. To recapitulate, the social survey with Japanese university students showed that the egalitarian ideology was endorsed more strongly by blue-collar, working-class citizens than by white-collar, middle-class citizens. The cross-cultural ethnographies of primordial societies revealed that variance in acquisition of a food resource is a key determinant of how much of the resource is shared with nonfamily members. The agent-based computer simulations showed that such a communal-sharing strategy is evolutionarily stable in uncertain environments. Last, both American and Japanese students showed the "windfall psychology" about uncertain resources, independent of their personal distributive ideologies.

What are the overall implications of these findings? Let us speculate. Given that uncertainty in resource supply was a recurrent adaptive problem in the Environments of Evolutionary Adaptedness (EEA) and that most humans have been unable to solve this problem individually until quite recently, our minds may have been built, by evolution, as egalitarian sharers. (Again, it should be emphasized that the image of the egalitarian sharer we propose here is not "moralistic," as envisaged by Karl Marx, John Rawls, and other social philosophers, but better described as "cautious," someone who reluctantly evinces an egalitarian tendency to preempt the implicit demands of others.) An egalitarian psychology could be an evolved adaptation to high uncertainty in the EEA and, if so, should be a pan-human universal.[1] Ethnographies of hunter-gatherer groups, evolutionary game analyses of communal sharing, and the operation of a windfall psychology all support this conjecture.

Yet, this cannot be the whole story. Recall, in our survey, that white-collar, middle-class citizens were less supportive for the egalitarian ideology than blue-collar, working-class citizens. How can such variations within the same society emerge and be sustained over time? We think that these within-society variations should be viewed as a consequence of adaptations to respective local ecologies, which have taken place over many generations. Recall that in the event that personal buffers are insufficient, a communal-sharing norm and its psychological counterpart (an egalitarian ideology) may represent the only viable adaptive solution available to ancestral humans when faced with uncertainty under poor ecological conditions. As pointed out by historians, however, modernization has freed a substantial portion of the population (the middle class) from this basic state through the accumulation of personal wealth and other buffers (e.g., education). Middle-class ideology can thus be seen as a new adaptation to the enriched ecology, inserted on top of a fundamentally egalitarian mind.

It is important to realize that macrosocial systems have played a vital role in emergence and persistence of related new adaptations over time. Merit-based education and employment systems, which are cornerstones of liberal democratic societies, provide a case in point. What longitudinal consequence do these social systems yield? As repeatedly pointed out by economists and sociologists, the merit-based systems perpetuate social classes across generations—because they have greater access to economic and cultural resources, children born into middle-class families are more likely to have white-collar jobs when they come of age than children born into working-class families (Bourdieu & Passeron, 1990; Bowles, Gintis, & Groves, 2005; Coleman, 1990). This means that material conditions of the parents' generation are largely inherited by the children; the degree of life uncertainties experienced by children is also essentially similar to those experienced by their parents. Thus, fixations of different distributive ideologies and related values across middle and working classes would result via the ecological inheritance of different material conditions, as well as via differential socialization processes.

We believe that such a "niche-construction" perspective, focusing on the complex relationships between microbehavior and macrosocial structure (Laland, Odling-Smee, & Feldman, 2000), is essential to explain different sociocultural models of distributive justice. Social justice systems that exist in our world may be highly complex and varied, yet they can all be understood as adaptations to respective social and natural ecologies where people live, that is, as adjustments to local ecologies on top of the fundamentally egalitarian mind.

ACKNOWLEDGMENTS

Writing of this chapter was supported by the Grant-in-Aid for Scientific Research 17330133 from the Ministry of Education, Culture, Sports, Science, and Technology of Japan to the first author. We are grateful to Paul Wehr, Ryo Tamura, Keigo Inukai, Nobuye Ishibashi, Megumi Iwabuchi, and Joanna Schug for their helpful comments.

NOTE

1. We conjecture that the type of psychological universality here falls into the category of "functional universality" (Norenzayan & Heine, 2005), where the focal psychological process operates in all cultures, serves an identical adaptive function, but can vary across cultures in the extent to which it is activated. The experimental result that resource sharing was enhanced under uncertainty in both Japan and the United States but to a different extent is in line with this speculation.

REFERENCES

Adams, J. S. (1965). Inequity in social exchange. In L. Berkowitz (Ed.), *Advances in experimental social psychology* (Vol. 2, pp. 267–299). New York: Academic Press.

Axelrod, R. (1986). An evolutionary approach to norms. *American Political Science Review*, *80*, 1095–1111.

Bliege Bird, R. L., & Bird, D. W. (1997). Delayed reciprocity and tolerated theft: The behavioral ecology of food-sharing strategies. *Current Anthropology*, *38*, 49–77.

Blurton Jones, N. (1987). Tolerated theft: Suggestions about the ecology and evolution of sharing, hoarding, and scrounging. *Social Science Information*, *26*, 131–144.

Bond, M. H., Leung, K., & Wan, K. C. (1982). How does cultural collectivism operate? The impact of task and maintenance contributions on reward distribution. *Journal of Cross Cultural Psychology*, *13*, 186–200.

Bourdieu, P., & Passeron, J. C. (1990). *Reproduction in education, society and culture* (2nd ed.). London: Sage.

Bowles, S., Gintis, H., & Groves, M. O. (Eds.). (2005). *Unequal chances: Family background and economic success*. Princeton, NJ: Princeton University Press.

Camerer, C. (2003). *Behavioral game theory: Experiments in strategic interaction*. Princeton, NJ: Princeton University Press.

Coleman, J. S. (1990). *Foundations of social theory*. Cambridge, MA: Harvard University Press.

Cosmides, L., & Tooby, J. (1992). Cognitive adaptations for social exchange. In J. H. Barkow, L. Cosmides, & J. Tooby (Eds.), *The adapted mind: Evolutionary psychology and the generation of culture* (pp. 163–228). Oxford, UK: Oxford University Press.

Dawes, R. M. (1980). Social dilemmas. *Annual Review of Psychology*, *31*, 169–193.

de Waal, F. B. M. (1996). *Good natured: The origins of right and wrong in primates and other animals*. Cambridge, MA: Harvard University Press.

Eckblad, G. F., & von der Lippe, A. L. (1995). Norwegian lottery winners: Cautious realists. *Journal of Gambling Studies*, *10*, 305–322.

Elster, J. (1992). *Local justice: How institutions allocate scarce goods and necessary burdens*. New York: Russell Sage Foundation.

Erdal, D., & Whiten, A. (1994). On human egalitarianism: An evolutionary product of Machiavellian status escalation? *Current Anthropology*, *35*, 175–178.

Fukuyama, F. (1992). *The end of history and the last man*. New York: Free Press.

Fukuyama, F. (2006). *America at the crossroads: Democracy, power, and the neoconservative legacy*. New Haven, CT: Yale University Press.

Gintis, H. (2000). *Game theory evolving: A problem-centered introduction to modeling strategic behavior*. Princeton, NJ: Princeton University Press.

Gurven, M. (2004). To give or not to give: An evolutionary ecology of human food transfers. *Behavioral and Brain Sciences*, *27*, 543–559.

Güth, W., Schmittberger, R., & Schwarze, B. (1982). An experimental analysis of ultimatum bargaining. *Journal of Economic Behavior and Organization*, *3*, 367–388.

Hamilton, W. D. (1963). The evolution of altruistic behavior. *American Naturalist*, *96*, 354–356.

Hawkes, K., O'Connell, J. F., & Blurton Jones, N. G. (2001). Hadza meat sharing. *Evolution and Human Behavior*, *22*, 113–142.

Henrich, J., Boyd, R., Bowles, S., Camerer, C., Fehr, E., & Gintis, H. (Eds.). (2004). *Foundations of human sociality. Economic experiments and ethnographic evidence from fifteen small-scale societies*. Oxford, UK: Oxford University Press.

Kameda, T., Takezawa, M., & Hastie, R. (2003). The logic of social sharing: An evolutionary game analysis of adaptive norm development. *Personality and Social Psychology Review*, *7*, 2–19.

Kameda, T., Takezawa, M., Tindale, R. S., & Smith, C. (2002). Social sharing and risk reduction: Exploring a computational algorithm for the psychology of windfall gains. *Evolution and Human Behavior*, *23*, 11–33.

Kaplan, H., & Hill, K. (1985). Food sharing among Ache foragers: Tests of explanatory hypotheses. *Current Anthropology*, *26*, 223–246.

Kelley, H. H., Holmes, J. G., Kerr, N. L., Reis, H. T., Rusbult, C. E., & Van Lange, P. A. M. (2003). *An atlas of interpersonal situations*. Cambridge, UK: Cambridge University Press.

Laland, K. N., Odling-Smee, J., & Feldman, M. W. (2000). Niche construction, biological evolution, and cultural change. *Behavioral and Brain Sciences*, *23*, 131–175.

Maynard Smith, J. (1982). *Evolution and the theory of games*. Cambridge, UK: Cambridge University Press.

Norenzayan, A., & Heine, S. J. (2005). Psychological universals: What are they and how can we know? *Psychological Bulletin*, *131*, 763–784.

Rawls, J. (1971). *A theory of justice*. Cambridge, MA: Harvard University Press.

Yamagishi, T. (1986). The provision of a sanctioning system as a public good. *Journal of Personality and Social Psychology*, *50*, 110–116.

Section *III*

Evolutionary Universals and Cross-Cultural Differences

12

Color in Mind, Culture, and Language

DEBI ROBERSON

For more than 50 years, philosophers, psychologists, linguists, and anthropologists have debated the relationship between culture, language, and the mind (our mental representation of the experienced world). Researchers have considered both whether cognition is affected by differences in the content of communication across different speech communities (what people choose to talk about) and whether it is affected by the structure of language itself (how the content is communicated). Moreover, differences at the level of either structure (e.g., whether verbs carry information about tense, whether objects are assigned grammatical gender) or content (whether a language has words for *shame*, *democracy*, or *purple*) might have either an acute online influence on cognition—just in the act of "thinking for speaking" (Slobin, 1996)—or a deeper, more profound, chronic influence on what an individual is able to think about.

BACKGROUND TO THE LANGUAGE AND THOUGHT DEBATE

The hypothesis that different cultures encode and represent different aspects of experience (the linguistic relativity hypothesis) was first proposed by Benjamin Lee Whorf (and his mentor Edward Sapir). Whorf suggested, "We cut nature up, organize it into concepts and ascribe significances as we do, largely because we are parties to an agreement to organize it in this way—an agreement that holds throughout our speech community and is codified in the patterns of our language" (Whorf, 1956, p. 214). This view that culture, through language, influences cognition has ignited heated debate across research fields and continues to be contentious. Many researchers take the view that linguistic categories and structures constitute a "straightforward mapping from a pre-existing conceptual space, programmed into our biological nature: Humans invent words that label their concepts" (Li & Gleitman, 2002, p. 166). According to this nativist view, language is no more than a transparent medium through which our preexisting mental representations of the world are communicated to others. People without language, prelinguistic infants, and closely related animal species could have the same mental representations as those possessed by normal adults. Their concepts might be either innate or learned, but these researchers are adamant that language-specific experience does not affect nonlinguistic representations (Munnich & Landau, 2003; Pinker, 1994). Researchers have proposed as evidence of the complete independence of thought and language that some cognitive domains are organized along universal principles, regardless of the ways in which different linguistic communities describe them. Domains such as spatial cognition, color, shape, and time are described in radically different ways in different languages, but nativists argue that common experience of the world should drive all communities to process information about the environment in "optimal" ways, regardless of the limitations of their language (Landau & Jackendoff, 1993; Regier, Kay, & Khetarpal, 2007; Talmy, 1983).

In recent years, however, researchers have reported studies from each of these domains in which differences in linguistic classification are associated with different cognitive and behavioral responses. For instance, Gordon (2004) found that members of an Amazonian culture, whose language lacks words for numbers greater than 2, also failed simple numerical reasoning tasks with three or more items, whereas Levinson (1996), Majid, Bowerman, Kita, Haun, and Levinson (2004), and Choi, McDonough, Bowerman, and Mandler (1999) found differences in spatial reasoning between cultures whose lexical terms for spatial relations differed. Lucy (1992) found that classification of objects by material or shape differed between speakers of languages with, or without, a mass–count grammatical distinction, and Boroditsky (2001) reported similar differences between English and Chinese speakers' judgments of time. Roberson, Davidoff, and Shapiro (2002) found differences in classification of basic shapes in speakers of languages with or without basic shape terms, whereas other researchers (Sera, Berge, & del Castillo Pintado, 1994; Sera et al., 2002; Vigliocco, Vinson, & Siri, 2005) have reported differing effects of grammatical gender on classification of items across languages.

THE DEBATE IN THE COLOR DOMAIN

One of the earliest reports of such effects of language was in the domain of color (Brown & Lenneberg, 1954), and it is in this domain that the largest volume of work concerned with linguistic relativity has been carried out. Color categorization has been particularly well researched, because the range of colors visible by humans is large (approximately 2 million just noticeable differences), but the range of cognitive and linguistic color categories available to describe them is generally small (between 2 and 22 basic terms).[1] Moreover, not only do some languages have gross differences in terminology (e.g., the use of a single term to refer to everything that an English speaker would call either blue, green, or purple) but even those languages with similar color vocabularies have some variations in the range of stimuli covered by a particular term (e.g., Russian and Turkish, which both have two basic terms for English *blue*; Özgen & Davies, 1998; Paramei, 2005; Winawer, Witthoft, Frank, Wu, & Boroditsky, 2007). The color domain may also be a particularly fruitful ground for investigating the relationships between culture, language, and thought because the acquisition of color vocabulary by children is typically rather slow, laborious, and error prone compared to other domains (Bornstein, 1985; Braisby & Dockrell, 1999; O'Hanlon & Roberson, 2006; Roberson, Davidoff, Davies, & Shapiro, 2004), so there is ample opportunity to study the process of color term acquisition.

Empirical investigations of human color categorization have approached this question in three distinct ways. Some studies have looked for two cultures that differ in the way they code experience in language and then sought evidence that these differences affect some behavioral outcome (e.g., the judged similarity, discrimination or memory of objects or events). Others have examined the behavior of adult English speakers in laboratory tasks, when verbal coding is prevented by the requirement to carry out a simultaneous secondary verbal task. A third raft of investigation has examined the developmental trajectory of concepts from prelinguistic infants through childhood to the attainment of full adult concepts.

In this chapter I discuss the issues raised by each strand of investigation in relation to color categorization. I examine the evidence from each of the three main fields of research just outlined, highlighting the benefits and drawbacks of each approach, before discussing whether the different fields of investigation provide any converging evidence about the relationship between language and thought. I argue that the role of culture and language is likely to operate from early in development and that categorization is an automatic process of which people are largely unaware, even though it affects the acquisition of knowledge and its transmission across generations.

THE EARLY DEBATE IN THE COLOR DOMAIN

Historically the debate concerning the linguistic relativity of color categories was sharply divided, and, at times, the same evidence for wide variability of color naming has been cited both as supporting the view that language influences one's worldview (Brown & Lenneberg, 1954; Ratner, 1989) and also in support of the view that color thought and language are completely independent (Bornstein, 1985; Rosch, 1973; Rosch Heider, 1972b).

Brown and Lenneberg (1954) found that colors that were easier to name (in English) were both easier to communicate to others and easier to remember (for English speakers). Several subsequent cross-cultural studies also found positive correlations, in other languages, between a range of measures of codability for colors (such as speed of naming, consistency of naming both across informants and within an informant across time, accuracy with which a color could be communicated to a third party, and accuracy of recognition memory for colors; Lantz & Stefflre, 1964; Stefflre, Castillo Vales, & Morley, 1966). The principal conclusion drawn from these studies was that the ease with which a color could be named would be a generally good predictor of memory accuracy in all languages and across a wide range of color stimuli. Because the number of color terms and the range of their referents differed across languages, this suggested that color categorization, in thought and language, could vary arbitrarily between cultures and, in consequence, must be learned in the context of a particular culture and language (Ratner, 1989). The major challenge facing such a theoretical viewpoint was that many, widely geographically separated, languages were found to have quite similar sets of color terms, so that approximate translations could easily be made between languages (e.g., *red* = *rouge*) (Berlin & Kay, 1969).

An opposing view was put forward by Berlin and Kay (1969) who, instead, suggested that, regardless of what terms were used to describe colors, the underlying cognitive representations of all humans were organized into exactly 11 basic categories (black, white, gray, red, yellow, green, blue, pink, purple, orange, and brown). One hypothesis put forward to explain such a possibility is that these categories might be innate and hardwired into the setup of human (and some primate) color vision pathways (Bornstein, Kessen, & Weiskopf, 1976; Kay & McDaniel, 1978; Sandell, Gross, & Bornstein, 1979). Saunders and van Brakel (1997) reviewed these arguments. This universalist view was largely based on a series of studies carried out by Rosch (Heider) in a small, remote community of hunter-gatherers, the Dugum Dani (Heider & Olivier, 1972; Rosch, 1973, 1975; Rosch Heider, 1972b). Members of this tribe appeared to have better recognition memory for colors that were named easily in English (good examples of red, green, blue, etc.), despite the fact that their own language used only two basic terms to label the whole range of visible colors (but see Rosch Heider, 1972a). Even though the results of these experiments were somewhat ambiguous, they were widely taken to support the proposal that a particular set of color categories might be panhuman cognitive universals, transcending differences in linguistic labeling across cultures. The universalist hypothesis thus implied that there could be such large differences between the "structure of the color space in memory" and the structure of the lexical categories used to describe it (Heider & Olivier, 1972, p. 351) that the two sets of categories (one in language, the other in thought) could (in the case of Dugum Dani) be effectively orthogonal.

Such an extreme nativist view of color cognition would mean that a child growing up in a culture that names all 11 universal categories would need to learn only which labels mapped onto each existing category. Speakers of languages that expressed fewer categories might have to overcome this universal organization to learn the appropriate reference set for their culture-appropriate set of terms (Bornstein, 1985). Such a hypothesis could not, however, explain why there is such diversity in the number of linguistic color categories across cultures with correspondingly diverse sets of referents, and it could not explain the widely reported finding that even English-speaking children, whose language codes all of the proposed universal categories, nevertheless found color terms very difficult to learn.

The explanatory limitations of either extreme view have led to the modification of both, as empirical investigation has further illuminated the issue. Where Brown and Lenneberg (1954) had proposed a simple and straightforward mapping between naming and memory for colors, more recent investigations (Agrillo & Roberson, 2009; Guest & Van Laar, 2002; Lucy, 1992) have demonstrated that the visual context in which colors are seen moderates this relationship. Lucy (1992) suggested that although categorical judgments of color may be made on the basis of name alone, discrimination judgments also employ nonverbal perceptual processing that may not be organized categorically. A number of similarly detailed empirical studies have moved the debate away from the polar extreme positions that either individuals have a full innate set of color categories at birth or they learn an entirely arbitrary set determined by language and then subsequently perceive color only through the lens of language (Gilbert, Regier, Kay, & Ivry, 2006; Guest & Van Laar, 2002; Jameson & Alvarado, 2003; Kay & Kempton, 1984; Moore, Romney, & Hsia, 1997; Özgen, 2004; Pilling, Wiggett, Özgen, & Davies, 2003). Instead, recent formulations of the relativist view of color categorization have emphasized the role of cultural needs in shaping both language and cognition: "The essential idea of linguistic relativity [is] the idea that culture, *through* language, affects the way we think, especially perhaps our classification of the experienced world" (Gumperz & Levinson, 1997, p. 613). Similar cultural needs, such as evolutionary pressure for successful frugivory (Sumner & Mollon, 2000), combined with the need to communicate efficiently about colors could thus cause some category divisions to be more likely than others, even among widely separated societies (Jameson, 2005; Levinson, 1997, 2005; Steels & Belpaeme, 2005; Wierzbicka, 1990).

Recent formulations of the universality hypothesis have acknowledged that "there is ample evidence that differences in color category boundaries between languages may influence color memory, learning or discrimination" (Kay & Regier, 2006, p. 297), while still maintaining that there are strong universal tendencies both in color naming (Kay & Regier, 2003) and in selection of the best examples of categories, which are believed to "cluster near the prototypes for English white, black, red, green, yellow, and blue" (Kay, Regier, & Cook, 2005, p. 8386). The number of proposed universal categories, however, has been reduced from the 11 expressed in English to 6 (red, green, yellow, blue, black, and white; Kay et al., 2005). Moreover, although some groupings may appear more likely than others, because they maximize perceptual similarity within categories and minimize it across categories (Regier et al., 2007), these authors also acknowledge a wide variety of languages whose categorization of color do not follow such principles of well-formedness. Thus, two once diametrically opposed viewpoints now share much common ground. Empirical cross-cultural investigations have contributed both to refinements of each position and to current views of the relationship between language and thought in the color domain.

One issue that contributes to the continued disagreement between researchers is the question of at what point of general similarity should sets of categories across languages be regarded as effectively equivalent (and thus universal) because they are more similar than any other sets of arbitrarily divided categories would be. Another is the question of whether there is a particular set of universally salient category centers, or foci, around which categories are formed (Kay et al., 2005). Cross-cultural studies often differ in the method of data collection used, and this may also contribute to differences in interpretation of results.

CROSS-CULTURAL INVESTIGATIONS

Studies that have sought universal tendencies in color naming (such as the World Color Survey) have collected a large quantity of naming data, using a standard range of highly saturated color examples, from a wide range of industrialized and traditional cultures. Subsequent data analysis has projected the proposed universal set of categories onto the sets of categories named in traditional cultures (Kay, Berlin, & Merrifield, 1991; Kay & Regier, 2006; Lindsey & Brown, 2006). Many of the cultures considered, however, have no overarching term for color as a domain, so researchers who wish to ask the question "What color is this?" must resort to some alternative, such as "What would you call this?" (Kuschel & Monberg, 1974; Lucy, 1992; Lucy & Gaskins, 2001). Despite this, responses are interpreted as referring solely to color (rather than texture, shape, artifact–natural distinction, or any other nameable property), because that is the domain of interest to the researcher. Moreover, the range of stimuli used to collect naming data contains only highly colorful (saturated) stimuli that are artificially produced. Most traditional communities, lacking printing and dyeing facilities, may see such colorful stimuli for the first time when naming them for an experimenter.

In behavioral studies, where larger sets of stimuli have been considered, the similarity between naming patterns for the most colorful stimuli does not extend to less colorful stimuli. For example, the similarity found between two languages, Berinmo and Himba, that both use five color terms drops from 61% interlanguage naming agreement for the standard, highly colorful set to 27% for less colorful stimuli (Roberson, Davidoff, Davies, & Shapiro, 2005). Himba and English speakers use a large number of secondary terms to label less colorful stimuli (e.g., maroon, dun, olive, khaki) whereas Berinmo speakers readily extend basic terms to such stimuli. Reliance on the naming of only the most colorful stimuli may have led, in the past, to overestimation of the similarity of divergent languages' color term systems. Roberson, Davidoff, et al. (2005) suggested that for any strong conclusions about universality to be made, at the very least the full set of items that would be named by any particular term should be considered, rather than just a subset or the placement of the category center. Whether there are any behavioral consequences of either the gross similarities or the small differences between the categories under consideration should also be investigated when possible.

To date, the available data suggest that there are few behavioral consequences of the similarities between different naming systems, although even small differences in name categories produce significant differences in behavioral response across languages in grouping behavior (free sorting a range of colors into groups; Roberson, Davies, Corbett, & Vandervyver, 2005) as well as in perceptual and/or memory judgments (Davidoff, Davies, & Roberson, 1999; Roberson, Davies, & Davidoff, 2000; Roberson, Davidoff et al., 2005; Webster et al., 2002; Winawer et al., 2007). Thus, although the observed similarities between naming patterns across cultures may be interesting, they should not, alone, be taken as evidence for a set of underlying cognitive universals.

An ongoing series of investigations has addressed the issue of whether speakers of any language can have more (and different) cognitive color categories than those that they label linguistically. In particular, evidence has been sought for a set of cognitively privileged focal points that might be universal prototypes around which new categories will inevitably form (Kay et al., 1991). One set of experiments has involved adult speakers of two languages with a very restricted set of basic color terms (Davidoff et al., 1999; Roberson, Davidoff et al., 2005; Roberson et al., 2000). Both the Berinmo language, spoken in Papua New Guinea, and the Himba language, spoken in northern Namibia, have only 5 basic color terms, compared to the 11 in English.

Both the traditional cultures investigated are remote peoples, living lifestyles that have changed little over the past 200 years, although they are widely separated geographically and live in very different visual environments. We examined recognition memory for a range of both good and poor examples of English color categories from an array of 120 distracters after a 30-second delay. Regardless of the status of stimuli in English color categories, within each language group, good examples of *their own* linguistic color categories were recognized better than poor examples. Paired-associate learning also failed to show any advantage for a set of proposed universal prototypes in either Berinmo or Himba speakers. A similar lack of preeminence for this particular set, either in naming or in categorization, was found by Jameson and Alvarado (2003) in Vietnamese speakers and in Korean speakers (Roberson, Pak, & Hanley, 2008). Taken together, these results point to the conclusion that there is no single set of prototypical colors that are universally cognitively privileged and are the basis around which categories are formed. Rather, those stimuli that are best examples of one's own linguistic categories are remembered more easily than those that are not. Ease of naming, in all these experiments, was a good predictor of memory performance for speakers of all languages.

In a further series of experiments, we compared perceptual and memory judgments at the boundaries of the English categories "green" and "blue" as well as at the Berinmo and Himba category boundaries (which fall toward the center of the English category "green"). For each language, discrimination and recognition memory were facilitated when target and distracter stimuli came from different name categories (e.g., blue target with green distracter). Critically, only the presence of a boundary in a speaker's own language facilitated discrimination and recognition. There was no effect for the proposed universal boundaries (e.g., for that between green and blue) in speakers of languages that did not make the distinction, a result also previously reported by Kay and Kempton (1984). It thus appears that boundary divisions between categories are heavily influenced by culture and language and not the result of perceptual category divisions at a deeper or more primitive level.

These and other findings (e.g., Gilbert et al., 2006; Pilling et al., 2003; Webster, Miyahara, Malkoc, & Raker, 2000; Webster et al., 2002) have led to moderations of the extreme versions of the universality hypothesis. It is now widely accepted that small differences in the placement of a linguistic category boundary produce measurable differences in behavioral response and that adult speakers of different languages, with the same number of color terms, show different patterns of discrimination and memory for the same set of colors.

Even if there are tight links in adult cognition between language and thought, there remains the question of whether learning a particular categorical language coding system can permanently alter the perceptual system so that between-category differences are genuinely perceived as greater than within-category differences. If the effect of language learning were to permanently alter the perceptual system in this way, then speakers of different languages could truly be said to see the world differently. If, however, the interaction between systems were habitual, but not at this deep level, then, although the use of verbal labeling might distort one's perception of the experienced world, free from this influence, our perceptions might still be truly universal. This issue has been considered in laboratory studies of adults performing recognition and discrimination tasks under verbal suppression as well as in developmental studies of children learning sets of color terms.

LABORATORY STUDIES OF ADULT COLOR MEMORY AND DISCRIMINATION

Several studies have investigated adult English speakers' discrimination of and memory for colors under articulatory suppression in an attempt to release the underlying perceptual representation of color from linguistic influence. When participants have carried out a concurrent verbal task, categorical perception of color categories such as green and blue (once thought to be the result of low level discontinuities, hardwired into the visual system) disappears. Instead participants appear to experience a smoothly graded continuum of color, without any abrupt perceptual change at category boundaries. This result has been found both for recognition memory judgments, where verbal coding of to-be-remembered colors is a likely strategy (Pilling et al., 2003; Roberson & Davidoff, 2000), and also for tasks that require only perceptual discrimination, such as visual search, where there is no reason to expect participants to code the stimuli verbally (Gilbert et al., 2006; Winawer et al., 2007).

Winawer et al.'s (2007) study took advantage of the fact that *siniy* (dark blue) and *goluboy* (light blue) are distinct basic color terms for speakers of Russian (Paramei, 2005). When asked to select which of two colors matched a *siniy* target, Russian participants were faster if the distracter was *goluboy* than if it was a different shade of *siniy*. These results were observed even though the physical difference between targets and distracters was equated. English speakers, who would call all the stimuli "blue," did not show the same cross-category advantage. When Russian participants had to perform a verbal interference task, the category advantage disappeared, suggesting that it arose because participants accessed their verbal labels for colors while performing the matching task.

In the visual-search task used by Gilbert et al. (2006), participants were required to fixate on a cross in the center of the computer screen. They were then asked to report the location of an oddball-colored target appearing among an array of identically colored distracters. English speakers were faster to locate a target from a category different from the distracters (e.g., *green* target, *blue* distracters) than a target from the same category (e.g., *blue* target, *blue* distracters), even though the degree of physical difference of targets from distracters was equated. This effect, however, was observed only when the target appeared in the right visual field; participants were no faster for cross-category targets than within-category targets in the left visual field. Because information presented to the right visual field has preferential access to lexical representations in the left hemisphere, whereas access to these representations from the left visual field would require transfer of information across the corpus callosum, this finding was interpreted as providing strong evidence for a linguistic influence on the cognitive representation of color. Once again, when a secondary verbal task was undertaken simultaneously, the advantage for cross-category discrimination disappeared.

To investigate the wider applicability of this finding, we recently used the same visual search task to compare discriminations by English and Korean participants around the boundary between *yeondu* (yellow-green) and *chorok* (green). This boundary is not marked in English, and English speakers called all the colors tested *green*. The results demonstrated that speakers of Korean also show categorical perception of two shades of green that are linguistically marked in Korean but not in English (Roberson et al., 2008). English speakers showed no such effect. Because this boundary is unique to Korean, these results, as with the findings of Winawer et al. (2007) for the Russian division of *siniy* and *goluboy*, suggest that perceptual categorization of color varies between cultures with different linguistic terminology.

Other researchers have shown that novel boundaries within existing categories can be induced with training for a variety of domains, including arbitrary color categories (Özgen & Davies, 2002). On one hand, this supports the view that language influences cognitive processing online and that the similarities and the differences observed between adult speakers of different languages reflect the influence of habitual linguistic coding but not underlying differences in how the world is perceived. On the other hand, it suggests that what is universal about color perception is that humans perceive a smoothly graded continuum across the visible range of colors, with no natural boundaries at all. Confirmatory evidence of this proposal has been sought in two ways.

If color categories are learned and culture specific, then the learning process should be observable in young children. Moreover, there would be no reason to expect that other species would display a particular set of categorical sensitivities, even though they may share the mechanisms of color vision with humans. If, however, there is a particular set of universally optimal categories that are innate in humans (and possibly other primates with a similar color vision system) but that cultural learning overrides, then color term learning should differ considerably between cultures that do and cultures that do not mark the universal set (Bornstein, 1985).

STUDIES OF CHILDREN'S COLOR TERM LEARNING

Another line of investigation has therefore compared children's discrimination and memory for colors before and after they acquired their color vocabulary. Individuals who have yet to learn the set of categories appropriate to their own culture and language might either behave as if they already have some cognitive organization of color categories or treat colors as a true perceptual continuum without fixed boundaries. Bornstein (1985) predicted that if there was an innate, hue-based, universal set of color categories, then acquiring color terms would be even more difficult for children learning a language in which those must be overwritten by a new set, even if the new set contained fewer terms to be learned. Bowerman and Choi (2003) suggested that the more robust and prepotent the prelinguistic organization of the perceived world is, the greater the resistance that language acquisition must overcome to restructure mental life. Thus, children acquiring a smaller set of named categories in a traditional language should show a different developmental pattern to that of English-speaking children.

Some studies have suggested that categorical divisions between red, green, blue, and yellow might be perceived categorically by infants (Bornstein, Kessen, & Weiskopf, 1976; Franklin & Davies, 2004). There were methodological issues, however, with Bornstein et al.'s early study (Banks & Salapatek, 1976; Werner & Wooten, 1985). Franklin and Davies (2004), using a preferential looking technique, found that 4-month-old infants showed categorical novelty preferences for a wide range of color categories, both across hue boundaries (such as that between blue and green) and across brightness boundaries (such as that between pink and red), but there are reasons to be cautious of interpreting infant categorization as resembling that acquired later in life.

Infants appear to easily form a variety of short-term dynamic online categories, within a preferential looking paradigm, for stimuli as diverse as cats, dogs, and lions (Quinn & Eimas, 1997), but these categorizations are labile and can change when the perceptual features of the input are changed (Rakison & Butterworth, 1998a, 1998b). Recent research has suggested that color and location information are processed separately in infants (Bremner, Bryant, & Mareschal, 2006; Mareschal & Bremner, 2006), because dorsal and ventral streams of visual processing are not integrated until much later in development. Such a finding is problematic for preferential looking paradigms, which measure looking time to a color in a particular location. This may be why children of 2 to 3 years of age often fail on other categorization tasks that infants appear to have passed, because it is around this age that children begin to try to integrate information about color, shape, texture, and location of stimuli.

Because it is difficult to interpret infant performance on preferential looking tasks, an alternative approach is to examine the performance of slightly older children as they acquire their color categories. We therefore examined the acquisition of color terms by children in two ways. In the first study, we examined naming and comprehension systematically over a 3-year period to establish a reliable measure of children's color term acquisition as well as to compare color term acquisition in speakers of different languages. Roberson et al. (2004) studied English 3-year-olds, who were tested initially before they entered preschool and, subsequently, through 3 years of formal education, and Himba 3-year-olds from northern Namibia, few of whom received any formal education during the period of the study. Children's color term knowledge and memory for colors were tested at 6-month intervals over 3 years.

Despite the considerable environmental, linguistic, and educational differences between the two groups, the process of color term learning appeared to be remarkably similar in the two groups, and there was no predictable order of category acquisition in either group. English children acquired their first color words earlier than Himba children, and it is likely that greater exposure to colored objects and the increased cultural salience of color in Western society contribute to an earlier conceptual understanding of color as a separable dimension. Both Himba and English children, however, showed similar patterns of memory errors before they learned any color terms, and, critically, this pattern did not resemble that derived from the 11 basic categories of English but appeared to be based on perceptual distance rather than a particular set of predetermined categories. In addition, for this group of children, there was no advantage in memory for the stimuli that were central (focal) to the basic terms of *either* language.

From an initial, shared reliance on perceptual similarity, an advantage for the (language-appropriate) set of focal colors became evident as soon as children acquired any color terms. Of those children knowing one or more terms at the first time of testing, Himba children showed superior memory performance for the items that are central to Himba but not to English categories, whereas English children showed the reverse pattern. Such rapid divergence in the cognitive organization of color for the two groups, from the time that the first terms are learned, suggests that cognitive color categories must be learned. Thus, these data, like those for adult English, Himba, and Berinmo speakers, argue against an innate origin for the proposed universalist tendencies in color naming. Considering the trajectory of color term acquisition in the two cultures, the longitudinal results suggested that children continue to refine their conceptual color categories for some years after they first show evidence of term knowledge for focal colors. For both populations, once color terms were acquired, memory performance was determined by the number of terms known. The advantage for items central (focal) to children's native language categories also increased throughout the longitudinal study, so the emphasis that has been given to category centers in establishing categories (Heider & Olivier, 1972; Kay et al., 2005) seems justified from the present data. Nevertheless, it is important to stress that the focality found here is not universal but, as shown for both children and adults, language dependent. Daoutis, Franklin, Riddett, Clifford, and Davies (2006) confirmed, cross-culturally, that the influence of language on color perception can be found in children as young as 4 years old (see also Goldstein, Davidoff, & Roberson, 2009). Cross-cultural studies of very young children have produced somewhat inconsistent results, and there are particular problems associated with testing young children in remote communities, where accurate records of children's ages are not available and their monolingual status cannot be confirmed.

In a further attempt to ascertain whether language contributes to color term acquisition, we investigated novel color term learning in 3-year-old English-speaking children (O'Hanlon & Roberson, 2006). Much research has suggested that even English-speaking children acquire accurate color naming relatively late (Andrick & Tager-Flusberg, 1986; Braisby & Dockrell, 1999; Mervis, Bertrand, & Pani, 1995; Rice, 1980; but see also Pitchford & Mullen, 2001). Given a contrastive task that makes learning additional category terms easy (Carey & Bartlett, 1978; Heibeck & Markman, 1987), 2-year-old children who knew no color names, but hundreds of other names, took on average 800 trials simply to respond "red" to red objects and "green" to green objects (Rice, 1980).

In our studies, groups of children, matched for chronological and vocabulary age and with similar levels of prior knowledge of color terms, were taught novel, low-frequency color terms using either corrective linguistic feedback or perceptual matching conditions. These learning studies replicated and extended previous findings that corrective linguistic contrast helps children to learn new color terms (Au & Laframboise, 1990; Gottfried & Tonks, 1996) and, in addition, showed that enhancing the perceptual salience of the color dimension, without linguistic feedback, also facilitates learning. Both attentional and linguistic factors contributed independently to the process of color term learning, and more was learned about the novel color terms when attentional and linguistic aspects of the learning context converged. The pattern of results for the individual color terms taught (beige, crimson, and teal) supported the hypothesis that children have difficulty learning color terms because they overextend the terms they know (e.g., by calling all colors "red") and then misapply a principle of mutual exclusivity (that no single object or object-property will have more than one name). Moreover, Italian children, who showed a similar pattern of learning for beige and crimson, learned an additional term (teal) in the blue-green region more easily than English children. Because Italian adults use a variety of terms in this region (*blu*, which covers the range English speakers would call dark blue; *celeste*, which is sky blue; and *azzurro*, which includes shades of blue toward turquoise), Italian children receive a more varied adult input in this region that should weaken their expectations that terms for colors in this area are mutually exclusive.

An existing color vocabulary indicates that the child has already learned to abstract color as a relevant dimension of stimuli and override initial attentional biases that mediate against attending to the color dimension (Soja, 1994). In our studies, more was learned about the novel color terms when both attentional and linguistic aspects of the learning context converged and, as for the Himba and English children in Roberson et al.'s (2004) study, adult linguistic input played a crucial role in enabling children to abstract color as a dimension of objects and to learn new color categories.

CONCLUSION

Put together, the results of these behavioral studies of color categorization by adults and children, across a variety of languages and cultures, suggest that segmentation of color categories, like that of other perceptual continua such as shape, musical tone, speech sounds, or facial expression, is the result of learning and experience within a particular speech community. Language appears to play a role both in guiding the initial learning of categories by children and in maintaining those categories in adults. When linguistic categories have not yet been learned in children, do not exist in the adult language, or are disrupted by a concurrent verbal task, perception and memory of colors in humans appears to follow a smooth, undifferentiated continuum rather than exhibit a predetermined set of discontinuities. Recent results from studies with baboons support this hypothesis (Fagot, Goldstein, Davidoff, & Pickering, 2006), and there are significant consequences of delayed or deviant language acquisition on color categorization in autism (Heaton, Ludlow, & Roberson, 2008).

There is thus converging evidence from studies of adults and children in a range of different communities that learning to use a set of color terms appropriately, within a particular speech community, affects the way that colors are distinguished and remembered. The accumulating evidence that the effects of categorization on both perception and memory disappear under verbal interference, however, suggests that these effects occur online rather than at a deep and permanent level. Categorization appears to be a cognitive process that directs attention to cross-boundary differences and within-category similarities without permanently altering the continuous nature of perceptual just noticeable differences across the range of visible colors. Categorization processes appear nevertheless to be highly automated, so that experimental participants are unaware of their use and maintain, when asked, that cross-boundary pairs of stimuli "just look more different." If verbal coding of all perceptual stimuli is automated and occurs without awareness from early childhood onward, then even the acute infection of experience by culture and language has profound effects on our representation of the experienced world.

Even with a converging body of evidence supporting the influence of culture and language on color cognition, however, any comprehensive theory of color categorization must accommodate both the differences and the similarities of segmentation across languages with similar numbers of terms (Bimler, 2005; Sayim, Jameson, Alvarado, & Szeszel, 2005). Kay and Regier (2006, p. 297) noted, in a comparison of nine five-term languages from nonindustrialized languages, "Berinmo color naming appears to be quite similar to that of other five-term languages from a range of genetically and geographically separated language families." Because communities with few color terms tend to be those without printing or dyeing facilities, where color is little discussed or attended to, similar needs or the overriding salience of some colorful things (such as blood or the ripeness of fruit) could have driven different cultures, living in very different environments, to arrive at very similar solutions to the problem of dividing the color continuum (Jameson, 2005; Roberson, 2005; Wierzbicka, 1990). With the rapid expansion of technology in developing countries, many languages acquire borrowed terms. If categories are initially formed based on the relative similarity of stimuli, as Dedrick (1996) and Roberson et al. (2000) argued, or because different category arrangements are optimal at different levels of categorization (Regier et al., 2007), then both the range of available stimuli in the environment and the variability in the need to communicate about color should affect the eventual set that a community arrives at.

NOTE

1. The criteria set by Kay, Berlin, and Merrifield (1991) for terms considered to be basic are that the terms be monolexemic, be present in the idiolect of every observer, and not be subsumed within the range of another term.

REFERENCES

Agrillo, C. & Roberson, D. (2009). Colour language and colour cognition: Brown and Lenneberg revisited. *Visual Cognition, 17*, 412–430.

Andrick, G. R., & Tager-Flusberg, H. (1986). The acquisition of color terms. *Journal of Child Language, 13*, 119–134.

Au, T. K., & Laframboise, D. E. (1990). Acquiring color names via linguistic contrast: The influence of contrasting terms. *Child Development, 61*, 1808–1823.

Banks, M. S., & Salapatek, P. (1976). Contrast sensitivity function of infant visual system. *Vision Research, 16*, 876–879.

Berlin, B., & Kay, P. (1969). *Basic color terms: Their universality and evolution*. Berkeley: University of California Press.

Bimler, D. (2005). Intimations of optimality: Extensions of simulation testing on color-language hypotheses. *Behavioral and Brain Sciences, 28*, 489–490.

Bornstein, M. H. (1985). On the development of color naming in young children: Data and theory. *Brain and Language, 26*, 72–93.

Bornstein, M. H., Kessen, W., & Weiskopf, S. (1976). Color vision and hue categorization in young human infants. *Journal of Experimental Psychology: Human Perception and Performance, 2*, 115–129.

Boroditsky, L. (2001). Does language shape thought? English and Mandarin speakers' conceptions of time. *Cognitive Psychology, 43*, 1–22.

Bowerman, M., & Choi, S. (2003). Space under construction: Language-specific categorization in first language acquisition. In D. Gentner & S. Goldin-Meadow (Eds.), *Language in mind: Advances in the study of language and thought* (pp. 387–498). Cambridge, MA: MIT Press.

Braisby, N., & Dockrell, J. (1999). Why is color naming difficult? *Journal of Child Language, 26*, 23–47.

Bremner, A. J., Bryant, P. E., & Mareschal, D. (2006). Object-centred spatial reference in 4-month-old infants. *Infant Behavior and Development, 29*, 1–10.

Brown, R., & Lenneberg, E. (1954). A study in language and cognition. *Journal of Abnormal and Social Psychology, 49*, 454–462.

Carey, S., & Bartlett, E. (1978). Acquiring a single new word. *Papers and Reports on Child Language Development, 15*, 17–29.

Choi, S., McDonough, L., Bowerman, M., & Mandler, J. M. (1999). Early sensitivity to language-specific spatial categories in English and Korean. *Cognitive Development, 14*, 241–268.

Daoutis, C. A., Franklin, A., Riddett, A., Clifford, A., & Davies, I. R. L. (2006). Categorical effects in children's colour search: A cross-linguistic comparison. *British Journal of Developmental Psychology, 24*, 373–400.

Davidoff, J., Davies, I., & Roberson, D. (1999). Colour categories in a stone-age tribe. *Nature, 402*, 604–604.

Dedrick, D. (1996). Color language, universality and evolution: On the explanation for basic color terms. *Philosophical Psychology, 9*, 497–524.

Fagot, J. Goldstein, J., Davidoff, J. & Pickering, A. (2006). Cross-species differences in color categorization. *Psychonomic Bulletin & Review, 13*, 275–280.

Franklin, A., & Davies, I. R. L. (2004). New evidence for infant colour categories. *British Journal of Developmental Psychology, 22*, 349–377.

Gilbert, A. L., Regier, T., Kay, P., & Ivry, R. B. (2006). Whorf is supported in the left visual field but not the right. *Proceedings of the National Academy of Science, USA, 103*, 489–494.

Goldstein, J., Davidoff, J. & Roberson, D. (2009). Knowing color terms enhances recognition: Further evidence from English and Himba. *Journal of Experimental Child Psychology, 102*, 219–238.

Gordon, P. (2004). Numerical cognition without words: Evidence from Amazonia. *Science, 15*, 496–499.

Gottfried, G. M., & Tonks, S. J. M. (1996). Specifying the relation between novel and known: Input affects the acquisition of novel color terms. *Child Development, 67*, 850–866.

Guest, S., & Van Laar, D. (2002). The effect of name category and discriminability on the search characteristics of colour sets. *Perception, 31*, 445–461.

Gumperz, J. J., & Levinson, S. C. (1997). Rethinking linguistic relativity. *Current Anthropology, 32*, 613–623.

Heaton, P., Ludlow, A. & Roberson, D. (2008). When less is more: poor discrimination but good colour memory in Autism. *Research in Autism Spectrum Disorders, 2*, 147–156.

Heibeck, T. H., & Markman, E. M. (1987). Word learning in children: An examination of fast mapping. *Child Development, 58*, 1021–1034.

Heider, E. R., & Olivier, D. C. (1972). The structure of the color space in naming and memory for two languages. *Cognitive Psychology, 3*, 337–354.

Jameson, K. A. (2005). Why GRUE? An interpoint-distance model of composite color categories. *Cross-Cultural Research, 39*, 159–204.

Jameson, K. A., & Alvarado, N. (2003). Differences in color naming and color salience in Vietnamese and English. *Color Research and Application, 28*, 113–138.

Kay, P., Berlin, B., & Merrifield, W. R. (1991). Biocultural implications of systems of color naming. *Journal of Linguistic Anthropology, 1*(1), 12–25.

Kay, P., & Kempton, W. (1984). What is the Sapir-Whorf hypothesis? *American Anthropologist, 86*, 65–78.

Kay, P. & McDaniel, C. (1978). The linguistic significance of the meanings of basic color terms. *Language, 54*, 610–646.

Kay, P., & Regier, T. (2003). Resolving the question of color naming universals. *Proceedings of the National Academy of Science, USA, 100*, 9085–9089.

Kay, P., & Regier, T. (2006). Color naming universals: The case of Berinmo. *Cognition, 102*, 289–298.

Kay, P., Regier, T., & Cook, S. (2005). Focal colors are universal after all. *Proceedings of the National Academy of Science, USA, 102*, 8386–8391.

Kuschel, R., & Monberg, T. (1974). "We don't talk much about color here": A study of color semantics on Bellona Island. *Man, 9*, 213–242.

Landau, B., & Jackendoff, R. (1993). "What" and "where" in spatial language and spatial cognition. *Behavioral and Brain Sciences, 16*, 217–238.

Lantz, D., & Stefflre, V. (1964). Language and cognition revisited. *Journal of Abnormal and Social Psychology, 69*, 472–481.

Levinson, S. C. (1996). Language and space. *Annual Review of Anthropology, 25*, 353–382.

Levinson, S. C. (1997, June). *Yélî dyne and the theory of basic color terms*. Paper presented at the Max Plank Institute for Psycholinguistics, Njimegen, The Netherlands.

Levinson, S. C. (2005). Living with Manny's dangerous idea. *Discourse Studies, 7*, 431.

Li, P., & Gleitman, L. (2002). Turning the tables: Language and spatial reasoning. *Cognition, 87*, 15–178.

Lindsey, D. T., & Brown, A. M. (2006). Universality of color names. *Proceedings of the National Academy of Science, USA, 103*, 16608–16613.

Lucy, J. A. (1992). *Language and diversity of thought: A reformulation of the linguistic relativity hypothesis*. Cambridge, UK: Cambridge University Press.

Lucy, J. A., & Gaskins, S. (2001). Grammatical categories and the development of classification preferences: A comparative approach. In M. Bowerman & S. C. Levinson (Eds.), *Language acquisition and conceptual development* (pp. 465–492). Cambridge, UK: Cambridge University Press.

Majid, A., Bowerman, M., Kita, S., Haun, D. B. M., & Levinson, S. C. (2004). Can language restructure cognition? The case for space. *Trends in Cognitive Sciences*, *8*, 108–114.

Mareschal, D., & Bremner, A. J. (2006). When do 4-month-olds remember the "what" and "where" of hidden objects? *Attention and Performance, 21*, 427–447.

Mervis, C. B., Bertrand, J., & Pani, J. R. (1995). Transaction of cognitive-linguistic abilities and adult input: A case-study of the acquisition of color terms and color-based subordinate object categories. *British Journal of Developmental Psychology*, *13*, 285–302.

Moore, C. C., Romney, A. K., & Hsia, T. (1997). Shared cognitive representations of perceptual and semantic structures of basic colors in Chinese and English, *PNAS USA*, *97*, 5007–5010.

Munnich, E., & Landau, B. (2003). The effects of spatial language on spatial representation: Setting some boundaries. In D. Gentner & S. Goldin-Meadow (Eds.), *Language in mind: Advances in the study of language and thought* (pp. 113–156). Cambridge, MA: MIT Press.

O'Hanlon, C. G., & Roberson, D. (2006). Learning in context: Linguistic and attentional constraints on children's color term learning. *Journal of Experimental Child Psychology*, *94*, 275–300.

Özgen, E. (2004). Language, learning, and color perception. *Current Directions in Psychological Science 13*, 95–98.

Özgen, E., & Davies, I. R. L. (1998). Turkish color terms: Tests of Berlin and Kay's theory of color universals and linguistic relativity. *Linguistics*, *36*, 919–956.

Özgen, E., & Davies, I. R. L. (2002). Acquisition of categorical color perception: A perceptual learning approach to the linguistic relativity hypothesis. *Journal of Experimental Psychology: General*, *131*, 477–493.

Paramei, G. V. (2005). Singing the Russian blues: An argument for culturally based color terms. *Cross-Cultural Research*, *39*, 10–38.

Pilling, M., Wiggett, A., Özgen, E., & Davies, I. R. L. (2003). Is color "categorical perception" really perceptual? *Memory and Cognition*, *31*, 538–551.

Pinker, S. (1994). *The language instinct.* New York: Morrow.

Pitchford, N. J., & Mullen, K. T. (2001). Conceptualization of perceptual attributes: A special case for color? *Journal of Experimental Child Psychology*, *80*, 289–314.

Quinn, P. C., & Eimas, P. D. (1997). A reexamination of the perceptual-to-conceptual shift in mental representations. *Review of General Psychology*, *1*, 271–287.

Rakison, D. H., & Butterworth, G. (1998a). Infant attention to object structure in early categorization. *Developmental Psychology*, *34*, 1310–1325.

Rakison, D. H., & Butterworth, G. (1998b). Infants' use of parts in early categorization. *Developmental Psychology*, *34*, 49–62.

Ratner, C. (1989). A sociohistorical critique of naturalistic theories of color perception. *Journal of Mind and Behavior*, *10*(4), 361–373.

Regier, T., Kay, P., & Khetarpal, N. (2007). Color naming reflects optimal partitions of color space. *Proceedings of the National Academy of Science, USA*, *104*(4), 1436–1441.

Rice, M. (1980). *Cognition to language: Categories, word meanings, and training*. Baltimore: University Park Press.

Roberson, D. (2005). Color categories are culturally diverse in cognition as well as in language. *Cross-Cultural Research*, *39*, 56–71.

Roberson, D., & Davidoff, J. (2000). The categorical perception of colors and facial expressions: The effect of verbal interference. *Memory and Cognition*, *28*, 977–986.

Roberson, D., Davidoff, J., Davies, I. R. L., & Shapiro, L. R. (2004). The development of color categories in two languages: A longitudinal study. *Journal of Experimental Psychology: General*, *133*, 554–571.

Roberson, D., Davidoff, J., Davies, I., & Shapiro, L. (2005). Colour categories in Himba: Evidence for the cultural relativity hypothesis. *Cognitive Psychology*, *50*, 378–411.

Roberson, D., Davidoff, J., & Shapiro, L. (2002). Squaring the circle: The cultural relativity of good shape. *Journal of Culture and Cognition*, *2*, 29–52.

Roberson, D., Davies, I. R. L., Corbett, G., & Vandervyver, M. (2005). Freesorting of colors across cultures: Are there universal grounds for grouping? *Journal of Cognition and Culture*, *5*, 349–386.

Roberson, D., Davies, I., & Davidoff, J. (2000). Color categories are not universal: Replications and new evidence from a Stone-age culture. *Journal of Experimental Psychology: General*, *129*, 369–398.

Roberson, D., Pak, H. S., & Hanley, J. R. (2008). Categorical perception of colour in the left and right visual field is verbally mediated: Evidence from Korean. *Cognition, 107*, 752–762.

Rosch, E. (1973). Natural categories. *Cognitive Psychology*, *4*, 328–350.

Rosch, E. (1975). The nature of mental codes for color categories. *Journal of Experimental Psychology: Human Perception and Performance*, *1*, 303–322.

Rosch Heider, E. (1972a). Probabilities, sampling, and ethnographic method: The case of Dani color names. *Man*, *7*, 448–466.

Rosch Heider, E. (1972b). Universals in color naming and memory. *Journal of Experimental Psychology*, *93*, 10–20.

Sandell, J. H., Gross, C. G., & Bornstein, M. H. (1979). Color categories in macaques. *Journal of Comparative and Physiological Psychology*, *93*, 626–635.

Saunders, B. A. C., & van Brakel, J. (1997). Are there non-trivial constraints on color categorization? *Behavioral and Brain Sciences*, *20*, 167–178.

Sayim, B., Jameson, K. A., Alvarado, N., & Szeszel, M. K. (2005). Semantic and perceptual representations of color: Evidence of a shared color-naming function. *Journal of Cognition and Culture*, 5(3/4), 427–486.

Sera, M. D., Berge, C., & del Castillo Pintado, J. (1994). Grammatical and conceptual forces in the attribution of gender by English and Spanish speakers. *Cognitive Development*, 9, 261–292.

Sera, M. D., Elieff, C., Forbes, J., Burch, M. C., Rodriguez, W., & Dubois, D. P. (2002). When language affects cognition and when it does not: An analysis of grammatical gender and classification. *Journal of Experimental Psychology: General*, *131*, 377–397.

Slobin, D. L. (1996). From "thought and language" to "thinking for speaking." In J. J. Gumperz & S. C. Levinson (Eds.), *Rethinking linguistic relativity* (pp. 70–96). Cambridge, UK: Cambridge University Press.

Soja, N. N. (1994). Young children's concept of color and its relation to the acquisition of color words. *Child Development*, *65*, 918–937.

Steels, L., & Belpaeme, T. (2005). Coordinating perceptually grounded categories through language: A case study for colour. *Behavioral and Brain Sciences*, *28*, 469–501.

Stefflre, V., Castillo Vales, V., & Morley, L. (1966). Language and cognition in Yucatan: A cross-cultural replication. *Journal of Personality and Social Psychology*, *4*, 112–115.

Sumner, P., & Mollon, J. D. (2000). Catarrhine photopigments are optimized for detecting targets against a foliage background. *Journal of Experimental Biology*, *203*, 1963–1986.

Talmy, L. (1983). How languages structure space. In H. Pick & L. Acredolo (Eds.), *Spatial orientation: Theory, research, and application* (pp. 225–282). New York: Plenum.

Vigliocco, G., Vinson, D. P., & Siri, S. (2005). Semantic and grammatical class effects in naming actions. *Cognition*, *94*, 91–100.

Webster, M. A., Miyahara, E., Malkoc, G., & Raker, V. E. (2000). Variations in normal color vision. II. Unique hues. *Journal of the Optical Society of America A*, *17*, 1545–1555.

Webster, M. A., Webster, S. M., Bharadwadj, S., Verma, R., Jaikumar, J., Madan, G., & Vaithilingam, E. (2002). Variations in normal color vision. III: Unique hues in Indian and United States observers. *Journal of the Optical Society of America A, 19*, 1951–1962.

Werner, J. S., & Wooten, B. R. (1985). Unsettled issues in infant color vision. *Infant Behavior and Development*, *8*, 99–107.

Whorf, B. L. (1956). The relation of habitual thought and behavior to language. In J. B. Carroll (Ed.), *Language, thought and reality: Essays by B. L. Whorf* (pp. 35–270). Cambridge, MA: MIT Press.

Wierzbicka, A. (1990). The meaning of color terms: Semantics, culture and cognition. *Cognitive Linguistics*, *1*, 99–150.

Winawer, J., Witthoft, N., Frank, M., Wu, L., & Boroditsky, L. (2007). Russian blues reveal effects of language on color discrimination. *Proceedings of the National Academy of Science, USA, 104*, 7780–7784.

13

An Institutional Approach to Culture

TOSHIO YAMAGISHI and NAOTO SUZUKI

The goal of this chapter is to advance an institutional approach to analyzing culture as a self-sustaining system of beliefs. The core of the institutional approach presented here lies in what Cohen (2001) called the "equilibrium perspective." An institution is a self-sustaining system of beliefs, behavior, and incentives among interdependent individuals. Individuals' behaviors, guided by their beliefs about other individuals' responses to their own behavior, collectively constitute the incentives for other individuals, and thus the set of incentives collectively sustain themselves through the behaviors of the individuals who follow them. We use the term *institution* instead of the more generic term *equilibrium* to emphasize that the continuity of behavior patterns in a culture is produced by the presence of a particular incentive structure rather than simple inertia. We adopted the use of this term in this way from comparative institutional economics, in particular Masahiko Aoki's (2001) work.[1]

Because the institutional approach we advance in this chapter is based on game theory, individuals' behaviors are conceptualized as strategies—that is, goal-oriented behavior guided by beliefs about how other interdependent individuals will respond. The difference of this approach to standard game theory resides in the new concept of "default strategies." A default strategy is a strategy in the sense used by biological game theorists, not by game theorists in economics. It is an ecologically rational (Gigerenzer, 2000) decision rule used by individuals (both nonhuman animals and human animals) without conscious calculation. Human individuals use a default strategy in many situations in the absence of cues suggesting ecological misfit of the strategy.

According to the institutional approach we advance in this chapter, a cultural mind is conceived as a set of adaptive strategies, each of which is dedicated to a particular task imposed by one's social as well as natural environment. Furthermore, social adaptation tasks are not exogenous to these strategies. It is not simply that a particular strategy is adopted by individuals by a given culture, unidirectionally imposing a particular adaptation task on its members, such as presenting oneself to others in such a way as to elicit high evaluations from them. Instead, we emphasize the aspect that the behavior guided by such a strategy collectively creates a distribution of contingent behavior in a society, to which individuals are advised to adapt.

Cultural psychologists similarly examine the mutual constitution of the mind and culture (cf. Markus & Kitayama, 1994). For example, Kim and Markus (1999) argued that preferences shared by a majority of people in a culture come to constitute social norms for that culture and that social norms in a culture are internalized as preferences. Although we agree with the idea of the mutual constitution of mind and culture, we argue in this chapter that this process is not simply an aggregation of individual preferences into social norms and subsequent internalization by individuals; rather, the process of mutual constitution of mind and culture is mediated by social institutions.

The core of the institutional approach is in the equilibrium analysis by which social adaptation tasks themselves are the product of actions of the individuals who apply adaptive strategies in an effort to solve the very tasks they have collectively created. Individuals who are endowed with a set of strategies collectively create a set of social adaptation tasks, the solutions of which are best achieved by applying the very strategies they use. In short, the strategy set becomes self-sustaining in the sense that a circular reinforcement relationship is established between strategies and adaptation tasks. Let us start with a simple example to show how this approach relates to interpreting cultural differences in psychological functioning, commonly studied by cultural psychologists.

PREFERENCES AND STRATEGIES

Cultural beliefs—beliefs shared by most members of a culture—about human nature and how society works differ greatly from culture to culture. A good example of this is found between Western and East Asian construals of the self, discussed by Markus and Kitayama (Kitayama & Markus, 1994; Markus & Kitayama, 1994). Markus and Kitayama argued that Westerners and East Asians have different beliefs about what human beings are like. Westerners share an independent construal of the self, a belief that human beings are internally driven agents operating independently from other agents, whereas East Asians share an interdependent construal of the self, a belief that human beings are a component of a larger system and are designed to accommodate their internal states and behavior to the system state. These general beliefs about human nature guide individuals' understanding of their own self. For example, Westerners are said to see themselves as a unique entity equipped with unique goals, desires, emotions, feelings, and so on, which are different from and often conflict with those of others. One implication of such a belief is that the best way to achieve one's goals, which generally compete with the goals of others, is to be actively assertive. These cultural beliefs and values, Kim and Markus (1999, p. 797) argued, shape individuals' preferences: "Whatever the culture says is right and good becomes what people like." Specifically, they argued that the general belief shared by Westerners—that humans are independent, internally driven agents—leads them to acquire a preference for uniqueness. In contrast, East Asians, who share the cultural belief that humans are components of a larger system, come to *prefer* conforming to others' expectations and behavior: "Once a person comes to understand that conformity is desirable, then the person will continue to try to be like others, not because he or she consciously thinks about the cultural values, but because being like others will feel good" (Kim & Markus, 1999, p. 797). Finally, Kim and Markus argued that such preferences and preference-based actions collectively create a social reality in which cultural beliefs coincide with actual behavior:

> Such preferences and actions are reflections of the values one adopted from the culture, but, at the same time, they constitute part of the social reality for oneself and others. … Social reality, then, is made up of numerous actions of participating individuals. At the individual level, one's actions do not seem to affect culture. However, when a critical mass of individuals are committed to similar values and act in concert in similar situations, the individuals will, as a collective, maintain or transform the culture. (Kim & Markus, 1999, p. 796)

To demonstrate that cultural beliefs engender cultural preferences, Kim and Markus (1999) conducted a series of experiments designed to measure the preference for uniqueness and conformity among Westerners and East Asians. In one of these experiments (Experiment 3), their research team distributed a questionnaire at an airport and asked travelers in the waiting lounges to fill it out. As a token of appreciation, they offered a pen to the traveler who participated in the study. An experimenter took out five pens from a bag that contained a large number of pens and asked a traveler to choose one of them to take home. The pens in the bag came in two external (not ink) colors. The color combination of the five pens was either 1–4 or 2–3, the first of the two numbers referring to the number of a "minority" color and the second to a "majority" color. The purpose of this field experiment was to demonstrate that Caucasians would prefer the minority color pen, whereas East Asians (judged by their appearance) would prefer the majority color pen. The result of this field experiment confirmed this prediction. Caucasians showed a preference for a pen in the minority color, whereas East Asians showed a preference for a pen in the majority color.

Yamagishi, Hashimoto, and Schug (2008) sought a different interpretation of this finding. They posited that the differential preferences for the two types of pens in the pen-choice experiment were caused by the use of a particular default strategy, which they termed the *not-offend-others strategy* (NOO strategy). People have numerous strategies appropriate for use in various situations. How they define a situation will lead them to select a particular strategy. A default strategy is one that people use when they do not deliberately choose a particular strategy, a strategy that is useful in most situations that people find themselves in everyday life, until they learn that its use is inappropriate. The particular strategy used as the default depends on its ecological fit, that is, the frequency of situations in a culture in which the strategy's use and nonuse produce desirable and undesirable outcomes, and the positive and negative impact of these outcomes (see Yamagishi,

Terai, Kiyonari, Mifune, & Kanazawa, 2007, for the logic of error management behind the choice of default strategies).

We speculate that the ecological fit of the strategy for avoiding a bad reputation varies between Western and East Asian societies. This is because the ecological fit of this strategy depends on the cost of being rejected by others. Ultimately, the cost is a function of the opportunities available to individuals in the event that they are excluded from their current relationships or groups. To the degree that one's social life is circumscribed by the boundaries of a group that one belongs to, a large cost is imposed on people who are excluded from that group. In a collectivist society in which groups or relationship are typically closed to outsiders, those who are ostracized from their group or relationship may find no alternatives to accept them. The cost of being excluded, therefore, is much higher in collectivistic societies than in individualistic societies, in which individuals can more easily replace lost relationships (Greif, 1994; Yamagishi, Cook, & Watabe, 1998; Yamagishi & Yamagishi, 1994). The NOO strategy thus has a higher ecological fit in a collectivist society than in an individualistic society.

Yamagishi and colleagues speculated that at least some of the East–West difference in preferences might be better explained as a reflection of differences in the default strategies, which were ultimately based on their ecological fit to differing social contexts. When travelers are presented with an opportunity to choose a pen, there is no strong need to use a particular strategy. In the absence of cues indicating a need to use a particular strategy appropriate to the situation, people should rely on a default strategy. The NOO strategy is a more likely default strategy for interpersonal situations in general among East Asians, who live in a collectivist society, than for Americans living in an individualist society, where "assert yourself when you want to have something" is a more likely candidate for a default strategy.

CHOOSING A PEN IN THE LABORATORY

In an experimental study by Yamagishi and colleagues (Yamagishi et al., 2008), American (50 students, 14 males and 36 females) and Japanese participants (55 students, 27 males and 28 females) were presented with a set of scenarios. In each scenario, participants were asked to imagine that they had been asked to fill out a questionnaire and had been offered a pen as a token of appreciation. They were also asked to imagine that they had been offered five pens to choose from. Four of the five pens were of one color, and the remaining one was of a different color. They were asked to indicate what color pen they would choose. There were four scenarios. In the first scenario, they were simply told that they had been provided with a choice of a pen. In the second scenario, they were told that they were the first out of the five people to choose a pen. In this scenario, participant's choice of a unique pen restricts the opportunity to select the unique pen. This manipulation was designed to exacerbate participants' concerns regarding the social implications of their actions and thus enhance their tendency to use the NOO strategy, as the choice of the minority pen might lead to resentment from others. In the third scenario, participants were told that they were the last person to choose a pen.

This manipulation was designed to eliminate concern for the social implications of their action and thus reduce their tendency to use the NOO strategy, because no one would be offended by either choice. Finally, participants were asked which color pen they would buy from a stationery shop. Because purchasing activities are generally regarded as solely private activities in which social implications are irrelevant, this manipulation is expected to reduce the use of the NOO strategy. In addition to these questions, they were asked how they thought other people would evaluate (on a 9-point scale ranging from *not so good* to *good*) a minority chooser (who chooses a minority pen) and a majority chooser (who chooses a majority pen).

The results indicated, first, that both Japanese and American participants expected that most people would evaluate the majority chooser more favorably than the minority chooser. The majority chooser received an average evaluation score (expected from other people) of 7.09 by Japanese and 6.63 by Americans, whereas the minority chooser received an average evaluation score of 4.11 by Japanese and 4.80 by Americans. The difference between the expected evaluations of the majority chooser and the minority chooser was significant among either Japanese or American participants. Furthermore, the cultural difference in the evaluations of majority and minority choosers between Japanese and American participants was not significant. Being unique and taking what one (presumably) wants was not regarded as a socially desirable deed for either Americans or Japanese.

Interestingly, however, both Japanese and American participants personally evaluated minority choosers as favorably as majority choosers. On average Japanese participants gave an evaluation score of 6.02 to the majority chooser and 6.16 to the minority chooser. Similarly, the average evaluation score of American participants was 6.45 for the majority chooser and 5.71 for the minority choosers. No difference turned out to be significant. It was thus shown that Japanese and American participants did not personally endorse the choice of the majority pen, and yet they shared the belief that other people would endorse the choice of the majority pen.

The proportion of participants who stated that they would choose the minority pen was significantly greater among American participants (71%) than Japanese participants (53%).[2] As we predicted, however, this difference existed only in the default choice condition. In the first-choice condition in which the participant was asked to report which pen they would select if they did so first among five potential choosers, the proportion of minority choosers did not differ between American (49%) and Japanese participants (45%). The proportion of minority choosers in the last-choice condition, in which the participants were asked to imagine what they would do if they chose a pen after all others had chosen, was higher than in the first-choice condition. Again, as predicted there was no significant difference between Japanese (71%) and American participants (72%). A similar pattern was observed in the purchase condition, in which there was no need for participants to worry about the social implications of their choice. The proportion of minority choosers was 73% among Japanese and 76% among American participants; again, the difference was not significant. In the default condition, the proportion of Japanese minority choosers (53%) was similar to that observed in the first-choice condition (49%), whereas the proportion of American minority choosers (71%) was similar to that observed in the last-choice condition (72%) and the purchase condition (76%).

These findings indicate, first, that the cultural difference emerged only in the default condition in which the nature of social constraints on the situation were unclear. In the first-choice condition, in which the social implications of choosing a minority pen are clear (remember that both Japanese and American participants believed that most people would give a low evaluation to the minority chooser), the cultural difference disappeared. The majority of American and Japanese participants refrained from choosing a minority pen under these circumstances. In the last-choice and purchase conditions, it was clear that participants' choices posed no negative consequences for others, and as such it was possible for participants to freely act on their preferences without having to consider the social implications of their actions. As a result, Japanese participants chose the minority pen just as frequently as did American participants.

The findings further suggest that the cultural difference in the default condition resulted from the way that the no-context situation was regarded—as being similar to either the first-choice condition or the last-choice condition. Japanese participants' propensity to choose a minority pen in the default condition was equivalent to that in the first-choice condition, in which social implications of their choice was obvious, despite the fact that Japanese chose a minority pen when they could freely do so. From this, we can conclude that the Japanese propensity to avoid the minority pen cannot be explained by preferences—Japanese participants chose the minority pen just as often as Americans participants did in the last-choice and purchase conditions. When the nature of the social constraints present in the situation was not clear (i.e., in the default condition), Japanese participants applied the NOO strategy. In contrast, American participants seemed to treat the default condition similarly to the last-choice condition, in which the NOO strategy was irrelevant: Although they applied the NOO strategy when its relevance was clear (first-choice condition), they did not apply it when nature of social constraints was not apparent (default condition). Thus, it is clear that the NOO strategy was not the default strategy for Americans.

ACTUAL CHOICE OF A PEN IN THE PRESENCE AND ABSENCE OF AN EXPERIMENTER

Because the study just presented involved hypothetical rather than real choices, Yamagishi and colleagues (2008) further attempted to see if Japanese participants' actual choice of majority and unique-colored pens would be affected by the nature of the situation in which the choice occurred. To do so, they asked students participating in various experiments (which had nothing to do with the pen choice) to select one of five pens. Of the five pens, four were of a randomly determined majority color, and one was a unique color. They first compared the 30 American participants with 126 Japanese participants who had experienced the same experiment as the Americans, and they successfully replicated Kim and Markus's (1999) study. Nineteen of the 30

(63%) American participants selected the unique pen, whereas only 44 of 126 (35%) Japanese participants did so, and the difference was significant.

They then offered the pen to Japanese participants at the beginning of experiments they participated in, in two differing situations. Seventy-eight participants chose a pen in the experimenter-present condition, in which an experimenter presented the participant with five pens in a cup and asked the participant to pick up one both for use in the experiment and to take home when the experiment was over. In the experimenter-absent condition, the experimenter placed a cup with five pens in front of the participant and asked him or her to take one, then left before the participant made his or her selection. Thus, participants were able to make their selection in privacy. The proportion of the participants who chose a minority pen was much higher in the experimenter-absent condition (52% or 14 of 27) than in the experimenter-present condition (23% or 18 of 78). The presence of another person (or a cue suggesting the social nature of the situation) seems to be required for the NOO strategy to be applied by Japanese participants.

THE EFFECT OF THE EXPERIMENT TYPE

There were 12 experiments (or experimental treatments) in which participants chose a pen after they went through the experimental task. These experiments are broken down into two general types: those in which participants' rewards were determined through participation with other players in the context of experimental games, and experiments that involved no interactions with other participants. The game-based experiments are further broken down into two types: ones in which participants' actions were monitored by other players (monitored game condition), and ones in which players' actions were completely anonymous (anonymous game condition). There were 175 participants in the nongame experiments (nongame condition), 272 participants in the anonymous game condition, and 111 participants in the monitored game condition. The selection of unique pens varied greatly depending on the type of experiment participants experienced prior to selecting the pen; participants were less likely to choose the unique pen after experiments in which players monitored each other's behaviors. The proportion of the unique pens was 22% after experiencing a game experiment with monitoring, 44% after a game experiment without monitoring, and 41% after a nongame experiment. The difference was more pronounced when participants chose a pen in the absence of the experimenter: 17% after experiencing a game experiment with monitoring, 49% after a game experiment without monitoring, and 57% after a nongame experiment. Because, as discussed previously, the presence of the experimenter reduces the choice of unique pens, the difference was not as pronounced when the experimenter was present: 22% after experiencing a game experiment with monitoring, 45% after a game experiment without monitoring, and 45% after a nongame experiment. These findings demonstrate that the presence of the experimenter and the experience of being monitored by other participants promoted the use of the NOO strategy.

I WILL PAY YOU IF YOU CORRECTLY JUDGE YOUR OWN ABILITY

Let us present another example of the NOO strategy as a default strategy. Cross-cultural differences in self-enhancement have been reported in many studies (Heine, Takata, & Lehman, 2000; see Heine, Lehman, Markus, & Kitayama, 1999), showing that Westerners have a much stronger tendency to perceive themselves in a positive light than do East Asians. Furthermore, East Asians are oftentimes shown to be self-effacing rather than self-enhancing and appear to focus attention to their weakness rather than their strengths (Heine et al., 2001). Cultural psychologists have explained this cultural difference in terms of the difference in self-construal discussed earlier (Heine et al., 1999).

The self-effacement tendency among East Asians, and especially among the Japanese, has also been interpreted as resulting from a conscious application of the NOO strategy (Bond, Leung, & Wan, 1982; Murakami & Ishiguro, 2005; Muramoto, 2003; Yoshida, Kojo, & Kaku, 1982). One difficulty, however, with this alternative interpretation of self-effacement among the Japanese is the fact that the self-effacing tendency has been observed in completely anonymous settings. Because their responses are unknown to anyone, they cannot be produced by the NOO strategy. This difficulty, however, does not prevent us from interpreting self-effacement as a *default* NOO strategy for self-presentation. We argue that the NOO strategy can be used as a default strategy when the situation lacks any clear indication that the strategy is unsuitable. The default strategy should be used until proved irrelevant or inappropriate. When a Japanese is asked to reveal his or her self-evaluation to someone else in a social setting—either directly to another person or in response to questionnaire items—he or she should automatically use the NOO strategy unless there is an obvious reason not to. Anonymity should be insufficient to override default strategies. Thus, Suzuki and Yamagishi (2004) conducted an experiment to demonstrate that Japanese use the NOO strategy as a default strategy in self-presentation.

First, participants in the experiment (Suzuki & Yamagishi, 2004) took a bogus intelligence test called the *Integrated Cognitive Ability Test*. Then, they were asked to judge whether their performance was above or below the average performance in their school (Hokkaido University). In the control condition, in which no additional instructions were provided, 72% (79 of 110 participants) judged that their performance was below the school average. This result is in sharp contrast to the better-than-average effect often observed among Americans (Alicke, 1985; Alicke, Klotz, Breitenbecher, Yurak, & Vredenburg, 1995; Dunning, Meyerowitz, & Holzberg, 1989). A very clear self-effacement tendency emerged in this condition, despite the fact that participants were assured of complete anonymity when making their judgment. To eliminate the possibility that participants were consciously concerned with the possibility that others might evaluate their self-judgment, we carefully constructed the situation to assure participants complete anonymity.

This self-effacement tendency, which is supposedly a reflection of the self, however, was completely reversed in the other condition, that is, the bonus condition. Before making a judgment, participants in the bonus condition were told that they would be paid an extra 100 yen (approximately $1.00) if their judgment was correct, in addition to the promised fixed payment of 300 yen. In the bonus condition, participants were provided with a reason for making the judgment—accurate judgments earn extra money. In contrast, participants in the control condition were provided with no reason for making the judgment. This difference, Suzuki and Yamagishi (2004) predicted, would affect the use of default strategies for self-presentation; Japanese should use the NOO strategy by default in the control condition but not in the bonus condition where it was apparent that the NOO strategy was inappropriate or, in other words, that an accurate judgment was required. Sixty-nine percent (36 of 52) of the participants judged their performance above, not below, the school average in the bonus condition, and the difference between the two conditions in the proportion of the above-average judgments was statistically significant.

The strong tendency toward self-effacement as observed in the control condition under strict anonymity was not a straightforward reflection of the participants' self-assessment of their ability. It was only when they lacked a reason to report their self-judgments that Japanese participants exhibited self-effacement. When it was made clear that the reason for the judgment was not for evaluating how they present themselves but to earn money, the self-effacing tendency completely disappeared, and Japanese participants showed a very clear tendency for self-enhancement. Although a great deal of psychological research on biases reveals that biases can be overridden with appropriate care, the results of this study go one step further, because one bias—self-effacing tendency—was replaced by another bias—self-enhancing tendency—rather than by the absence of any bias.

INSTITUTIONS

What difference does it make whether cultural differences in the way people perceive their world and behave accordingly are caused by different default strategies or different preferences and beliefs about human nature? From the cultural psychological perspective, the difference may appear small, if not trivial. Most cultural psychologists (e.g., Cohen, 2001; Heine et al., 1999; Nisbett, 2003; Nisbett, Peng, Choi, & Norenzayan, 2001) would not disagree with the general view that cultural beliefs and culture-specific cognition styles somehow serve as a means for adapting to the social–cultural reality surrounding individuals, and the default strategies are one specific means by which individuals adapt to their personal social–cultural reality. The fundamental question to cultural psychologists, we suspect, is how cultural beliefs and an individual's psychological functioning permeate one another. The stability of a culture is a matter of *self-perpetuation*. On the one hand, cultural beliefs permeate psychological functioning in the form of preferences. The preferences affect individuals' behavior, which in turn collectively creates a social reality. Finally, the beliefs are formed as a reflection of the social reality. Together, this cycle carries on as a stable state of the cultural reality.

In contrast, the fundamental question that we address in this chapter, which is shared by some economists who are interested in culture, is how a shared system of beliefs becomes self-sustaining (cf. Aoki, 2001). The stability of culture from this perspective is a matter of self-sustaining incentives; individuals' strategic behavior collectively produces a set of incentives that make particular strategies more adaptive than others. We use the term *institution* to describe a state in which a set of incentives encourages people to behave in ways that maintain the state.

Understanding institutions in this sense requires a social definition of incentives. Incentives are something that an individual desires to obtain. As humans are a social species, it is often the case that others provide what humans desire to obtain, such as money, love, prestige, and so on. In this sense, incentives are inseparable from the actions taken by others to provide the desired object. An incentive-driven behavior, then, is a strategy designed to elicit a behavior by other individuals that provides the object. We use the term *incentives* in this chapter to refer to the responses of other people to an individual's eliciting strategy. Suppose person *P* desires to obtain object *x*, and it is frequently the case that taking action *Pa* will induce person *O* to take action *Ox*, thereby providing *P* with the desired object. Conversely, taking action *Pb* will cause *O* to take a different action (*Oy*), which provides *P* with a different object (*y*), one that *P* does not desire to obtain. (The objects *x* and *y* do not need to be material; they can be psychological objects such as love or social objects such as prestige.) In this example, *O*'s behavior *Ox* is an incentive to instigate *P* to take action *Pa*.

A PRIMITIVE FORM OF AN INSTITUTION AS A SELF-SUSTAINING SYSTEM OF INCENTIVES

A primitive form of an institution is a situation in which *Pa* constitutes an incentive for *O* to perform *Ox*, and *Ox* constitutes an incentive for *P* to perform *Pa*. *O* provides the incentive (*Ox*) for *P* to perform *Pa*, which in turn provides the incentive for *O* to perform *Ox*. The incentives in this mutually reinforcing relationship are thus self-sustaining; each party's response to an incentive provides an incentive for the other party to respond in a certain way. This is what is meant by a self-sustaining system of incentives in its most primitive form. A voluntary-based system of social exchange is an example of a self-sustaining system of incentives, insofar as each provides what the other desires, and the relationship is voluntarily maintained.

A primitive form of an institution as a self-sustaining system of incentives is in fact an equilibrium in which a particular pattern of behavior is maintained because other parties' behavior provides the incentive to maintain a current behavior, and this holds true for all members of the institution. A simple example is a pattern of pedestrians in a crowded street. Pedestrians avoid collision with oncoming pedestrians by walking toward areas where there are fewer people. Because all pedestrians behave according to the same principle of the least resistance, they cannot help but to congregate to form a pattern, walking either on the left side, on the right side, in the middle, or on both sides, leaving the center for oncoming pedestrians. Once a pattern is formed, not too many people venture to break the pattern. Which side is better to walk on is determined by the flow of other pedestrians who also seek the path of least resistance. That is, the very behavior of walking on the left side, for example, provides an incentive for others to walk on the left side. Once walking behavior reaches a critical mass, it provides the incentive for other pedestrians to conform, thereby maintaining equilibrium (see Sugden, 1989).

Another example of this type of a primitive institution is a collectivist institution produced by in-group favoritism. In-group favoritism represents conditional cooperation, in which group members treat each other more favorably than they treat nonmembers, with the expectation that other group members will do the same (Yamagishi, Jin, & Kiyonari, 1999). Once enough individuals in the group adopt this strategy, individuals are better off adhering to this strategy than defecting from it. This is because only those who use this strategy can receive favorable treatment from similar others; those who fail to favor similar others fail to benefit from similar others. Once threshold is achieved, in-group favoritism reaches equilibrium; collectively, the practice of in-group favoritism constitutes a collectivistic institution.

AN INSTITUTION AS A SHARED SYSTEM OF SELF-SUSTAINING BELIEFS

The choice of an equilibrium state from multiple equilibria is often affected by players' beliefs. For example, among pedestrians who share the belief that people generally walk on the left side of the street, an equilibrium will form in which people walk on the left side. Different beliefs form different equilibria. In these examples, beliefs are self-sustaining in the sense that any behavior derived from said beliefs collectively creates an incentive system in which the beliefs correctly describe which strategies produce desirable outcomes. The example of a collectivistic institution mentioned earlier may often be mediated and facilitated by a belief that other individuals favor in-group members. For example, Hirai (2000) found that Japanese respondents to her questionnaire survey believed that they themselves were individualistic but that people around them were collectivistic. Similarly, in the vignette study of pen choice presented earlier, participants believed that most people would evaluate minority choosers negatively, whereas they themselves were indifferent. These findings suggest that the Japanese belief in the collectivistic nature of Japanese culture is maintained through pluralistic ignorance. Although they would prefer not to conform to social pressure, Japanese assume that other Japanese are collectivistic and would be disposed to enforce sanctions against individualistic behavior. This leads Japanese to behave in a collectivistic manner, without necessarily being collectivist themselves. Observers cite this behavior, however, as evidence that the Japanese are in fact collectivists. Ohbuchi and Saito (2007) provided evidence for this argument, showing that the collectivistic behavior pattern of conflict avoidance is maintained among Japanese people as a type of self-fulfilling prophecy. In short, in the words of Dov Cohen, a "considerable amount of people's public behavior and publicly stated performances is dependent on how they believe others will behave, whose behavior is, in turn, partially regulated by their own perceptions of public opinion" (Cohen, 2001, p. 454).

The sociological version of a self-fulfilling prophecy, originally developed by Merton (1968) and distinct from its psychological version, is in fact an institution as a self-sustaining system of shared beliefs. The difference between the sociological and psychological versions of the self-fulfilling prophecy is in the role incentives play. In the psychological version, of which the most famous example is "Pygmalion in the classroom" (Rosenthal & Jacobson, 1968), the teacher's expectation provides subtle cues to students, encouraging their progress. In contrast to this subtle cue theory of the self-fulfilling prophecy, the original sociological version, concerned at the time with racial discrimination in the northern industrial areas of the United States, centers on incentives. White workers in the north during the '40s and '50s expected that recent black immigrants from the nonindustrial south were "traitors to the working class," because they undercut their wage level and were often employed as strikebreakers. Thus, white workers denied blacks membership to their trade union. The lack of union membership kept blacks out of regular jobs, leaving them to seek any job including jobs as strikebreakers. The belief that blacks were traitors of the labor movement thus became the reality. In this example, the beliefs about black workers led to their exclusion, which in turn provided the incentive for black workers to replace striking workers.

Another good example of an institution as a self-sustaining system of shared beliefs is discrimination and differential investment in human capital. Investment in human capital, such as higher education, yields differential returns to those who are and are not discriminated against, as shown by the relative disparity in the increase to annual income resulting from greater education between men and women. This differential return rate from investment in human capital provides differential incentives for human capital development among those who are discriminated against and those who are not. When a group of people is believed to be less qualified and, consequently, excluded from high-paying jobs, individual members of said group are provided with fewer incentives to invest in the development of proper human capital. That is, the act of discrimination itself creates an incentive system that discourages those discriminated against from developing proper human capital. This will cause the discriminated group to actually become less qualified in terms of the human capital required for high-paying jobs. The belief among the majority, thus, is sustained by creating an incentive system, as discrimination based on a belief collectively creates an incentive system that discourages discriminated groups to acquire human capital, thereby making the belief into reality.

MAGHRIBI AND GENOESE TRADERS

The NOO strategy discussed earlier is adaptive in a collectivist institution in which most people share the same beliefs and favor in-group members. A collectivistic institution is a social setting in which conditional in-group favoritism constitutes an equilibrium. That is, individuals receive favors from other in-group members when and only when they give favors to other in-group members, and they fail to receive favors when they fail to give favors to other members of their group. Insofar as the benefits received exceed the costs incurred, it is in the individual's self-interest to give favors to other group members. In other words, a system of generalized exchange operates within a group, and an individual benefits from being a part of the system. In this collectivistic institution, the condition for receiving favors from other members of the system is a reputation as a compliant member of the system who is willing to perform his or her part in providing favors to others (see the literature on indirect reciprocity for more formal treatments of this discussion; Milinski, Semmann, Bakker, & Krambeck, 2001; Nowak & Sigmund, 1998a, 1998b; Panchanathan & Boyd, 2003; Takahashi, 2000). Having a reputation as a noncompliant member of the system limits the favors he or she will receive from other members of the group whose behaviors are guided by the conditional in-group favoring strategy. A critical condition for survival in such an institution is that others regard one as a compliant member. Applying the NOO strategy as a default strategy minimizes the risk of receiving a bad reputation as a noncompliant member.

Avner Greif, a renowned economic historian, analyzed the relationship between cultural beliefs and institutions by comparing the two hegemonic powers in medieval Mediterranean trade, Maghrib and Genoa (Greif, 1989, 1994). Medieval traders across the Mediterranean Sea faced a serious agency problem—a problem of how to control agents and curtail cheating by agents who operated across the sea. Finding a viable solution to the problem was critical to the trader's success. Maghribi traders and Genoese traders developed different solutions to this problem, the former being a collectivistic solution and the latter being an individualistic solution. The solution adopted by Maghribi traders was to close their trading circle to outsiders, hire only their own kind as agents, spread reputation about cheaters, and refrain from hiring an agent who developed a bad reputation. This is a collectivistic institution. Once this institution was established, they became very successful and dominated the Mediterranean trade during the 11th century. The success of similar trading institutions was observed in many places and many historical periods of the world; this is a standard practice for traders when central authority either does not exist or exists but does not enforce trading laws. According to Okazaki (1999), the Tokugawa regime of the 17th through 19th centuries in Japan did not intervene in civil matters, and the civil issues were left to the merchants themselves who did not have their own government to enforce civil laws. To solve the agency problem, they formed *kabu-nakama*, guildlike groups of merchants, and traded only within the circle. *Kabu-nakama* was banned twice by the central government during two and half centuries of Tokugawa regime, because they were perceived as cartels conspiring together to raise prices. Each time they were banned, trading activities throughout Japan dwindled, and the government was forced to lift the ban.

The solution adopted by Genoese traders was different. They established formal organizations to control cheating activities, such as city government and legal authorities. This is a more expensive alternative compared to the collectivistic solution adopted by Maghribi traders. Despite the disadvantage of higher maintenance costs, Genoese traders overtook Maghribi traders as the dominant power in the Mediterranean by the 12th century. Universal laws enforced by formal organizations reduced the risks for traders from other areas of the Mediterranean, and they began to trade with Genoese citizens who were accountable for their transactions. Conversely, the collectivistic institution of Maghribi traders kept outsiders from engaging in trading activities with them. As such, Maghribi traders had to bear an opportunity cost by restricting their trades to within their own circle, and that put them at a disadvantage in competing with Genoese traders.

Greif argued that "rational cultural beliefs that capture individuals' expectations with respect to actions that others will take in various contingencies" (Greif, 1994, p. 915) contributed to the adoption of the two types of solutions in the two societies. Collectivist cultural beliefs among the Maghribi people "restricted economic and social interactions to a small group and further facilitated in-group communication and economic and social collective punishments" (Greif, 1994, p. 942). In contrast, individualistic cultural beliefs—such as the belief that people mind their own business and do not care about other people—among the Genoese "weakened the dependence of each individual on any specific group, thereby limiting each group's ability to use economic, social, and moral sanctions against individual members" (Greif, 1994, p. 942). The absence of a collective punishment system (i.e., ostracism against exploitative agents) led the Genoese to adopt another, more expensive solution. "To support collective actions and to facilitate exchange, an individualist society needs to develop formal legal and political enforcement organizations" (Greif, 1994, p. 936). On the one hand, the two institutions were contingent on two separate cultural beliefs (collectivistic beliefs and individualistic beliefs), which in turn reinforced their respective cultural beliefs. Individuals who live under a collectivistic institution face a reality in which their behavior is constantly monitored and sanctioned by their peers. Collectivistic beliefs are reflections of this reality. Those who live under an individualistic institution, characterized by a legal system and other formal organizations, face another reality in which their behavior is judged against an objective criterion of behavior (i.e., the code of law) rather than by the informal approval of their peers.

SETTING UP AN INSTITUTION IN THE LABORATORY

Let us conclude this chapter with a brief description of our study that experimentally demonstrates the effect of institutions on cultural beliefs. This experiment was a conceptual reproduction of Maghribi and Genoese institutions to control the agency problem. A total of 91 participants were divided into two groups. Each group represented a total society and consisted of five communities of 9 or 10 participants each. Participants traded abstract commodities either within their communities or across communities. Trading of commodities involved information asymmetry and an accompanying agency problem. That is, only the seller knew the true quality of the commodity before it was sold; the buyer discovered its true quality only after the commodity was paid for. Payoffs were designed so that trading across communities was more profitable than trading within communities.

In addition to engaging in trading activities, each member of the community was provided with an opportunity to contribute money to establish a formal system of monitoring and sanctioning. The efficiency of this system depended on how much community members contributed for its establishment. In addition, community members of one society—the Maghribi society—were given the opportunity to get together and exchange information regarding their experience of trading with other members of their own community—whom they traded with and whether or not they were cheated. Members were identified only by ID numbers, not personally. Members of the other society, Genoese, were not given this opportunity to engage in reputation exchange within the community. We predicted that most trades in the Maghribi society would be restricted to within-community trades, because mutual monitoring and informal sanctioning through reputation sharing effectively controls the agency problem in each community, making within-community trades safe. This control mechanism did not exist for across-community trading, making these potentially very lucrative trades far riskier to pursue. In the Genoese society, the advantage of engaging in within-community trades did not exist because within-community reputation sharing was not allowed. The only means to control the agency problem was to bear the cost with other community members and establish an expensive formal system to supervise trading. Once one community established an effective system of social control, all members of the society would want to trade with its members, because they were forced to be honest by the system. The pattern of trading—within-community trades in the Maghribi society and across-community trades in the Genoese society—generally conformed to this prediction.

What we were interested in was whether the two institutional arrangements engendered different cultural self-construals. Specifically, we predicted that the Maghribi society would promote the interdependent self-construal and the Genoese society would promote the independent self-construal. As shown in Figure 13.1, the result of the experiment confirmed this prediction. Participants who had experienced the Maghribi type of collectivistic institution for controlling the agency problem had a higher score on the interdependent self-construal scale (Takata, Ohmoto, & Seike, 1996) than those who experienced the Genoese type of individualistic institution, and the direction of the difference was reversed on the independent self-construal scale.

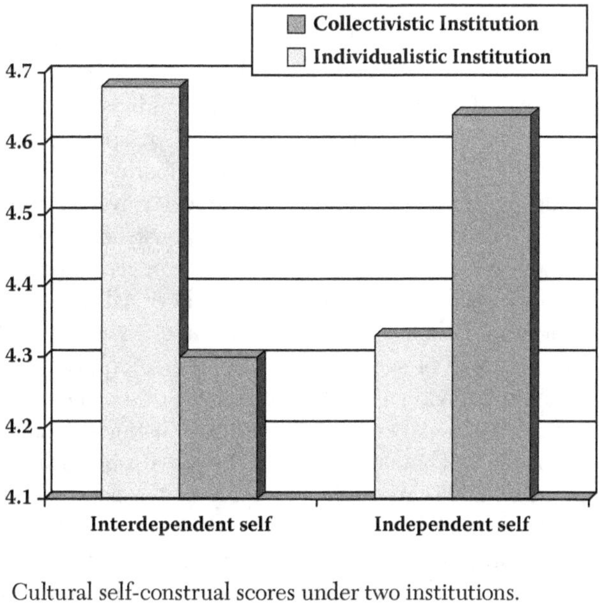

Figure 13.1 Cultural self-construal scores under two institutions.

CONCLUSION

Many scholars and pundits have traced cultural beliefs and cultural practices to the ecology of the area in which a particular culture originated (Umesao, 1967; Watsuji, 1935/1991). Nisbett (2003), for example, traced the Western way of thinking to the Greek tradition, which originated in a rocky fractured geological region where fishing and trading were the major industries. Success in fishing and trading depended greatly on individual skills and performance rather than on successful collaboration among individuals. Their way of life required individual decision making and initiative. This encouraged development of individual autonomy and analytic thinking. In contrast, the East Asian way of thinking can be traced to ancient China where rice cultivation, which requires collective efforts such as building and maintaining an irrigation system, was the main industry. Organizing collaborative groups was essential to their success. This encouraged conflict avoidance strategies within a group and dialectical thinking. Many of the contemporary differences in the way Westerners and Easterners perceive and interpret their world are extensions of belief systems formed centuries and millenniums ago, which originated from local ecological differences.

The institutional approach to culture we advanced in this chapter also attributes the ultimate source of cultural beliefs and practices to ecological conditions. On the other hand, it is different from the unidirectional (i.e., environment → culture) argument mentioned previously in that it emphasizes the coevolution of cultural beliefs and institutions. What our approach emphasizes is the fact that the social (and socially altered natural) environment itself is a product of cultural practices guided by cultural beliefs. The focus of our analysis is to identify how cultural beliefs become self-sustained; that is, to identify how cultural practices and the use of culturally sanctioned strategies create an incentive system in which the very strategies become adaptive.

Cohen (2001) listed four niches that play important roles for cultural diversity: physical, social, intracultural, and intercultural. Among those four, our approach focuses on social niches. Social niches are created by individuals' behavior, on the one hand, and shape individuals' behavior, on the other. Specifying the nature of interdependence among individuals and how the distribution of individuals' behaviors and individuals' beliefs about the distribution are converted to incentives for each individual is at the core of the social niche construction. Construction of social niches, we believe, is at the core of cultural diversity and cultural continuity. At the same time, social niches may have played critical roles in the evolution of some aspects of human nature. The "cheater detection module" (Cosmides & Tooby, 1992) is an example of such a default strategy. Acquiring such a module enhanced adaptation of individuals to the social niche of social exchange (or, using our term, the institution of social exchange), on the one hand, and made the practice of social exchange possible in human society, on the other. Yamagishi and colleagues (Yamagishi et al., 1999, 2007) are explicit in applying the notion of default strategies and the logic of social niche construction in their account of the institution of social exchange (including generalized exchange). The institution as a self-sustaining system of shared beliefs and incentives, we hope, thus provides a useful venue for evolutionary psychology and cultural psychology to collaborate in our effort to understand how humans create and maintain divergent patterns of behavior.

The studies presented in this chapter touch on only some of the vast collection of cultural differences in psychological functioning known to cultural and cross-cultural psychologists. We do not dare to claim that our approach is capable of explaining all of such cultural differences, which is an impossible task for any single approach. Future studies should address the issue of what types of cultural elements are best explained and predicted from the institutional model and what types are most difficult to do so. We hope this chapter and the studies presented in this chapter provide an impetus for further research to identify the range and limits of the institutional approach.

NOTES

1. Readers of this chapter are advised to keep in mind that the term *institution* is used in a more general manner than its more reified use in our everyday conversation, such as schools, hospitals, and the police, as examples of institutions.
2. There was a significant gender difference such that males chose a minority pen more frequently than females. The Gender × Nationality interaction was not significant. The gender composition favored females among American participants, whereas it was roughly balanced among Japanese participants. This reduces the nationality difference unless gender is controlled. We report here the simple percentages without adjusting for the difference in gender composition. When the gender composition was adjusted for, the nationality difference was more pronounced.

REFERENCES

Alicke, M. D. (1985). Global self-evaluation as determined by the desirability and controllability of trait adjectives. *Journal of Personality and Social Psychology*, *49*, 1621–1630.

Alicke, M. D., Klotz, M. L., Breitenbecher, D. L., Yurak, T. J., & Vredenburg, D. S. (1995). Personal contact, individuation, and the better-than-average effect. *Journal of Personality and Social Psychology*, *68*, 804–825.

Aoki, M. (2001). *Toward a comparative institutional analysis*. Cambridge, MA: MIT Press.

Bond, M. H., Leung, K., & Wan, K. C. (1982). The social impact of self-effacing attributions: The Chinese case. *Journal of Social Psychology*, *118*, 157–166.

Cohen, D. (2001). Cultural variation: Considerations and implications. *Psychological Bulletin*, *127*(4), 451–471.

Cosmides, L., & Tooby, J. (1992). Cognitive adaptations for social exchange. In J. Barkow, L. Cosmides, & J. Tooby (Eds.), *The adapted mind: Evolutionary psychology and the generation of culture*. New York: Oxford University Press.

Dunning, D., Meyerowitz, J. A., & Holzberg, A. D. (1989). Ambiguity and self-evaluation: The role of idiosyncratic trait definitions in self-serving assessments of ability. *Journal of Personality and Social Psychology*, *57*, 1082–1090.

Gigerenzer, G. (2000). *Adaptive thinking: Rationality in the real world*. Oxford, UK: Oxford University Press.

Greif, A. (1989). Reputation and coalitions in medieval trade: Evidence on the Maghribi traders. *Journal of Economic History*, *49*, 857–882.

Greif, A. (1994). Cultural beliefs and the organization of society: A historical and theoretical reflection on collectivist and individualist societies. *Journal of Political Economy, 102*, 912–950.

Heine, S. J., Kitayama, S., Lehman, D. R., Takata, T., Ide, E., Leung, C., & Matsumoto, H. (2001). Divergent consequences of success and failure in Japan and North America: An investigation of self-improving motivations and malleable selves. *Journal of Personality and Social Psychology, 81*, 599–615.

Heine, S. J., Lehman, D. R., Markus, H. R., & Kitayama, S. (1999). Is there a universal need for positive self-regard? *Psychological Review, 106*, 766–794.

Heine, S. J., Takata, T., & Lehman, D. R. (2000). Beyond self-presentation: Evidence for self-criticism among Japanese. *Personality and Social Psychology Bulletin, 26*, 71–78.

Hirai, M. (2000). Stereotypes about the Japanese: Differences in evaluations between "the Japanese" and "myself." *Japanese Journal of Experimental Social Psychology, 39*, 103–113.

Kim, H., & Markus, H. R. (1999). Deviance or uniqueness, harmony or conformity? A cultural analysis. *Journal of Personality and Social Psychology, 77*, 785–800.

Kitayama, S., & Markus, H. R. (1994). *Emotion and culture: Empirical studies of mutual influence*. Washington, DC: American Psychological Association.

Markus, H. R., & Kitayama, S. (1994). The cultural construction of self and emotion: Implications for social behavior. In S. Kitayama & H. Markus (Eds.), *Emotion and culture* (pp. 89–130). Washington, DC: American Psychological Association.

Merton, R. K. (1968). *Social theory and social structure* (enlarged ed.). New York: Free Press.

Milinski, M., Semmann, D., Bakker, T., & Krambeck, H. (2001). Cooperation through indirect reciprocity: Image scoring or standing strategy? *Proceedings of the Royal Society of London, Series B, Biological Sciences, 268*, 2495–2501.

Murakami, F., & Ishiguro, I. (2005). The effects of the communication target on self-effacement. *Japanese Journal of Social Psychology, 21*, 1–11.

Muramoto, Y. (2003). An indirect self-enhancement in relationship among Japanese. *Journal of Cross-Cultural Psychology, 34*, 552–566.

Nisbett, R. E. (2003). *The geography of thought: How Asians and Westerners think differently … and why*. New York: Free Press.

Nisbett, R. E., Peng, K., Choi, I., & Norenzayan, A. (2001). Culture and systems of thought: Holistic versus analytic cognition. *Psychological Review, 108*, 291–310.

Nowak, M., & Sigmund, K. (1998a). The dynamics of indirect reciprocity. *Journal of Theoretical Biology, 194*, 561–574.

Nowak, M., & Sigmund, K. (1998b). Evolution of indirect reciprocity by image scoring. *Nature, 393*, 573–577.

Ohbuchi, K., & Saito, T. (2007). Cognitive causes of conflict avoidance among Japanese: An approach from pluralistic ignorance. *Progress in Asian Social Psychology, 6*, 83–97.

Okazaki, T. (1999). *Edo no shijou keizai* [Market economies in Edo period]. Tokyo: Kodansha.

Panchanathan, K., & Boyd, R. (2003). A tale of two defectors: The importance of standing in the evolution of indirect reciprocity. *Journal of Theoretical Biology, 224*, 115–126.

Rosenthal, R., & Jacobson, L. (1968). *Pygmalion in the classroom: Teacher expectation and pupils' intellectual development*. New York: Holt, Rinehart and Winston.

Sugden, R. (1989). Spontaneous order. *Journal of Economic Perspective, 3*, 85–97.

Suzuki, N., & Yamagishi, T. (2004). An experimental study of self-effacement and self-enhancement among the Japanese. *Japanese Journal of Social Psychology, 20*, 17–25.

Takahashi, N. (2000). The emergence of generalized exchange. *American Journal of Sociology, 105*, 1105–1134.

Takata, T., Ohmoto, M., & Seike, M. (1996). Development of an independent–interdependent self-construal scale (rev. ed.). *Nara University Bulletin, 24*, 157–173.

Umesao, T. (1967). *Bunmeino seitai shikann* [An ecological view of civilization]. Tokyo: Chuo Koronsha.

Watsuji, T. (1991). *Fudo: Ningengakuteki kousatsu* [Climate and culture: A philosophical study]. Tokyo: Iwanami Shoten. (Original work published 1935)

Yamagishi, T., Cook, K. S., & Watabe, M. (1998). Uncertainty, trust and commitment formation in the United States and Japan. *American Journal of Sociology*, *104*, 165–194.

Yamagishi, T., Hashimoto, H. & Schug, J. (2008). Preferences versus strategies as explanations for culture-specific behavior. *Psychological Science, 19*, 579–584.

Yamagishi, T., Jin, N., & Kiyonari, T. (1999). Bounded generalized reciprocity: In-group favoritism and in-group boasting. *Advances in Group Processes*, *16*, 161–197.

Yamagishi, T., Terai, S., Kiyonari, T., Mifune, N., & Kanazawa, S. (2007). The social exchange heuristic: Managing errors in social exchange. *Rationality and Society*, *19*, 259–298.

Yamagishi, T., & Yamagishi, M. (1994). Trust and commitment in the United States and Japan. *Motivation and Emotion*, *18*, 129–166.

Yoshida, T., Kojo, K., & Kaku, H. (1982). A study on the development of self-presentation in children. *Japanese Journal of Educational Psychology*, *30*, 120–127.

14

Cultural Consequences of Voluntary Settlement in the Frontier *Evidence and Implications*

SHINOBU KITAYAMA and NICHOLAS A. BOWMAN

Cultural psychology has made substantial progress in the past two decades. The body of research accumulated over the period was taken stock of in the first handbook for the field (Kitayama & Cohen, 2007). Moreover, two comprehensive textbooks are now available (Chiu & Hong, 2006; Heine, 2008). There are numerous monographs and edited volumes on culture (Benson, 2001; Cole, 1996; Mahalingam, 2006; Matsumoto, 2001; Nisbett, 2003; Nisbett & Cohen, 1996; Schaller & Crandall, 2004; Shweder, 2003; Sperber, 1996).

One major contribution of this literature is the extent to which some important psychological processes can vary as a function of sociocultural contexts has been drawn out. It has done so by making systematic macroscopic comparisons across broadly defined cultural regions of East (e.g., East Asia) and West (e.g., North America). More than once or twice, entirely unexpected and thus surprising degrees of variation have been uncovered. For example, self-enhancing biases in social judgment, which are one of the most robust and replicable effects in North America, are often very hard to obtain in East Asia (Heine, Lehman, Markus, & Kitayama, 1999; Kitayama, Markus, Matsumoto, & Norasakkunkit, 1997). Another highly robust effect in North America involves a dispositional bias in attribution of another person's behavior. This bias, called the fundamental attribution error, is typically much weaker or even nonexistent in East Asian cultures (Choi, Nisbett, & Norenzayan, 1999; Masuda & Kitayama, 2004; Miyamoto & Kitayama, 2002; Morris & Peng, 1994).

The same degree of cultural variation is also evident in yet another staple effect in social psychology—namely, cognitive dissonance. A recent series of cross-cultural and cross-ethnic experiments has provided abundant evidence that this effect occurs in both North America and Asia, and, accordingly, at first glance it appears universal. One important caveat, however, is that the specific conditions in which the effect is observed are very different across cultures (Hoshino-Browne et al., 2005; Imada & Kitayama, 2004; Kitayama, Snibbe, Markus, & Suzuki, 2004). Findings such as these, in turn, suggest that the initial appearance of identity is deceptive. The seemingly identical effect is mediated by very different psychological mechanisms. In terms of psychological mechanisms, the universality is more apparent than real.

So far, however, this literature is largely confined to macroscopic comparisons between globally demarcated areas such as East and West. One issue that is largely overlooked concerns possible within-culture variations. The issue of subcultural variation can be easily ignored as secondary if this variation represents noise in observation. Perhaps some individuals may acquire the dominant culture more than some others, or some instruments measure the relevant construct with greater precision for some people than for others. And, of course, some participants are more conscientious in taking psychological tests than are some others. All these sources of variation can contribute to the within-culture variability. One could argue, however, that if one's ultimate aim is to understand sociocultural processes, these factors do not bear much theoretical significance.

This argument, however, may be seriously challenged for several reasons. First, despite a moderate to large East–West effect size on a variety of explicit and implicit psychological measures (for meta-analyses, see Heine & Hamamura, 2007; Miyamoto, Kitayama, & Talhelm, 2006; Oyserman, Coon, & Kemmelmeier, 2002), there is substantial variance within these broad cultural groups. These within-culture variations are sometimes so great that they are difficult to dismiss as measurement error. Second, broadly grouping nations into a single "East" or "West" category ignores the important influences of modernization, political structure, and religion that differentiate nations or subcultural groups within these categories (e.g., Inglehart & Baker, 2000). Third, there are significant regional variations in personality and values within countries that can predict social behaviors including voting patterns (Caprara, Schwartz, Capanna, Vecchione, & Barbaranelli, 2005; Rentfrow, Jost, Gosling, & Potter, 2007) and well-being (Plaut, Markus, & Lachman, 2002). And fourth, most relevant to our current discussion, there are important historical events that can have a pronounced impact on present-day psychological processes and cultural practices (e.g., Vandello & Cohen, 1999).

Furthermore, we believe that by examining regional variations, it is possible to test theories of processes underlying the maintenance and change of culture. For example, since Max Weber (1904–1906/1958), it has been suggested that Protestantism is an important element of what constitutes the present-day capitalism and an associated version of individualism. This historical hypothesis is difficult to test in any direct fashion by observing history itself. Nevertheless, because this history has produced a number of contemporary consequences, it would be possible to derive predictions about systematic subgroup differences within the United States regarding certain outcome variables. In fact, a series of studies by Sanchez-Burks (2005) showed that as compared to European Americans with

Catholic backgrounds, those with Protestant backgrounds become actively disengaged from personal relations, especially in work settings. Findings such as these provide important clues about the nature of influences Protestantism has historically exerted on American individualism.

In this chapter we propose yet another factor that we believe has had pivotal influences on the contemporary American culture: the history of voluntary settlement in frontiers. In testing this idea, we rely on an examination of within-culture variations in certain aspects of independence and individualism. In what follows, we first sketch basic ideas underlying this hypothesis. We then describe two lines of work that lend support to the hypothesis. We finish with a discussion of further implications of the hypothesis in understanding some aspects of the contemporary cultural dynamics.

THE VOLUNTARY SETTLEMENT HYPOTHESIS

Drawing on earlier sociological theorizing on origins of the contemporary American culture (Turner, 1920), Kitayama, Ishii, Imada, Takemura, and Ramaswamy (2006) argued that the history of voluntary settlement in frontiers gives rise to independent agency. To delineate the hypothesis, we need to be specific about what we mean by independent agency.

Self-in-Action and Culture: Independence and Interdependence

Agency refers to a psychological system of active self-regulation. It therefore implies self-in-action (Kitayama, Duffy, & Uchida, 2007; Kitayama & Uchida, 2005; Markus & Kitayama, 1991, 2004). We thus define independent agency as self-in-action that is grounded in a cultural belief of independence, self-reliance, and separation from others. Independent agency is typically found in Western, individualistic societies.

Independent agency may be contrasted with interdependent agency, which implies self-in-action that is based on a cultural belief of interdependence, mutual interpersonal attunement and harmony, and connectedness with others (Kitayama & Uchida, 2005). This latter mode of agency is more commonly found in Eastern, collectivist cultures. Each of these modes of self-in-action is composed of numerous partially independent elements. We suggest that these elements can be organized into three global components: epistemic, normative, and motivational.

Epistemic component. First, the cultural views of self as independent or interdependent offer a set of beliefs about the nature of self. Because of this descriptive or epistemic function of cultural self, once individuals are socialized in a given cultural context, they are often inculcated with beliefs of either independence or interdependence. These beliefs provide an epistemic basis for the corresponding modes of agency.

There is evidence that independent beliefs are more dominant in Western cultures, and interdependent beliefs are more dominant in Eastern contexts. Numerous empirical studies have demonstrated that the cross-culturally contrasting epistemic views of self have profound influences on social perception. That is, those with an independent view are more likely than those with an interdependent view to emphasize dispositional factors in accounting for another person's behavior.

Miller (1984), in her classic work, demonstrated that the dispositional bias is highly robust among North Americans, thus replicating the existing evidence (Ross, 1977). Importantly, however, Miller also found that this tendency is largely nonexistent and even reversed among Indians in India. For example, in accounting for a socially desirable behavior of their acquaintances, Americans tended to report that the acquaintances were kind and well intentioned, whereas Indians tended to report that there were strong social norms or role expectations on the acquaintances.

This initial finding was followed by numerous studies that demonstrated similar cross-cultural variations (e.g., Kitayama, Ishii, et al., 2006; Morris & Peng, 1994). Moreover, the same cross-cultural contrast has been suggested in studies of correspondence bias. When observing someone's behavior (e.g., after stating that President George W. Bush's policy in Iraq is justifiable), Americans are highly prone to infer an attitude that corresponds to the behavior (a pro-Bush attitude) even when the behavior is constrained in very obvious ways (e.g., Gilbert & Malone, 1995; Jones, 1979). Asians, however, are much more likely to discount the behavior, especially when the situational constraint is made salient (Choi & Nisbett, 1998; Masuda & Kitayama, 2004) or when the behavior at issue is minimally diagnostic of a strong attitude (Miyamoto & Kitayama, 2002). Similar cross-cultural variations have been shown in the analysis of media contents and other cultural artifacts (Lee, Hallahan, & Herzog, 1996; Markus, Uchida, Omoregie, Townsend, & Kitayama, 2006; Morris & Peng, 1994).

Normative component. The cultural views of self as independent or interdependent have a normative function as well—that is, the prescriptive function of specifying what behaviors and actions are desirable and what are not desirable. Specifically, once individuals who share beliefs in either independence or interdependence are brought together to form a community, they begin to develop corresponding normative expectations. That is, in societies or communities where independence is widely shared and thus taken for granted, independent behaviors (e.g., self-expression, personal goal pursuit) are more likely to be encouraged and rewarded than interdependent behaviors (e.g., modesty, communal goal pursuit). Conversely, in societies or communities where interdependence is more widely shared and taken for granted, interdependent behaviors are more likely to be normatively sanctioned than independent behaviors.

As may be expected, acts of independence, including self-expression and personal goal pursuit, are likely to be rewarded in Western cultural contexts more than in Eastern contexts. But acts of interdependence, including modesty and communal goal pursuit, are more likely to be rewarded in Eastern contexts than in Western contexts. One way of assessing such societal reward contingencies is to examine who is most likely to experience happiness and well-being over time. Relatively independent people should be happier than relatively interdependent people in Western contexts, but the reverse should be the case in Eastern contexts. This general expectation has been strongly borne out in several cross-cultural studies (Kitayama, Markus, & Kurokawa, 2000; Kwan, Bond, & Singelis, 1997).

For example, Kitayama, Markus, and Kurokawa (2000) had both American and Japanese undergraduates report how frequently they experienced a variety of positive emotions. Some positive emotions, such as happiness, joy, and calmness, were general, indicating global happiness and well-being. Some others, such as pride in self and self-confidence, were socially disengaging, meaning that they are based on success in achieving independence. Finally, some other emotions, such as friendly feelings and feelings of connection with others, were socially engaging, grounded on success in achieving interdependence. It was found that for Americans, happiness and well-being were predicted more strongly by socially disengaging positive emotions, but for Japanese, they were predicted more strongly by socially engaging positive emotions.

Motivational component. The third function of the views of self as independent or interdependent is motivational. These cultural views provide a variety of goal states: Whereas the independent view offers a number of personally oriented goals, the interdependent view supplies numerous communally oriented goals. Through socialization, individuals internalize these goal states, thereby forming their intrinsic motivational propensities. As a consequence, in independent cultural contexts, people will be intrinsically motivated to set personal goals, such as personal achievement and need satisfaction, and seek to achieve them. In contrast, in interdependent cultural contexts, people will be intrinsically motivated to accomplish group or interpersonal goals such as fitting in and meeting social and interpersonal expectations and norms.

The hypothesis that one's personal goals are highly central in organizing a person's intrinsic motivation in North America is supported by evidence that personal choice is quite motivating and influences one's subsequent cognitions and actions. For example, North Americans typically work much harder on a task when allowed to choose the task than when assigned to it (Deci & Ryan, 1985). Likewise, when having chosen certain commodities such as CDs, people are strongly motivated to justify the choice (Brehm, 1956). These findings indicate that for North Americans, one's motivation is grounded in personally chosen goals and agendas.

Importantly, however, these effects are often much weaker or nonexistent among Asians or even among Americans with Asian cultural backgrounds. For example, Iyengar and Lepper (1999) found that Asian American children were highly motivated when the choice was made by trusted authority figures or peers, but they were not as motivated by personal choice. Similar cross-cultural differences in the impact of choice on motivation have also been found among college students (Na & Kitayama, 2006). Likewise, the justification effect following a personal choice has been shown to be largely nonexistent among Asians (Heine & Lehman, 1997; Kitayama et al., 2004) and North Americans with Asian heritage (Hoshino-Browne et al., 2005). Importantly, however, when choice is perceived as public (i.e., under the scrutiny of others), it does enhance motivation (Na & Kitayama, 2006). Moreover, under such conditions, choice motivates Asians to engage in self-justification (Hoshino-Browne et al., 2005; Imada & Kitayama, 2004; Kitayama et al., 2004). This emerging evidence shows that the guiding motivational theme of Asians is much more social and interdependent, revolving around ever-important relational concerns such as face-saving, public reputation, approval by others, and social harmony.

Kitayama, Mesquita, and Karasawa (2006) suggested that motivational propensities toward independence and interdependence can be assessed by testing when each person is most likely to experience happiness and well-being. That is to say, if one's chronic goal is that of independence, the person's happiness and well-being should be enhanced especially when some form of independence is achieved, whereas if his or her chronic goal is that of interdependence, his or her happiness and well-being should be particularly enhanced when some form of interdependence is achieved. Using a diary method, the researchers showed that there is a substantial individual difference in the extent to which individuals experience happiness when they experience socially disengaging emotions (which result from successful independence) or socially engaging emotions (which result from successful interdependence). Interestingly, in support of the hypothesis that Asians are more oriented toward interdependence (vis-à-vis independence) than North Americans, happiness depended far more on socially engaging (vis-à-vis disengaging) emotions for Asians than for North Americans.

Notice that the analysis here focuses on the within-subjects correlations between happiness and socially engaging or disengaging emotions. This is a measure of when one experiences happiness and, as such, is an indicator of the person's motivational propensity toward independence or interdependence. In contrast, one can use the same data to compute between-subject correlations between happiness and the two types of positive emotions. This is a measure of who is most likely to experience happiness over time. This chronic measure serves as an indicator of societal reward contingencies. Reanalyzing the data from Kitayama, Mesquita, et al. (2006), we found, indeed, that chronic interdependence predicts chronic happiness for Japanese, but chronic independence predicts chronic happiness for Americans. Of importance, there is no logical reason to expect, a priori, that the two types of correlations show the same pattern. In our theoretical analysis, the two measures are indicators of distinct, partially dissociable processes of normative independence and interdependence and their motivational counterparts.

Conditions of the Frontier and Independent Agency

We propose that the conditions of the frontier contribute to various elements of independent agency. By definition, frontiers are ecologically harsh and relatively unpopulated, and thus social mobility is bound to be quite high. In part because of scarce population, the infrastructure of any community that exists remains quite primitive. Moreover, the economic base of the community tends to be weak, and often heavily dependent on more populated areas. Hence, living standards in frontiers tend to be quite suboptimal. Not surprising, mere survival tends to be the primary concern of life. Moreover, because of the lack of sufficient resources, one's own survival will depend ultimately on the individual. Despite these negative conditions, frontiers offer freedom, the possibility of acquiring lands that are (imagined to be) limitless, and the corresponding promise of success and attendant wealth—or, if not, at least the hope for them.

Self-selection. Because of this promise of—or at least hope for—success and wealth, some people choose to move to frontiers. They do so despite the fact that many others instead choose to stay in populated areas of origin. There is a distinct possibility, then, that self-selection is a major factor that breeds independent ethos in the frontier. That is, the people who are most likely to move to the frontier are those who believe in the importance of independence (i.e., who are normatively independent) and/or those who enjoy being independent (i.e., who are intrinsically motivated toward independence). Nevertheless, it is important to note that immigration often takes place in a variety of social units, such as family and more extended kin groups, as well as nonkin social units, such as village and tribal groups. Thus, it is unlikely that self-selection alone is sufficient to account for independent agency that is likely to emerge in frontiers.

Adaptation to local conditions. Ecological conditions of the frontier are often extremely harsh, the infrastructure of the society is primitive at best, and, moreover, living conditions are substandard at best. These ecological and social structural conditions are likely to foster behaviors that are designed to promote and protect the self in lieu of communities or interpersonal ties. Therefore, a strong value for independence, self-reliance, and autonomy is likely to emerge. That is to say, independent behaviors will be normatively sanctioned and reinforced.

Institutionalization. Once people who either value or enjoy independence gather to form a community, they begin to develop social mores, informal rules and norms, and even more formal social institutions that are grounded in the value of independence. Because people begin to behave in a more consistently independent fashion, they may also begin to internalize the corresponding generic schematic understanding of person as highly independent—that is, as being guided by his or her internal attributes, such as attitudes, personal goals, and traits. In other words, a consensual belief about the nature of person and self as independent may become more common.

Summary

In short, the voluntary settlement hypothesis implies that because of ecological and social structural conditions of frontiers, settlers in the frontier are likely to develop three partially overlapping components of independent agency. The three components are epistemic, normative, and motivational. Moreover, the development of independent agency is likely to be mediated by three distinct processes of self-selection, adaptation to local conditions, and institutionalization. This general conceptualization of the voluntary settlement hypothesis is illustrated in Figure 14.1.

By itself, the hypothesis neither says much about which of the three components (i.e., epistemic, normative, and motivational) is most likely to emerge in any given circumstances nor specifies which of the three underlying processes (i.e., self-selection, adaptation to local conditions, and institutionalization) is most important. We believe that issues such as these are highly path dependent. That is, they depend on social and psychological conditions that exist prior to settlement

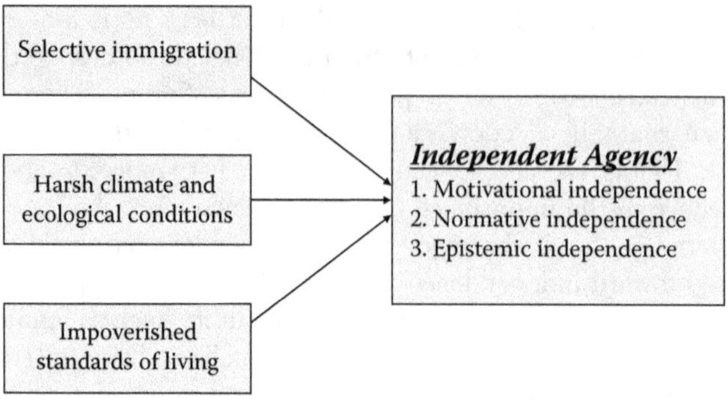

Figure 14.1 The voluntary settlement hypothesis: How conditions of the frontier foster and reinforce the ethos of independence, which is composed of three partially distinct facets of motivational independence, normative independence, and epistemic independence.

and immigration. As such, to resolve these issues, we need to take into account specific historical, societal, and economic circumstances in which settlement and immigration occurred. Nevertheless, the hypothesis asserts that some combination of components of independent agency listed in Figure 14.1 is likely to be fostered and reinforced in frontiers through one or more of the three processes described earlier.

UNITED STATES: CONSEQUENCES OF SETTLEMENT WITH PRIOR CULTURAL CONDITIONS OF INDEPENDENCE

Western European Heritage of North America

The voluntary settlement hypothesis has been informed by an earlier sociological theorizing to account for contemporary American culture as strongly influenced and even shaped by its history of both immigration to the "new" continent by a large number of people of mostly European descent and subsequent settlement in its Western frontiers (Turner, 1920). Hence, one obvious way to test the hypothesis involves a systematic comparison between matched samples from the United States and Western Europe (Kitayama, Park, Sevincer, & Karasawa, in press).

Even a cursory look at the history of the modern West makes it very clear that North America owes important cultural heritages to Western Europe. North America was initially settled by Puritans, followed by a large number of immigrants from Protestant countries in Western Europe. These individuals literally carried with them both practices and public meanings associated with independence. Even though a variety of ethnic groups subsequently came to the United States from all over the world, the foundation of the country was greatly shaped by discourses and social institutions implanted from Western Europe. Accordingly, it seems quite reasonable to expect that American culture is at least as individualistic and independently oriented as any Western European culture today. This may be true in respect to all the three facets of independence discussed previously.

Settlement History of North America

In one key respect, however, American culture is distinct from European cultures. Only American culture has undergone a history of settlement in a new continent and its western frontiers. According to the voluntary settlement hypothesis, this unique history must have bolstered some aspects of individualism even further (Turner, 1920). As argued earlier, three distinct processes might underlie this transformation.

First, only those Europeans who feel personally aspired to take challenges and to seek wealth and freedom may choose to engage in such settlement (self-selection). Second, because frontiers are often harsh in ecology and primitive in social institutions and infrastructures, sheer survival becomes an issue. Given the substantial social mobility of such a society, there may be minimal communal constraints on people's behaviors. As a consequence, personal initiatives and promotion of personal goals are likely to be not only tolerated but also highly encouraged (adaptation to local conditions). That is, both normative independence and motivational independence must have been amplified. Finally, once people who aspire for personal goals and personal promotion are gathered to form a community, the corresponding ethos and mores may develop (institutionalization). The resulting social system is likely to further bolster both motivational and normative independence.

Indeed, in the United States, a cultural discourse that emphasizes personal initiatives and hard work was eventually codified explicitly and shared collectively. This discourse is now known as the American Dream (Hochschild, 1995). It highlights strong personal drives toward tangible success. When we asked American undergraduates to write any ideas associated with the American Dream, they often referred to these general themes. For example, one female student said that for her, the American Dream includes a "lucrative career that I love, a wonderful husband, two beautiful kids and a nice house in an expensive neighborhood." For a male student in the same sample, the American Dream meant "an ability to make of yourself whatever you want."

In short, there is good reason to expect that as compared to Western Europeans, North Americans would be higher in both motivational independence and normative independence. These two aspects of independence are directly instrumental in promoting individual initiative, hard work, and motivational drive. Both people with stronger motivational independence and communities with greater normative independence must have had a better prospect for success and thus survival than those without it. In contrast, epistemic independence does not offer such functional values for success and survival. One who believes in independence of the self may or may not act in an independent fashion, much less show a strong drive for success or survival. Admittedly, both motivational independence and normative independence are likely to foster independent behaviors, which can, in turn, breed epistemic independence. Nevertheless, Western European culture had already entertained the idea of independence. Thus, it seems unlikely that settlement in North American contexts fosters any stronger degree of epistemic independence as it does for the other two facets of independence.

Data from a recently completed study (Kitayama et al., in press) are informative. This study used several measures that are likely to capture the three aspects of independence that are postulated here. Moreover, the study included matched samples of North American college students, German college students, and British college students. Furthermore, it may be expected that both of these Western groups would be more independent or less interdependent than Asians. To test this prediction, the researchers included another matched college student sample from mainland Japan. The inclusion of the two different West European groups is important. To the extent that the two groups are no different from one another and, yet, different from the U.S. sample, it will be all the more plausible to attribute the U.S.–Western European difference to settlement.

Motivational Independence in the United States, Western Europe, and Mainland Japan

Motivational independence was assessed with two measures based on an *Implicit Social Orientation Questionnaire* (ISOQ) (Kitayama & Park, 2007). In this newly developed questionnaire, participants are presented with 10 mundane social situations (e.g., "having a positive interaction with friends," "being overloaded with work"). Participants are asked to remember the latest event pertaining to each of the 10 situations. Then they are to report the extent to which they experienced a series of emotions in each of the situations on a rating scale. The list of emotions contains six theoretically derived types of emotion terms, some of which were described earlier. Some emotions are general in the sense that they are not specifically associated with either independence or interdependence and can be positive (happiness and joy) or negative (unhappiness). Other emotions are socially disengaging, resulting from either success or failure of independence (e.g., pride in self versus frustration). Still others are socially engaging, stemming from either success or failure of interdependence (e.g., friendly feelings versus shame).

Three measures are derived from the ISOQ, and two of these are relevant here (the remaining one is discussed in the section of normative independence). First, for each participant, the reported intensity of experiencing general positive emotions was regressed on both the reported intensity for disengaging positive emotions and the reported intensity for engaging positive emotions (Kitayama, Mesquita, et al., 2006). The resulting betas are indicators of motivational orientations toward independence and interdependence, respectively.

As predicted, Japanese are quite interdependent in this account, with their happiness strongly contingent on engaging rather than disengaging experiences. Replicating earlier studies (Kitayama, Mesquita, et al., 2006), Americans are relatively more independent. Their happiness was contingent more on disengaging rather than engaging experiences. Importantly, the West European data fell right in between, with no difference between Germans and British. This pattern suggests that Western groups are, overall, more independent—or less interdependent—than an Eastern group. Yet, among the three Western groups, Americans tend to be more independent.

A second measure derived from the ISOQ assesses the extent to which disengaging (vis-à-vis engaging) emotions were experienced in the 10 situations

(Kitayama, Mesquita, et al., 2006). Simply, the researchers subtracted the reported intensity of experiencing engaging emotions from the reported intensity of experiencing disengaging emotions. The expectation was that people who are motivated toward independent or interdependent goals ought to experience emotions that are relevant to the respective types of goals, namely, disengaging emotions and engaging emotions, respectively.

As with the predictor of happiness measure, the intensity of emotional experience measure showed that Americans are the most independent, with disengaging emotions reportedly experienced more strongly than engaging emotions. In contrast, Japanese were the most interdependent, showing the reverse pattern. Finally, and most important for the sake of testing the voluntary settlement hypothesis, West Europeans (Germans and British) were significantly different from both of the two groups. Clearly, Westerners are more independent (or less interdependent) overall compared to Easterners, yet within the Western culture, there is substantial regional variation. Thus, North Americans are more independent than both Germans and British on the measures of motivational independence. This pattern is precisely what the voluntary settlement hypothesis would predict.

Normative Independence in the United States, Western Europe, and Mainland Japan

In assessing the degree to which independence (vis-à-vis interdependence) is rewarded in a given society, data from Kitayama, Park, Servincer, Karasawa, and Uskul (in press) were reanalyzed for a third measure derived from the ISOQ. The assumption is that a given society is normatively independent if people who are happy tend to experience positive disengaging (vis-à-vis positive engaging) emotions (Kitayama et al., 2000). For each participant, across all the situations, the average intensity was computed for general positive emotions, disengaging positive emotions, and engaging positive emotions. The average intensity for general positive emotions was then regressed on the average intensity for disengaging positive emotions and that for engaging positive emotions. The relative size of the standardized regression coefficients for disengagement and engagement is the crucial measure. A society can be considered normatively independent if the regression coefficient is larger for disengagement than for engagement, whereas it is said to be normatively interdependent if the reverse is the case. As may be predicted, Kitayama et al. (in press) observed that the two Western European groups also were less independent in this measure than North Americans. As before, Japanese were least independent (or most interdependent).

Epistemic Independence in the United States, West Europe, and Mainland Japan

To assess a tacitly held belief of self as independent, Kitayama et al. (in press) assessed a dispositional bias in causal attribution. Following Kitayama, Ishii, et al.

(2006), the researchers presented participants with vignettes involving target individuals engaging in either a socially desirable behavior or a socially undesirable behavior. Participants were then asked to estimate the extent to which the behavior was influenced by either internal factors, such as the target person's attitudes, personality traits, and temperament, or external factors, such as social atmosphere and social norms. Replicating numerous studies on the topic, Americans estimated the internal factors to be far more important than the external factors as the cause of the behavior at issue. This dispositional bias in causal attribution was substantially weaker among Japanese, hence replicating a growing number of cross-cultural studies on the topic. Importantly, however, both Germans and British were no different from Americans on this measure of epistemic independence.

Regional Variation in Independence Within the United States

Given the general logic of the present chapter, we might expect to see greater independence in parts of the United States that most closely resemble the modern-day frontier, such as the Rocky Mountains (or, using the terminology of the U.S. Census, the Mountain West). These states, such as Montana, Wyoming, and Colorado, should exemplify the frontier spirit within the United States. Although no research has specifically examined regional differences in the components of motivational, normative, and epistemic independence, two studies may be informative. First, Plaut et al. (2002) examined regional differences in well-being and self in the United States. They found that the Mountain West region had the highest levels of environmental mastery and autonomy-based well-being, along with the second-lowest levels of perceived constraint. In addition, the Mountain West region was about average (relative to other U.S. regions) on other measures of well-being, including positive relations with others and self-acceptance. That is, high levels of well-being in the Mountain West were found only for dimensions that emphasized independence and individualism. This well-being pattern was consistent with participants' self-described traits: Mountain West residents reported particularly high levels of dominance, outspokenness, assertiveness, and self-confidence. Clearly, there is a tendency toward independence and individualistic behaviors among Americans who live in the modern-day frontier.

In another relevant study, Vandello and Cohen (1999) studied state-level differences in American behavior. They examined a variety of behaviors that are associated with individualism and collectivism (e.g., divorce rates, percentage of people living alone, and percentage of self-employed workers). One might expect that people who live in relatively harsh conditions would need to be more collectivistic to survive, whereas people living in relatively comfortable conditions would be afforded the luxury of independence. The voluntary settlement hypothesis, however, suggests the exact opposite pattern: Those who would choose to live in harsh conditions would be the most independent and would likely become even more independent over time. Using the collectivism index mentioned earlier, Vandello and Cohen (1999) ranked all 50 states from most to least collectivistic. As the voluntary settlement hypothesis predicts, the four states in the continental United States that are arguably the most frontierlike in terms of inclement weather and low population density—Montana, Wyoming, North Dakota, and South Dakota—were among the seven most individualistic (or least collectivistic) in this ranking. Other Rocky Mountain states were also quite high on individualism, including Colorado (#6), Washington (#8), and Idaho (#13) (note that Utah, with its large Mormon population, is considerably more collectivistic than other states in the region). These data do not provide specific information about the three types of independence, but they do suggest the prevalence of independent behavior in the modern-day frontier.

Summary

Overall, the Kitayama et al. (in press) study demonstrated that there is a consistent and quite sizable difference between Japan and the Western countries (Germany, Great Britain, and the United States). Thus, there is solid evidence indicating that Western cultures are much more independent or less interdependent than Japanese culture. Importantly, however, there were consistent differences between the two Western cultures. Specifically, Americans were more independent or less interdependent than Western Europeans in the two ISOQ-based measures of motivational independence as well as the measure of normative independence. In sum, the evidence is supportive of the prediction that the history of settlement in North America bolstered the motivational aspect of independence. In fact, there was virtually no difference between the United States and Western Europe in the measures of both epistemic independence and normative independence.

HOKKAIDO: CONSEQUENCES OF SETTLEMENT WITH PRIOR CULTURAL CONDITIONS OF INTERDEPENDENCE

Is Hokkaido the Wild North of Japan?

The comparison between North America and Western Europe, reported earlier, provided encouraging evidence for the voluntary settlement hypothesis. However, examining the hypothesis via only this single comparison is inherently problematic and limiting, for many other factors are confounded with both the initial immigration to the new continent and the territorial expansion of its western frontiers during the 17th through 19th centuries. These factors are interconnected and, therefore, hard to disentangle.

In another recent study, Kitayama, Ishii et al. (2006) proposed that one powerful method for testing the voluntary settlement hypothesis is to examine another place that does not share any elements of American or West European history, yet shares one theoretically crucial element: a relatively recent history of systematic voluntary settlement and subsequent struggle in the frontier in the absence of much social, political, and economic infrastructure or other resources for survival. For this purpose, Kitayama, Ishii et al. (2006) chose Hokkaido—Japan's northern island (see Figure 14.2 for a map).

Hokkaido was wilderness until the middle of the 18th century. Around that time, the central feudal government of Tokugawa collapsed. As a consequence,

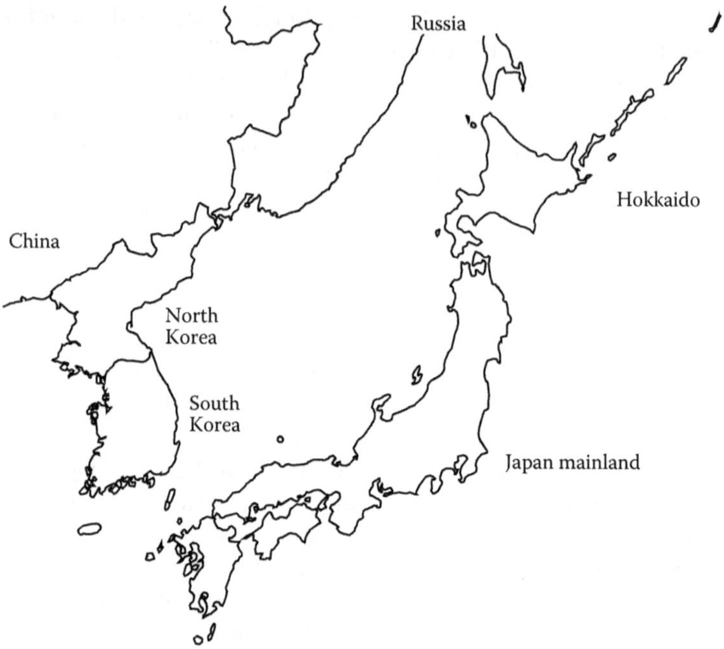

Figure 14.2 Japan and its neighboring countries: The northern island of Japan—called Hokkaido—is located between mainland Japan and Russia.

numerous samurai warriors lost their jobs. At the same time, Russia had expanded its territory to the Far East and become a major threat to Japan's northern territory, especially to Hokkaido and its neighboring islands. The new Meiji government decided to solve the two problems simultaneously by sending the jobless ex-samurai to settle in Hokkaido. This initial governmental initiative led numerous others to follow the ex-samurai; these settlers were mostly farmers and peasants, who voluntarily moved to Hokkaido to seek land and wealth. Over the several decades afterward, the immigration continued, and in 1920 when the first census became available in Japan, Hokkaido had already been transformed into an integral part of Japan, with a population of nearly 3 million.

Considerations Unique to Hokkaido

The contrast between Hokkaido and mainland Japan is simultaneously both similar and dissimilar from the contrast between North America and West Europe. First, these two contrasts are similar because in both cases, the first region of each pair underwent a substantial new settlement of their areas. Indeed, Hokkaido and the western frontiers of North America have extremely harsh ecology, with sparse population during the initial years of settlement. Although this settlement was initially encouraged and even organized by the central government, it was largely voluntary, and, moreover, it was motivated by personal economic aspirations.

There are also important differences, however, between these two histories. In particular, whereas Western Europe already had an individualistic ethos when North American settlement occurred, the predominant ethos of mainland Japan was highly collectivistic and interdependent. Hence, it is very important to explore exactly which aspects of independence might be fostered and even reinforced through the settlement history in Hokkaido.

Remember that in our comparison between North America and Western Europe, we discovered that Americans are more independent than Western Europeans in motivational and normative independence. In the epistemic aspect of independence, Americans were no higher than Western Europeans. The explanation we offered was that because Western Europeans already have a fairly strong degree of motivational and normative independence, they are likely to make full use of these psychological and collective propensities when deciding to move (self-selection) and, subsequently, when struggling to survive in the frontier (adaptation to local conditions).

Hokkaido is different. Among other factors, people in mainland Japan are relatively low in motivational independence. Instead, they are quite high in motivation toward interdependent values, such as social harmony, interpersonal connectedness, and fitting in (Kitayama, Mesquita, et al., 2006; Morling, Kitayama, & Miyamoto, 2002). Hence, it would be unlikely for them to make use of motivational independence either in choosing to move to Hokkaido or in struggling to survive in the land of frontier, for intrinsic motivation is hardly available for a vast majority. This raises an important question of what motivated settlers to immigrate in the first place. Moreover, it also begs the question of what it is that motivates people to struggle and to seek to survive in the frontier environment that is harsh in ecology, primitive in social infrastructure, and meager in socioeconomic resources.

We suggest that even when motivation toward independence is low, people may have normative beliefs in strong independence ("I think I have to be independent"). In interdependent cultural contexts, independence is a struggle. People do not necessarily enjoy practicing independence, because they are often not socialized to do so. Nevertheless, some of them may strive to achieve independence. It requires discipline, hard work, and resulting motivation. The motivation, however, is not intrinsic—they do not seek to achieve independence because they enjoy it. Instead, they seek to achieve it, because it is important in their environment.

It may then be suggested that a form of normative independence is an integral part of antecedents or consequences of settlement that take place in the context of interdependent culture. People with especially strong beliefs in independence may choose to move to Hokkaido when opportunities present themselves. Moreover, it may be these cognitive beliefs and values in independence on which people rely while struggling to survive and establish a living in the land of the frontier.

It is important to keep in mind that there is little or no cultural heritage of epistemic independence in mainland Japan. Nevertheless, once people who endorse independent values settle in Hokkaido and begin to form communities, they are likely to nurture a new belief of individuals as more independent than interdependent, simply because they are independent at least in a normative sense. Finally, once the newly formed culture is developed and elaborated in the direction of independence, those who are raised in it may begin to develop an intrinsic motivation toward independence. For this to occur, however, it may take generations, and, therefore, it is an empirical question to what degree motivational independence is evident in contemporary Hokkaido.

In sum, we believe that the voluntary settlement hypothesis is likely to be valid when applied to Hokkaido. Nevertheless, the actual social psychological dynamics underlying the settlement effect may be very different in this case than in the case of North America, because Hokkaido is settled by those brought up and socialized in an interdependent cultural context—namely, mainland Japan. It may be expected, in particular, that the psychological transformation would be most prominent in normative independence, and it may be also evident in epistemic independence. However, it might take generations of cultural transformation before the effect of settlement is manifest in motivational independence.

Normative and Epistemic Independence

The data reported in Kitayama, Ishii, et al. (2006) are consistent with the foregoing analysis. The researchers tested matched samples of college students in mainland Japan, Hokkaido, and the United States. In Study 1, they used the between-subject correlation measure of societal norm toward independence and interdependence. They first replicated earlier studies (e.g., Kitayama et al., 2000) to show that their American sample is normatively independent, in that people who frequently experience disengaging positive emotions (e.g., pride in self) are happier than those who frequently experience engaging positive emotions (e.g., friendly feelings). In contrast, their mainland Japanese sample was normatively interdependent, in that those who frequently experienced engaging positive emotions were happier than those who frequently experienced disengaging positive emotions. Importantly, the Hokkaido sample was more similar to the American sample than to the mainland Japanese sample, especially when the sample was limited only to those who were born and raised in Hokkaido.

In their second study, they tested cognitive dissonance (Brehm, 1956; Festinger, 1957). It has been shown that a threat to personal self-images (e.g., "Am I stupid?") often follows a personal choice, because the choice is almost always suboptimal (i.e., there are some valid reasons for choosing the rejected course of action). This threat, often called dissonance, in turn causes an effort to justify the choice so that the negative arousal can be reduced (Steele, 1988). Thus, people end up liking an item that has been chosen, while disliking an item that has been rejected.

Importantly, however, this effect is relatively unique to those who have strong beliefs in independence and, thus, who regard their internal attributes, such as competence and moral integrity, as centrally defining the self (Kitayama et al., 2004). Moreover, the self-justification occurs only when the choice is personal and private. When the choice is made public, it is construed as socially constrained, and, as a consequence, any threat to personal self-images tends to be defrayed. In contrast, for those who have strong beliefs in interdependence and who therefore regard their relationships as centrally defining the self, this effect occurs only when the choice is public such that the choice becomes a target of social evaluation. In a series of studies, Kitayama and colleagues (Hoshino-Browne et al., 2005; Imada & Kitayama, 2004; Kitayama, Snibbe, Markus, & Suzuki, 2004; Na & Kitayama, 2006) used a subtle manipulation to induce public scrutiny during choice and demonstrated that European Americans show this postchoice self-justification effect only when the choice is perceived as personal, but Asians and Asian Americans show the effect only when the choice is perceived as being under public scrutiny.

Kitayama, Ishii et al. (2006, Study 2) used the level of postchoice self-justification as a measure of an orientation toward either independence or interdependence in Hokkaido. The finding was quite straightforward: Hokkaido Japanese were no different from Americans. In stark contrast to their mainland counterparts, they showed a strong justification effect only when the choice is perceived as private, and they did not show such an effect when they were led to believe that the choice is under public scrutiny. This may be taken as an added piece of evidence for the hypothesis that normative independence is strongly nurtured in Hokkaido.

In Study 3, Kitayama, Ishii et al. (2006) went a step further and examined epistemic independence with an attribution measure described earlier. Remember that in a recent study by Kitayama et al. (in press), both Americans and Western Europeans showed a substantially stronger dispositional bias than did mainland Japanese. Kitayama, Ishii et al. (2006) first established the overall West–East difference, which was consistent with the Kitayama et al. (in press) study and numerous previous cross-cultural studies. The dispositional bias was very strong for Americans, but it was virtually nonexistent for mainland Japanese. However, in support for our analysis based on the voluntary settlement hypothesis, Hokkaido Japanese—especially those born and brought up in Hokkaido—were strongly dispositional. Importantly, their pattern of responses was no different from that of Americans.

Motivational Independence Among Hokkaido Japanese

Recently, Ishii and Kitayama (2007) addressed the question of whether Hokkaido Japanese might show greater intrinsic motivation toward independence than mainland Japanese. They used two ISOQ measures of motivational independence. One assesses the degree to which happiness is experienced when one achieves independence (as revealed in the reported experience of disengaging positive emotions) rather than interdependence (as revealed in the reported experience of engaging positive emotions). Another examines the relative salience of independent rather than interdependent goals (as revealed in chronically greater degree of experience of disengaging emotions versus engaging emotions).

As noted earlier, with these measures, Americans are high in independence or low in interdependence, even compared to West Europeans. Moreover, mainland Japanese are quite high in interdependence or low in independence. The results from the new study by Ishii and Kitayama (2007) are very clear. Hokkaido Japanese are no different from mainland Japanese on these specific measures. We may tentatively conclude, then, that Hokkaido Japanese are highly independent in terms of both normative independence and epistemic independence but quite low in motivational independence.

The conclusion reached here for Hokkaido, of course, is in stark contrast to the American case. As compared to Western Europeans, Americans are highly independent in terms of motivational independence, but no different from West Europeans in terms of epistemic independence and normative independence. The contrast here between Hokkaido and North America offers far-reaching implications for the very notion of independence and the corresponding social ideology of individualism. Whereas both Hokkaido and North America are quite independent in ethos, they are remarkably different in some specifics. We believe that the difference can be understood when historical contingencies unique to each location are fully articulated and taken into account. We turn to this issue in the next section.

TWO FORMS OF INDIVIDUALISM

Individualism in the United States

Research summarized here suggests that individualism in the frontier can take at least two different forms depending on initial conditions for settlement. When settlement occurs with prior cultural conditions of independence (North America), intrinsic motivation toward independence, such as intrinsic interest in seeking novelty, challenges, and opportunities, along with the sense of pleasure one gets from such activities of independence, becomes the primary guiding force underlying settlement and subsequent effort toward survival in the frontier. The culture of independence that emerges under these conditions is grounded in strong ethos of personal achievement, success, and happiness. Furthermore, this individual motivation toward independence can spread to cultural narratives and values, such as that of the American Dream.

It may be suggested that motivation toward independence is so strong that it may influence all aspects of life, even those pertaining to social relationships and interdependence. Social relations may become yet another domain in which personal motivation toward independence can be applied. In support of this conjecture, past research demonstrates that Americans are often extremely social, especially when social relations are conducive to the maintenance and enhancement of independent images of the self, such as self-esteem, self-efficacy, and personal sense of control (Baumeister & Leary, 1995). Indeed, even friendship choice is dominated by these independent concerns (Tesser, Crepaz, Collins, Cornell, & Beach, 2000). Along the same line, Uchida, Kitayama, Mesquita, Reyes, and Morling (2008) postulated that for North Americans, social support from close others enhances one's well-being only if the support enhances the person's self-esteem or his or her sense of control and efficacy and, thus, his or her sense of independence. What this means is that a commonly observed positive effect of social support on well-being might be fully mediated by the effect of social support to boost one's esteem, efficacy or control, and independence. The resulting individualism, then, takes on extremely social, proactive, and highly expansive tones.

Individualism in Hokkaido

These patterns do not seem to hold in Hokkaido. The research summarized here suggests that when settlement occurs with prior cultural conditions of interdependence (Hokkaido), normative beliefs in independent values, such as autonomy and self-reliance, become the primary guiding force underlying settlement and subsequent effort in survival in the frontier. The culture of independence that emerges under these conditions is grounded in strong ethos of societal norms for independence.

It may be suggested that normative demands for independence are so strong that people may avoid any risk of violating such norms. If true, this may lead to diametrically opposite consequences in the domain of social relationships. People in Hokkaido may actively seek to avoid social relationships, because relationship avoidance is an effective way of abiding by the societal norm for independence.

One implication of the foregoing analysis is that Hokkaido Japanese might prove to be less social than either mainland Japanese (who may be expected to be highly social because of the interdependent cultural ethos) or even Americans (who may be expected to be highly social because of the reason articulated earlier). According to one preliminary piece of evidence, this may not be too far-fetched a prediction to make. Ishii and Kitayama (2007) recently found that Hokkaido Japanese reported a significantly smaller number of friends in their social network than either mainland Japanese or Americans do. We believe that further research following up this initial lead is well warranted.

CONCLUDING REMARKS

The general thesis explored in this chapter holds that settlement in the frontier fosters the cultural ethos of independence. This cultural transformation may happen because of self-selection, adaptation to local conditions, and/or institutionalization. Moreover, it may have an impact on any one or more of the three facets of independence, namely, epistemic, motivational, and normative. We argued that this process is mediated by three general classes of factors, that is, self-selection, adaptation to new local environments, and institutionalization. Future work should examine each of these mechanisms in detail. Although we focused primarily on sociocultural factors, evolutionary considerations are likely to be increasingly important in understanding the psychological consequences of the frontier conditions.

Above and beyond these general points, the systematic empirical work in two regions of frontier highlighted specific processes that are highly path dependent. Thus, specific effects of settlement are likely to depend on prior sociohistorical conditions in which the settlement takes place. Specifically, independence founded on prior conditions of independence (the United States) is quite different in kind from independence based on prior conditions of interdependence (Hokkaido).

We believe that the significance of the present work stems from the fact that it begins to shed a new light on a particular configuration of social, ecological, and demographic conditions called the frontier as one major source of formative influence on cultural ethos of independence. Admittedly, the frontier is a specific condition that is associated with the Wild West in North America. To that extent, the current effort might appear to be highly limited in implications. It may reveal something unique about the history and culture of the United States, especially in comparison with West Europe. It might seem dubious, however, if there is any general significance in the notion of the frontier as a factor influencing cultural persistence and change in general.

Keep in mind, however, that conditions of the frontier do exist in fringes of all dynamically changing cultures. Sometimes, such conditions manifest themselves in demographic shifts (e.g., immigration to Hokkaido). But this need not be the case. Thus, conditions of the frontier may exist in cutting-edge sciences. Likewise, they may also exist among Chinese seasonal workers who come to big cities such as Beijing and Shanghai from nearby rural villages to seek wealth and opportunities. Or the same could be true for college students who are attracted to big cities for intellectual stimulations. Indeed, one may even be tempted to hypothesize that all cultural changes are initiated and mobilized at the fringes of any given cultural group.

We are therefore hopeful that the voluntary settlement hypothesis, broadly conceptualized in terms of both actual and symbolic forms of settlement, has the potential of going beyond the United States and Hokkaido to suggest a general principle involved in all forms of cultural change and persistence. Whether this will prove to be true will depend on further work, of course. We nonetheless believe that the current formulation can serve as an invaluable pathfinder for future work targeted on the mechanisms of cultural change and persistence.

REFERENCES

Baumeister, R. F., & Leary, M. R. (1995). The need to belong: Desire for interpersonal attachments as a fundamental human motivation. *Psychological Bulletin*, *117*, 497–529.

Benson, C. (2001). *Cultural psychology of self: Place, morality, and art in human worlds*. New York: Routledge.

Brehm, J. W. (1956). Postdecision changes in the desirability of alternatives. *Journal of Abnormal and Social Psychology*, 52, 384–389.

Caprara, J. V., Schwartz, S., Capanna, C., Vecchione, M., & Barbaranelli, C. (2005). Personality and politics: Values, traits, and political choice. *Political Psychology*, 27, 1–28.

Chiu, C.-Y., & Hong, Y. (2006). Cultural competence: Dynamic processes. In A. J. Elliot & C. S. Dweck (Eds.), *Handbook of competence and motivation* (pp. 489–505). New York: Guilford.

Choi, I., & Nisbett, R. E. (1998). Situational salience and cultural differences in the correspondence bias and the actor-observer bias. *Personality and Social Psychology Bulletin*, *24*, 949–960.

Choi, I., Nisbett, R. E., & Norenzayan, A. (1999). Causal attribution across cultures: Variation and universality. *Psychological Bulletin*, *125*, 47–63.

Cole, M. (1996). *Cultural psychology*. Cambridge, MA: Harvard University Press.

Deci, E. L., & Ryan, R. M. (1985). *Intrinsic motivation and self-determination in human behavior.* New York: Plenum.

Festinger, L. (1957). *A theory of cognitive dissonance*. Stanford, CA: Stanford University Press.

Gilbert, D. T., & Malone, P. S. (1995). The correspondence bias. *Psychological Bulletin, 117*, 21–38.
Heine, S. J. (2008). *Cultural psychology*. New York: Norton.
Heine, S. J., & Hamamura, T. (2007). In search of East Asian self-enhancement. *Personality and Social Psychology Review, 11*, 1–24.
Heine, S. J., & Lehman, D. R. (1997). Culture, dissonance, and self-affirmation. *Personality and Social Psychology Bulletin, 23*, 389–400.
Heine, S. J., Lehman, D. R., Markus, H. R., & Kitayama, S. (1999). Is there a universal need for positive self-regard? *Psychological Review, 106*, 766–794.
Hochschild, J. L. (1995). *Facing up to the American Dream*. Princeton, NJ: Princeton University Press.
Hoshino-Browne, E., Zanna, A. S., Spencer, S. J., Zanna, M. P., Kitayama, S., & Lackenbauer, S. (2005). On the cultural guises of cognitive dissonance: The case of Easterners and Westerners. *Journal of Personality and Social Psychology, 89*, 294–310.
Imada, T., & Kitayama, S. (2004). *Dissonance, self, and eyes of others in Japan and the U.S.* Unpublished manuscript, University of Michigan.
Inglehart, R., & Baker, W. E. (2000). Modernization, cultural change, and the persistence of traditional values. *American Sociological Review, 65*, 19–51.
Ishii, K., & Kitayama, S. (2007). [Socially avoidant individualism in Hokkaido: Making an independence out of an interdependent past]. Unpublished raw data, Hokkaido University, Sapporo, Japan.
Iyengar, S. S., & Lepper, M. R. (1999). Rethinking the value of choice: A cultural perspective on intrinsic motivation. *Journal of Personality and Social Psychology, 76*, 349–366.
Jones, E. E. (1979). The rocky road from acts to disposition. *American Psychologist, 34*, 107–117.
Kitayama, S., & Cohen, D. (2007). *Handbook of cultural psychology*. New York: Guilford.
Kitayama, S., Duffy, S., & Uchida, Y. (2007). Self as cultural mode of being. In S. Kitayama & D. Cohen (Eds.), *Handbook of cultural psychology* (pp. 136–174). New York: Guilford.
Kitayama, S., Ishii, K., Imada, T., Takemura, K., & Ramaswamy, J. (2006). Voluntary settlement and the spirit of independence: Evidence from Japan's "Northern Frontier." *Journal of Personality and Social Psychology, 91*, 369–384.
Kitayama, S., Markus, H. R., & Kurokawa, M. (2000). Culture, emotion, and well-being: Good feelings in Japan and the United States. *Cognition and Emotion, 14*, 93–124.
Kitayama, S., Markus, H. R., Matsumoto, H., & Norasakkunkit, V. (1997). Individual and collective processes in the construction of the self: Self-enhancement in the United States and self-criticism in Japan. *Journal of Personality and Social Psychology, 72*, 1245–1267.
Kitayama, S., Mesquita, B., & Karasawa, M. (2006). Cultural affordances and emotional experience: Socially engaging and disengaging emotions in Japan and the United States. *Journal of Personality and Social Psychology, 91*, 890–903.
Kitayama, S. & Park, H. (2007). Cultural shaping of emotion and well-being: How does it work? *Social and Personality Psychology Compass, 1*, 202–222.
Kitayama, S., Park, H., Servincer, A. T., Karasawa, M., & Uskul, A. K. (in press). A cultural task analysis of implicit independence: Comparing North America, West Europe, and East Asia. *Journal of Personality and Social Psychology.*
Kitayama, S., Snibbe, A. C., Markus, H. R., & Suzuki, T. (2004). Is there any "free" choice? Self and dissonance in two cultures. *Psychological Science, 15*, 527–533.
Kitayama, S., & Uchida, Y. (2005). Interdependent agency: An alternative system for action. In R. Sorrentino, D. Cohen, J. M. lson, & M. P. Zanna (eds.), *Culture and social behaviour: The Ontario symposium* (Vol. 10, pp. 165–198). Mahwah, NJ: Erlbaum.

Kwan, V. S. Y., Bond, M. H., & Singelis, T. M. (1997). Pancultural explanations for life satisfaction: Adding relationship harmony to self-esteem. *Journal of Personality and Social Psychology*, *73*, 1038–1051.

Lee, F., Hallahan, M., & Herzog, T. (1996). Explaining real-time events: How culture and domain shape attributions. *Personality and Social Psychology Bulletin*, *22*, 732–741.

Mahalingam, R. (2006). *Cultural psychology of immigrants*. Mahwah, NJ: Lawrence Erlbaum.

Markus, H. R., & Kitayama, S. (1991). Culture and the self: Implications for cognition, emotion, and motivation. *Psychological Review*, *98*, 224–253.

Markus, H. R., & Kitayama, S. (2004). Models of agency: Sociocultural diversity in the construction of action. In G. Berman & J. Berman (Eds.), *The 49th annual Nebraska symposium on motivation: Cross-cultural differences in perspectives on self* (pp. 18–74). Lincoln: University of Nebraska Press.

Markus, H. R., Uchida, Y., Omoregie, H., Townsend, S. S. M., & Kitayama, S. (2006). Models of agency in Japanese and American contexts. *Psychological Science*, *17*, 103–112.

Masuda, T., & Kitayama, S. (2004). Perceiver-induced constraint and attitude attribution in Japan and the U.S.: A case for the cultural dependence of the correspondence bias. *Journal of Experimental Social Psychology*, *40*, 409–416.

Matsumoto, D. R. (2001). *The handbook of culture and psychology*. Oxford, UK: Oxford University Press.

Miller, J. G. (1984). Culture and the development of everyday social explanation. *Journal of Personality and Social Psychology*, *46*, 961–978.

Miyamoto, Y., & Kitayama, S. (2002). Cultural variation in correspondence bias: The critical role of attitude diagnosticity of socially constrained behavior. *Journal of Personality and Social Psychology*, *83*, 1239–1248.

Miyamoto, Y., Kitayama, S., & Talhelm, T. (2006, January). *A meta-analytic review of cultural differences in cognitive processes*. Paper presented at the annual meeting of the Society for Personality and Social Psychology, Palm Springs, CA.

Morling, B., Kitayama, S., & Miyamoto, Y. (2002). Cultural practices emphasize influence in the U.S. and adjustment in Japan. *Personality and Social Psychology Bulletin*, *28*, 311–323.

Morris, M. W., & Peng, K. (1994). Culture and cause: American and Chinese attributions for social and physical events. *Journal of Personality and Social Psychology*, *67*, 949–971.

Na, J., & Kitayama, S. (2006). [Culture and intrinsic motivation]. Unpublished raw data, University of Michigan.

Nisbett, R. E. (2003). *The geography of thought: How Asians and Westerners think differently … and why*. New York: Free Press.

Nisbett, R., & Cohen, D. (1996). *Culture of honor: Psychology of violence in the South.* Boulder, CO: Westview.

Oyserman, D., Coon, H. M., & Kemmelmeier, M. (2002). Rethinking individualism and collectivism: Evaluation of theoretical assumptions and meta-analyses. *Psychological Bulletin*, *128*, 3–72.

Plaut, V. C., Markus, H. R., & Lachman, M. E. (2002). Place matters: Consensual features and regional variation in American well-being and self. *Journal of Personality and Social Psychology*, *83*, 160–184.

Rentfrow, P. J., Jost, J. T., Gosling, S. D., & Potter, J. (2007). *Regional differences in personality predict voting patterns in 1996–2004 U.S. presidential elections*. Unpublished manuscript, University of Cambridge.

Ross, L. (1977). The intuitive psychologist and his shortcomings: Distortions in the attribution process. In L. Berkowitz (Ed.), *Advances in experimental social psychology* (Vol. 10, pp. 174–220). New York: Academic Press.

Sanchez-Burks, J. (2005). Protestant relational ideology: The cognitive underpinnings and organizational implications of an American anomaly. In B. Staw & R. Kramer (Eds.), *Research in organizational behavior* (Vol. 26, pp. 267–308). New York: Elsevier.
Schaller, M., & Crandall, C. S. (2004). *The psychological foundations of culture.* Mahwah, NJ: Lawrence Erlbaum.
Shweder, R. (2003). *Why do men barbeque?* Cambridge, MA: Harvard University Press.
Sperber, D. (1996). *Explaining culture: A naturalistic approach.* Oxford, UK: Blackwell.
Steele, C. M. (1988). The psychology of self-affirmation: Sustaining the integrity of the self. In L. Berkowitz (Ed.), *Advances in experimental social psychology* (Vol. 21, pp. 261–302). San Diego, CA: Academic Press.
Tesser, A., Crepaz, N., Collins, J. C., Cornell, D., & Beach, S. R. H. (2000). Confluence of self-esteem regulation mechanisms: On integrating the self-zoo. *Personality and Social Psychology Bulletin*, *26*, 1476–1489.
Turner, F. J. (1920). *The frontier in American history*. New York: Henry Holt.
Uchida, Y., Kitayama, S., Mesquita, B., Reyes, J. A. S., & Morling, B. (2008). Is perceived emotional support beneficial? Well-being and health in independent and interdependent cultures. *Personality and Social Psychology Bulletin, 34*, 741–754.
Vandello, J. A., & Cohen, D. (1999). Patterns of individualism and collectivism across the United States. *Journal of Personality and Social Psychology*, *77*, 279–292.
Weber, M. (1958). *The Protestant ethic and the spirit of capitalism* (T. Parsons, Trans.). New York: Charles Scribner's Sons. (Original work published 1904–1906)
Witter, R. A., Okun, M. A., Stock, W. A., & Haring, M. J. (1984). Education and subjective well-being: A meta-analysis. *Educational Evaluation and Policy Analysis*, *6*, 165–173.

15

Cultural Inertia, Economic Incentives, and the Persistence of "Southern Violence"

MARTIN DALY and MARGO WILSON

HISTORY AND ECOLOGY; IDEALISM AND MATERIALISM

Cultural diversity challenges those wishing to apply insights from evolutionary theory to the study of human behavior. When we see that different human groups hunt with different technologies, raise different cultivars, and exclude different foodstuffs from the realm of the edible, for example, we can take the behavioral ecological approach and ask whether the people we observe are behaving like "optimal foragers" (e.g., Kaplan & Hill, 1992) only within the constraints of their culturally specific preferences, technical knowledge, and taboos. The cross-cultural diversity itself seems to require a different sort of explanation.

One popular response to this problem has been to insist that each culture can be "explained" only in terms of its unique history. According to Harris (1968), "By far the most sophisticated advocate … and most effective defender" of "the historical–particularist position" was the anthropologist R. H. Lowie, who famously asserted, "Culture is a thing *sui generis* which can only be explained in terms of itself. … *Omnis cultura ex cultura*" (Lowie, 1917, pp. 25–26). Such an extreme statement seems almost antiscientific, denying the possibility of principled cross-cultural analysis, but many anthropologists and sociologists have been attracted to such positions, perhaps as a way of defending their turf from the dual threats of psychological reductionism and Marxist materialism. And of course historical particularism has an undeniable grain of truth: Idiosyncratic historical trajectories surely *do* create cross-cultural differences that are "arbitrary" in the sense that nothing extrinsic to the cultural phenomenon itself presently favors maintaining it in one form rather than another. Lexicon is the prototypical example: Why we say *dog* whereas our neighbors say *chien* has a strictly historical answer.

Cultural materialists such as Marvin Harris have taken a different tack, maintaining, "Cultures on the whole have evolved along parallel and convergent paths which are highly predictable from a knowledge of the processes of production, reproduction, intensification, and depletion" (Harris, 1977, p. xii). And indeed, many cross-cultural analyses (some of which have incorporated good phylogenetic controls) indicate that particular ecologies favor particular social practices, which, in conjunction with panhuman cognitive processes and emotions, lead to convergent cultural "syndromes" of similar institutions, ideologies, and values among peoples with distinct histories. The pastoral way of life, for example, seems conducive to a cultural syndrome entailing patrilocal polygynous marriage, bridewealth, patrilineal inheritance, blood feud, and a "culture of honor" whereby personal affronts warrant violent responses (Nisbett & Cohen, 1996; Peristiany, 1965).

Historical particularists like Lowie, cultural materialists like Harris, and even contemporary human behavioral ecologists like Smith, Borgerhoff Mulder, and Hill (2001) eschew explicit psychologizing. Cultural phenomena, however, require human actors on whom historical and/or ecological factors have their effects, so evolutionary and cultural psychologists have been revisiting some of these issues in their discussions of "evoked" culture versus "transmitted" culture (Gangestad, Haselton, & Buss, 2006). This distinction echoes an even older polarity, namely, that of materialism versus idealism (Harris, 1968): Are cultural differences primarily to be understood as responses to the incentive structure of the environment or as the consequences of different ways of thinking? The proposition that cultural phenomena are evoked is a materialist claim that they result from responses of a universal human nature to ecological and economic determinants. The proposition that cultural phenomena are transmitted implies the idealist claim that the attitudes, values, and construals of reality that one learns from others are the causes of one's behavior.

The fact that there is truth to both of these positions has not rendered debate moot. The question of how *much* truth each perspective contains arises with reference to each specific cultural phenomenon under investigation, and the issue is of more than academic interest. To the extent that a cultural phenomenon is evoked, it can be changed by manipulating economic incentives, but if it is more a matter of transmitted attitudes and values that cohere in a socially valued belief system, then it may persist despite changing conditions that make it functionally anachronistic. People's positions in these controversies, which are perhaps a reflection—or even a partial determinant—of their politics, have policy implications: Idealists tend to favor educational and "consciousness-raising" remedies for social problems, whereas those with more materialist views favor attacking economic "root causes."

HOMICIDE AND THE CULTURE OF HONOR IN THE SOUTHERN UNITED STATES

Homicide can usefully be viewed as a cultural phenomenon, because there are large cross-cultural differences in relevant behavior, attitudes, and values (Daly & Wilson,

1989; Nisbett & Cohen, 1996). Anthropological studies tell us that a lethal response to certain transgressions is justified—even obligatory—in some societies but deplored and punished severely in others. In the modern world, killing a personal antagonist (as distinct from killing in warfare or otherwise "in the line of duty") is generally stigmatized, but in some small-scale societies lacking centralized government and courts of law, it is instead valorized (e.g., Chagnon, 1988; Patton, 2000).

Large cross-cultural differences persist in the modern world. According to data from the World Health Organization (2002), per capita rates of homicide vary by a factor of more than one-hundred-fold across contemporary nation-states, and even in the developed world, the differentials are remarkable. The homicide rate in the United States, for example, has for decades been more than 5 times that prevailing in the European Union and more than 10 times that of Japan. Moreover, cross-national variability in the prevalence of homicide is demonstrably associated with variability in attitudes, such that homicide rates tend to be highest where personal violence is most legitimized (e.g., McAlister, 2006).

Indeed, cross-cultural diversity in both the attitudinal supports of homicidal violence and its prevalence can be striking even within a single modern nation-state. Criminologists have long been aware that homicide rates are substantially higher in the southern United States than in the northern states, and there has long been debate about whether this difference should be attributed to structural (economics, racial politics) or cultural (attitudes, values) factors (see, e.g., Lee, Bankston, Hayes, & Thomas, 2007). Perhaps the most popular explanation for this pattern attributes it to the poisonous legacy of slavery, but a fresh perspective has emerged from the ingenious multimethod research program of Richard Nisbett, Dov Cohen, and their collaborators (Cohen & Nisbett, 1994; Cohen, Nisbett, Bowdle, & Schwarz, 1996; Nisbett, 1993; Nisbett & Cohen, 1996; Nisbett, Polly, & Lang, 1995). By focusing on the behavior and attitudes of white men, these researchers have circumvented some of the complications engendered by U.S. race relations and have assembled a strong case that "white" southerners, but not northerners, adhere to a version of the "culture of honor" that we mentioned earlier.

The prototypical culture of honor (e.g., Peristiany, 1965) consists of a suite of interrelated attitudes, values, and norms, whereby personal and familial honor reside largely in being respected as one who will brook no trespass, regardless of the costs and risks that a response may entail. Such a value system tends to characterize patrilineal societies with strong fraternal interest groups, institutionalized feuding, bridewealth, a heavy emphasis on female chastity, and, most important, "self-help justice." It is particularly characteristic of peoples who rely primarily on livestock rather than on crops, especially those whose herds graze on relatively unproductive pasture, and it is in this way of life that the apparent rationale for such a value system resides. Because of both their mobility and their low density, pastoralists cannot rely on central authorities for policing and protection, and their herds tempt thievery, which must be deterred by sincere, credible threats of a violent response by those victimized or their kinsmen. How better to advertise such a threat than by manifesting a willingness to respond dangerously to even minor affronts?

The tradition of dueling is the classic manifestation of the culture of honor in the U.S. South, but contemporary behavior and attitudes reflect it, too. Cohen and Nisbett (1994) showed, for example, that although southerners do not endorse violence more than northerners in all contexts, they are more approving of violent responses to trespassing, insults, and sexual advances against one's female partner or relatives. That this is not mere lip service was shown by an experiment in which undergraduate men at the University of Michigan who had been raised in the South responded very differently to a minor insult than those raised in the North, not just behaviorally and in questionnaire responses but physiologically as well: The southern men exhibited rapid cortisol and testosterone responses, whereas northerners were unaffected (Cohen, Nisbett, Bowdle, & Schwarz, 1996). And with regard to homicide, analyses by Nisbett (1993) and Lee et al. (2007) indicate that it is specifically in the context of "arguments" rather than "felonies" that white men from the southern states are more likely than their northern counterparts to kill.

What is the origin of these regional differences? Nisbett and his collaborators proposed that the answer resides in the ecological consideration that we discussed earlier: pastoralism. Citing historical evidence, they maintained that the North received its predominant economic and cultural influences from Puritans, Quakers, and Dutch farmers and townsmen with a strong ethic of neighborly cooperation, whereas those who settled the southern states were mostly "Scotch Irish" folk from the fringes of Britain who had made their living as herdsmen and to some extent continued to do so in the new world. These people brought with them an ethic of self-help justice, with its attendant notion that intimidatory capacity is central to manhood, and herding remained economically important in the southeast into the 19th century and in the southwest even longer. In short, white southerners and white northerners brought their cultural differences from Europe centuries ago and have transmitted those differences with sufficient fidelity that they persist to this day, even though their ecological foundations have now largely evaporated.

CULTURAL INERTIA IN THE ABSENCE OF MATERIAL FOUNDATIONS?

In our view, Nisbett and Cohen have unequivocally demonstrated that non-Hispanic, white U.S. southerners differ statistically from northerners in their attitudes and in their behavior, and we also agree that these differences are well characterized by saying that southerners adhere to a sort of culture of honor. The proposition that the culture's historical roots can be traced in this case, as in others, to the cost-benefit situation faced by pastoralists is more speculative, as Nisbett and Cohen acknowledged, but we have no quarrel with this idea either. The suggestion that we *do* want to challenge is this: that after its ecological origins have faded away, the culture of honor has nevertheless persisted for generations, in the absence of any material basis.

In their book's concluding section, Nisbett and Cohen reviewed their reasons for rejecting temperature, poverty, and slavery as useful explanations for southern violence and instead attributed its contemporary existence to more or less pure cultural transmission. Moreover, although they granted that the southern culture of honor may yet fade away for want of an enduring materialistic rationale, that's not where they're placing their own bets:

> Because culture is taken in without reflection, because we acquire it more by *absorbing* it than by studying it, the ultimate reason for why we do things (or why a behavior is functional) is often hidden from us. We do not reexamine cultural rules every generation or analyze how functionally adaptive they are before we internalize them. So, as long as they do not get us in too much trouble in some way that is manifest and as long as there is no far more attractive alternative, they will continue. Indeed, as long as there is social enforcement of the norms, it may be profitable to continue to behave in accord with such norms and costly to defy them even when one consciously, personally rejects them.
>
> An important implication of that analysis is that one should speculate not merely on whether the culture of honor will wither when material circumstances cease to make it rational but also whether the culture of honor could maintain itself, or even grow, for nonmaterial reasons. (Nisbett & Cohen, 1996, pp. 93–94)

This proposal that the culture of honor may persist "for nonmaterial reasons" has caught the fancy of many readers, and despite Nisbett and Cohen's acknowledgment that it is speculative, it has become an iconic example of culture's alleged inertia in the absence of material supports. Evolutionists seem to be especially fond of citing it, perhaps to defuse naive "nature versus culture" objections. In an essay aimed at convincing sociologists that they need to become selectionists, for example, Runciman (1998, p. 183) cited Nisbett and Cohen (1996) as having shown that "a 'culture of honour' may persist even after the environment that gave rise to it has changed and there has evolved a set of institutions which deprive it of its previous function." Cosmides and Tooby (1999, p. 461) cited the same work very similarly in an essay aimed at clinical psychologists. And in an essay on "The adaptive nature of culture," Alvard (2003, p. 138) stated that the norms adhered to by southerners are

> unlikely to be optimal in modern twenty-first century America. … Granted that the difference between farmers and herders discussed in Nisbett and Cohen's book can be construed as ecological in origin, it is difficult to understand without invoking cultural processes why such differences persist generations after the subjects' ancestors ceased to engage in their respective subsistence tasks.

Gangestad et al. (2006, p. 91) also accepted this example as an instance of "transmitted culture" to suggest that the social transmission process may work by changing thresholds for the elicitation of "evoked culture." Norenzayan (2006, p. 126) replied that the real role of the evoked culture concept in this case is in Nisbett and Cohen's "origin story" and concluded,

> The best explanation for the persistence of honor cultures is social transmission. … Perhaps a common scenario across cultures is that ecological differences evoke an initial cultural response that adaptively varies but then is

picked up by processes of transmitted culture, amplified, and perpetuated even when the initial conditions are no longer present.

Finally, it was Nisbett and Cohen's proposal that a culture "could maintain itself, or even grow, for nonmaterial reasons" that inspired Richerson and Boyd (2005) to use this example as their opening hook in *Not by Genes Alone: How Culture Transformed Human Evolution*, a book that argued that evolutionists must pay more attention to the social transmission process. After summarizing Nisbett and Cohen's main results and arguments, these authors elaborated on how the story illustrates their own thesis, as follows:

> *Culture is crucial for understanding human behavior.* ... Murder is more common in the South than in the North. If Nisbett and Cohen are right, this difference can't be explained in terms of contemporary economics, climate, or any other external factor. Their explanation is that people in the South have acquired a complex set of beliefs and attitudes about personal honor that make them more polite, but also more quick to take offense than people in the North. This complex persists because the beliefs of one generation are learned by the next. (Richerson & Boyd, 2005, p. 3)

INCOME INEQUALITY AND HOMICIDE RATES

What's wrong with these seemingly unexceptionable claims? Our only quarrel is with the suggestion that contemporary economics has been disposed of as a candidate explanation for the elevated rates of homicide in the U.S. South. It is certainly true that poverty and mean white per capita income are poor predictors of the variability in homicide rates that Nisbett and Cohen discussed, but there was never much reason to expect otherwise. Empirical studies of homicide and theoretical considerations have both suggested that it is not the average level of wealth that determines local homicide rates but *the degree to which that wealth is inequitably distributed*.

Gartner (1993, p. 205) summarized what homicide researchers have learned from multivariate, cross-national studies as follows: "Of the wide variety of political, economic, cultural, and social indicators included in these analyses, only one—income inequality—has shown a consistent (and positive) association with homicide rates." More recent cross-national analyses continue to tell the same story (Fajnzylber, Lederman, & Loayza, 2002; Messner, Raffalovich, & Schrock, 2002), and so do analyses within countries (Daly, Wilson, & Vasdev, 2001). Moreover, it is easy to understand why this should be so: Homicide rates reflect the intensity of male–male competition (Daly & Wilson, 2001; Eisner, 2003), and the intensity of competition is a function not of average well-being but of the *variability* among individuals in access to resources and hence of the (perceived) potential for gains from escalating one's competitive tactics.

How do these considerations apply to regional variation in U.S. homicide rates? Let us consider this question by following Nisbett and collaborators' dichotomization of the United States into 16 southern and 34 nonsouthern (henceforth called "northern") states. As one would expect, average income is not a significant predictor of homicide rates at the state level, and as Nisbett and Cohen's analysis suggests,

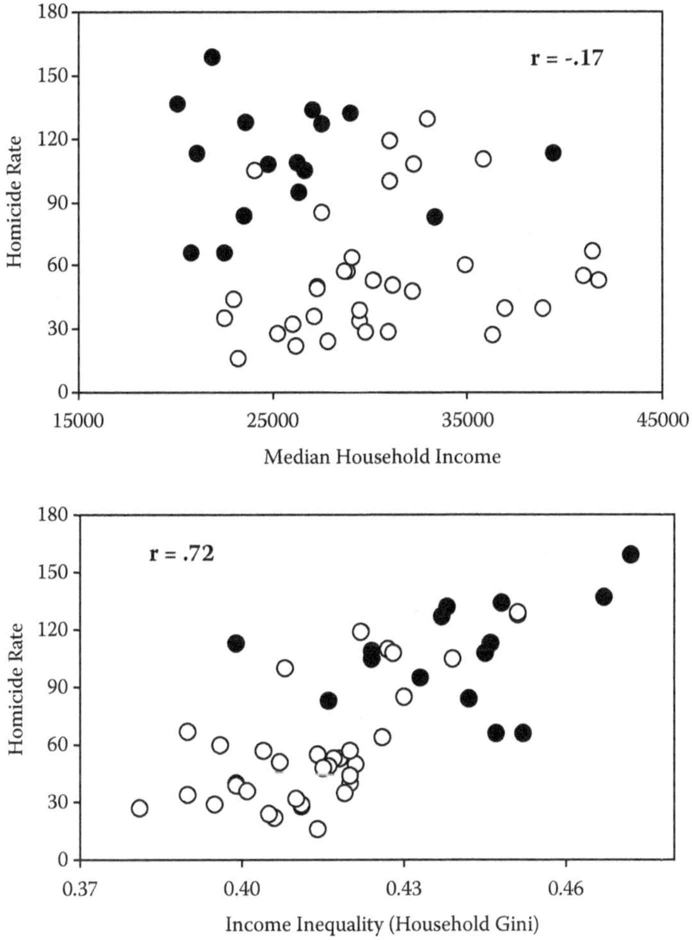

Figure 15.1 Mean homicide victimization rate per million persons per annum in southern (•) and northern (o) states, 1990, as a function of median household income in US$ (top panel) and of the Gini index of income inequality among households (bottom panel). Homicide data are from the FBI's Supplementary Homicide Reports: 1980–2006 (FBI, n.d.). Income data courtesy of Bruce Kennedy (Public Health, Harvard University); see Kennedy, Kawachi, and Prothrow-Stith (1996).

this variable cannot account for the large difference between south and north (see Figure 15.1). The standard measure of income inequality, the Gini index, however, performs much better. It accounts for slightly more than half of the variance in homicide rates among states (see Figure 15.1), and when its effect is partialled out, the difference in mean homicide rates between south and north is reduced from 56 deaths per million persons per annum to fewer than 20.

We presented these analyses at the "Mind, Culture, and Evolution" conference at the University of British Columbia in 2004, and Richard Nisbett immediately raised an objection. One of the strengths of the Nisbett and Cohen research program has been that by confining attention to white men, they have highlighted regional variations that cannot be artifacts of racial composition, and Nisbett argued that we must do the same if we wished to show that the effects of income inequality account for regional differences. Now, one might protest that insofar as male–male competition takes place within a multiracial pool, isolating racial categories could eliminate a relevant component of inequality, but let us accept Nisbett's argument and look at the situation for white men considered separately. As it happens, the results are essentially unchanged: Average income again fails to account for the variability between states or regions, whereas income inequality accounts for both (see Figure 15.2). If income inequality is ignored, the average southern state has a mean rate of white male homicide victimization that exceeds

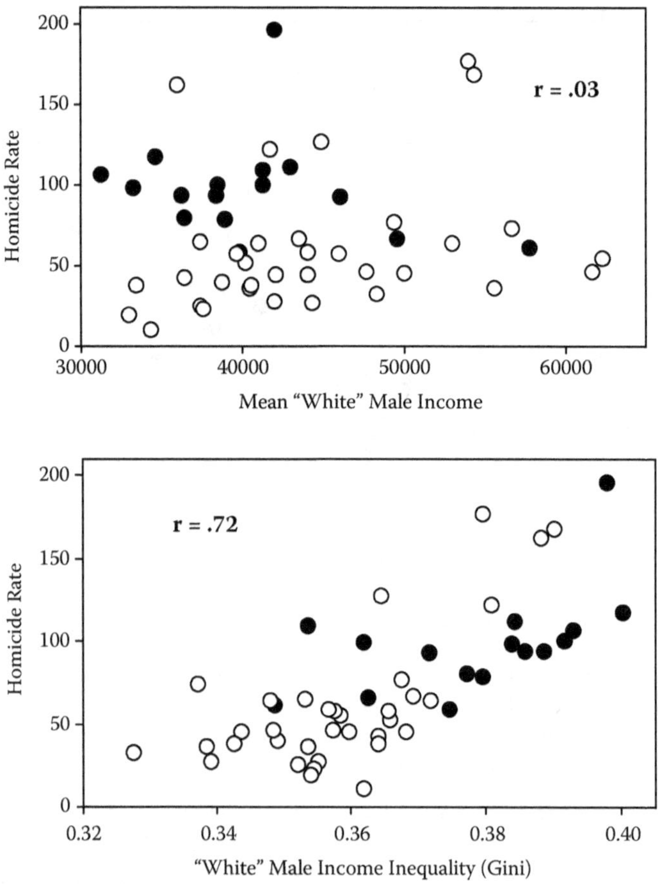

Figure 15.2. Mean homicide victimization rate per million white adult males per annum in southern (•) and northern (o) states, 1989–1991, as a function of 1990 mean white male income in US$ (top panel) and as a function of 1990 white male income inequality (Gini index). Homicide data are from the FBI's Supplementary Homicide Reports: 1980–2006 (FBI, n.d.). Income data courtesy of Angus Deaton (Economics, Princeton University); see Deaton and Lubotsky (2003).

that for northern states by 38 deaths per million men per annum, and the difference is highly significant (t_{48} = 3.11, p = .003). The Gini index, however, again accounts for just over half the between-state variance, and if this single variable's effect is partialled out, the south–north difference falls to fewer than 2 homicides per million white men per annum and is no longer statistically significant (t_{48} = 0.17, p = .87). It is particularly striking that the correlation between the Gini index and homicide is undiminished as a result of narrowing our focus from the entire population to white men, despite the fact that both the average level of inequality and its range of values across states are thereby reduced.

Cohen (personal communication, Aug. 8, 2007) drew our attention to the fact that the regional variable remains a significant predictor of residual variance, even in these state-level analyses, if one uses the Gastil index of "southernness" rather than a binary variable. Nevertheless, the Gini index is much the stronger predictor, greatly reducing the apparent importance of any such regional variable. In fairness to Nisbett and Cohen, it should be noted that although they do not include income inequality in their summary statements about "alternative explanations for southern violence," they have not totally ignored this variable. Nisbett et al. (1995; see also Nisbett & Cohen, 1996, Appendix A) used a city-level Gini index in certain analyses and found that although this measure was a significant univariate correlate of homicide rates, it dropped out in multivariate models. Arguably, however, this was not a good basis for dismissing income inequality from further consideration, because they did not follow Nisbett's own precept: Whereas the homicide data under consideration were for white men only, the Gini index that they used as a predictor was based on the cities' full multiracial income distributions.

Nisbett (personal communication, Aug. 12, 2007) furthermore suggested that we should be limiting these analyses to homicides that arose in the context of arguments rather than felonies. The theoretical rationale for making this distinction is that it is only in the former context that the defense of personal honor facilitates violence. But is this really so? "Thick descriptions" of homicides in the context of robberies and police action suggest that felony murders frequently entail a similar dynamic of insult, escalation, and a felt need to defend "face" (see, e.g., Toch, 1969). Moreover, the "circumstance" codes on which the argument versus felony distinction is based are problematic not only because potentially orthogonal case attributes are coded as alternatives but also because of missing data: 30% of male victim cases in 1990 are coded as "circumstances unknown," for example, and even the sex of the killer is unknown in 34% of cases. The upshot is that if we limit analysis to cases coded as "arguments," in which both victim and killer were known to be white males, we exclude 93% of the homicides in the U.S. national data archive for 1990, but we *still* find that Gini is a highly significant predictor and that controlling for Gini substantially reduces, but does not eliminate, the apparent effect of region.

Notwithstanding these caveats about the limitations of available data, Nisbett (1993) and Lee et al. (2007) *did* find that it was specifically in the context of arguments rather than felonies that white men from the southern states were more likely than their northern counterparts to kill, as noted earlier. This jibes with other sorts of data, such as Cohen et al.'s (1996) experimental demonstration of different responses by northern versus southern college students to a standard insult. It is difficult to prove, however, that the homicide data reflect regional differences in the relative prevalence of different types of lethal confrontations and not simply in the labels chosen to describe cases; and, in any event, the analyses incorporating income inequality indicate that this factor is a major contributor to regional variation in argument and felony homicides alike.

WHY DOES THIS MATTER?

Our point is not to cast doubt on the reality of the southern culture of honor, which seems to us well established. Neither is it to question whether having been enculturated in the South exacerbates violent responses to provocations. Our point is simply that Nisbett and Cohen, as well as others who have cited them, such as Richerson and Boyd, may be overstating the case that regional variations in U.S. homicide rates provide testimony to the inertial power of culture in the absence of material supports.

Homicide rates among white men are highest in those states in which white men's income is most inequitable, and the south–north variable appears to add little when this fact is taken into account. To suggest, even tentatively, that a cultural difference has no basis in *any* material factor is of necessity an argument by exclusion, and the trouble with such arguments is that eliminating one candidate material influence (such as average income) cannot rule out the possibility that another (such as income inequality) will turn out to be important. This is not to deny that many cross-culturally variable phenomena surely *are* devoid of material rationales. One can embrace this conclusion prematurely, however, and thus close the door on potentially fruitful further investigation.

It should be stressed that the difference of opinion that we are discussing is a small and subtle one and perhaps even a difference in emphasis rather than in basic claims. We do not doubt that "cultural inertia" plays some role in keeping homicide rates from fluctuating wildly (Daly & Wilson, 1988, 1989), and neither Nisbett and Cohen nor Richerson and Boyd doubt that individuals select among and manipulate the cultural messages that they "internalize." Nisbett and Cohen believe, as do we, that a panhuman evolved psychology underlies our capacity to develop cultural construals and practices, and they have stressed (as have we; Daly & Wilson, 1988) that cultures of honor arise in particular social and ecological settings that oblige self-help justice. Indeed, Nisbett and Cohen (1996) claimed, citing unpublished analyses by Nisbett's student Andrew Reaves, that the local prevalence of homicidal violence in the southern United States is *still* significantly associated with the contemporary economic importance of herding, and although these analyses have been challenged (Chu, Rivera, & Loftin, 2000), this claim illustrates the fact that Nisbett and Cohen are not arguing that ecology has lost *all* its relevance. Finally, we recognize that regional differences in income inequality may themselves reflect cultural differences, some of which (e.g., an emphasis on self-reliance) may be intimately linked to the ideological supports of the culture of honor.

That said, however, there are real implications of these differences of emphasis. Both Nisbett and Cohen and Richerson and Boyd apparently doubt the potential of economic policies as possible remedies for the problem of southern violence, on the grounds that its cultural supports are strong. We think their pessimism on this point is unfounded. People are demonstrably capable of jettisoning cultural baggage that they perceive as obsolete. Not only are the young often skeptical of the wisdom of their elders but an entire cultural complex can be abandoned in the face of evidence that alternative belief systems are the road to success, as witnessed in the phenomenon of cargo cults (Trompf, 1990).

As for homicide rates, they can change remarkably rapidly, in either direction. In the United States as a whole, the homicide rate fell by more than 30% between 1991 and 1997 (Blumstein & Rosenfeld, 1998); in New York, it fell by more than 60% and in Boston by more than 70%. Meanwhile, the Russian homicide rate doubled in the 5 years prior to the collapse of the Soviet Union, and it doubled again in the ensuing 5 years (Pridemore, 2006). Such rapid changes in homicide rates tend to consist primarily of increases and reductions in precisely the sorts of killings that are Nisbett and Cohen's focus: social conflicts between unrelated men, in a context of competition over status and the maintenance of respect (Daly & Wilson, 1988).

At least two important ecological supports of such honor killings have been identified: inequitable access to resources, which inspires escalated and sometimes dangerous competition, and a lack of access to law enforcement, which obliges self-help justice. Nisbett and Cohen have implied that policy directed at ameliorating these factors is likely to be ineffectual against the southern culture of honor, which might instead even spread, but in the same breath, they also have proposed that this culture has "allure" specifically for "those who have little to gain by playing by society's rules and little to lose by standing outside of them" (Nisbett & Cohen, 1996, p. 94). In other words, the compelling nature of the culture of honor's value system is affected by incentives. And policy can change incentives.

Chagnon (1988, p. 990) described a society in which men must be ready to use retaliatory violence to maintain respect and concluded with an anecdote that speaks to the possibility of change:

> A particularly acute insight into the power of law to thwart killing for revenge was provided to me by a young Yanomamö man in 1987. He had been taught Spanish by missionaries and sent to the territorial capital for training in practical nursing. There he discovered police and laws. He excitedly told me that he had visited the town's largest *pata* (the territorial governor) and urged him to make law and police available to his people so that they would not have to engage any longer in their wars of revenge and have to live in constant fear. Many of his close kinsmen had died violently and had, in turn, exacted lethal revenge; he worried about being a potential target of retaliations and made it known to all that he would have nothing to do with raiding.

We are powerfully shaped by cultural conditioning. But we are not its unthinking slaves.

ACKNOWLEDGMENTS

We thank Dov Cohen and Dick Nisbett for critical comments on prior drafts of this chapter and Bruce Kennedy and Angus Deaton for data. For research support, we thank the Natural Sciences and Engineering Research Council of Canada and the Social Sciences and Humanities Research Council of Canada.

REFERENCES

Alvard, M. S. (2003). The adaptive nature of culture. *Evolutionary Anthropology, 12*, 136–149.

Blumstein, A., & Rosenfeld, R. (1998). Explaining recent trends in U.S. homicide rates. *Journal of Criminal Law and Criminology, 88*, 1175–1216.

Chagnon, N. A. (1988). Life histories, blood revenge, and warfare in a tribal population. *Science, 239*, 985–992.

Chu, R., Rivera, C., & Loftin, C. (2000). Herding and homicide: An examination of the Nisbett–Reaves hypothesis. *Social Forces, 78*, 971–987.

Cohen, D., & Nisbett, R. E. (1994). Self-protection and the culture of honor: Explaining southern violence. *Personality and Social Psychology Bulletin, 20*, 551–567.

Cohen, D., Nisbett, R. E., Bowdle, B., & Schwarz, N. (1996). Insult, aggression, and the southern culture of honor: An "experimental ethnography." *Journal of Personality and Social Psychology, 70*, 945–960.

Cosmides, L., & Tooby, J. (1999). Toward an evolutionary taxonomy of treatable conditions. *Journal of Abnormal Psychology, 108*, 453–464.

Daly, M., & Wilson, M. (1988). *Homicide*. Hawthorne, NY: Aldine de Gruyter.

Daly, M., & Wilson, M. (1989). Homicide and cultural evolution. *Ethology and Sociobiology, 10*, 99–110.

Daly, M., & Wilson, M. (2001). Risk-taking, intrasexual competition, and homicide. *Nebraska Symposium on Motivation, 47*, 1–36.

Daly, M., Wilson, M., & Vasdev, S. (2001). Income inequality and homicide rates in Canada and the United States. *Canadian Journal of Criminology, 43*, 219–236.

Deaton, A., & Lubotsky, D. (2003). Mortality, inequality, and race in American cities and states. *Social Science and Medicine, 56*, 1139–1153.

Eisner, M. (2003). Longterm historical trends in violent crime. *Crime and Justice, 30*, 83–142.

FBI. (n.d.). The FBI's Supplementary Homicide Reports: 1980-2006. Retrieved January 2007, from the Office of Juvenile Justice and Delinquency Prevention. Access: http://ojjdp.ncjrs.org/ojstatbb/ezashr/

Fajnzylber, P., Lederman, D., & Loayza, N. (2002). Inequality and violent crime. *Journal of Law and Economics, 45*, 1–40.

Gangestad, S. W., Haselton, M. W., & Buss, D. M. (2006). Evolutionary foundations of cultural variation: Evoked culture and mate preferences. *Psychological Inquiry, 17*, 75–95.

Gartner, R. (1993). Methodological issues in cross-cultural large-scale survey research on violence. *Violence and Victims, 8*, 199–215.

Harris, M. (1968). *The rise of anthropological theory*. New York: T. Y. Crowell.

Harris, M. (1977). *Cannibals and kings: The origins of cultures*. New York: Random House.

Kaplan, H., & Hill, K. (1992). The evolutionary ecology of food acquisition. In E. A. Smith & B. Winterhalder (Eds.), *Evolutionary ecology and human behavior* (pp. 167–201). Hawthorne, NY: Aldine de Gruyter.

Kennedy, B. P., Kawachi, I., & Prothrow-Stith, D. (1996). Income distribution and mortality: Cross-sectional ecological study of the Robin Hood Index in the United States. *British Medical Journal, 312*, 1004–1007, 1194.

Lee, M. R., Bankston, W. B., Hayes, T. C., & Thomas, S. A. (2007). Revisiting the southern culture of violence. *Sociological Quarterly, 48*, 253–275.

Lowie, R. H. (1917). *Culture and ethnology.* New York: Basic Books.

McAlister, A. L. (2006). Acceptance of killing and homicide rates in nineteen nations. *European Journal of Public Health, 16*, 259–265.

Messner, S. F., Raffalovich, L. E., & Schrock, P. (2002). Reassessing the cross-national relationship between income inequality and homicide rates: Implications of data quality control in the measurement of income distribution. *Journal of Quantitative Criminology, 18*, 377–395.

Nisbett, R. E. (1993). Violence and U.S. regional culture. *American Psychologist, 48*, 441–449.

Nisbett, R. E., & Cohen, D. (1996). *Culture of honor: The psychology of violence in the south*. Boulder, CO: Westview.

Nisbett, R. E., Polly, G., & Lang, S. (1995). Homicide and U.S. regional culture. In R. B. Ruback & N. A. Weiner (Eds.), *Interpersonal violent behaviors: Social and cultural aspects* (pp. 135–151). New York: Springer.

Norenzayan, A. (2006). Evolution and transmitted culture. *Psychological Inquiry, 17*, 123–128.

Patton, J. Q. (2000). Reciprocal altruism and warfare: A case from the Ecuadorian Amazon. In L. Cronk, N. Chagnon, & W. Irons (Eds.), *Adaptation and human behavior: An anthropological perspective* (pp. 417–436). Hawthorne, NY: Aldine de Gruyter.

Peristiany, J. G. (Ed.). (1965). *Honour and shame: The values of Mediterranean society*. London: Weidenfeld & Nicolson.

Pridemore, W. A. (2006). Demographic, temporal, and spatial patterns of homicide rates in Russia. *European Sociological Review, 19*, 41–59.

Richerson, P. J., & Boyd, R. (2005). *Not by genes alone: How culture transformed human evolution*. Chicago: University of Chicago Press.

Runciman, W. G. (1998). The selectionist paradigm and its implications for sociology. *Sociology, 32*, 163–188.

Smith, E. A., Borgerhoff Mulder, M., & Hill, K. (2001). Controversies in the evolutionary social sciences: A guide for the perplexed. *Trends in Ecology and Evolution, 16*, 128–135.

Toch, H. (1969). *Violent men: An inquiry into the psychology of violence*. Chicago: Aldine.

Trompf, G. W. (1990). *Cargo cults and millenarian movements: Transoceanic comparisons of new religious movements.* Berlin: Mouton de Gruyter.

World Health Organization. (2002). *World report on violence and health*. Geneva: Author.

16

Infectious Diseases and the Evolution of Cross-Cultural Differences

MARK SCHALLER and DAMIAN R. MURRAY

We'll begin with a prediction: Years from now it will become obvious to observe that cross-cultural differences result substantially from regional differences in the prevalence of infectious diseases.

Does that sound presumptuous? Perhaps even preposterous? Maybe. For many readers, our prediction may seem like a provocation that we've made up out of thin air. That's not quite so. The scientific literature has, for years, documented cultural differences that are predicted by the prevalence of pathogens (e.g., Gangestad & Buss, 1993; Low, 1990; Quinlan, 2007; Sherman & Billing, 1999). But these findings tend to fly under the radar of the vast majority of social scientists who concern themselves with culture and cultural variation.

Why has there been so little attention paid to the potentially profound role that infectious diseases might play in the creation of cross-cultural differences? One reason, perhaps, is that much cultural scholarship (e.g., cultural anthropology, cultural psychology) is concerned primarily with *descriptions* of cultural differences or with the *consequences* that these cultural differences have for individual behavior. Less attention has been paid to the *origins* of cultural differences in the first place. Another reason may lie in the fact that, in the occasional articles that do link pathogen prevalence to cultural outcomes, those outcomes have been relatively narrow in scope (pertaining specifically to food preparation, for instance, or to mating behavior). These findings may strike readers as interesting curiosities but perhaps not diagnostic of cultural differences more broadly. A third reason may be that there really hasn't been much reason to expect that infectious diseases should have any sort of wide-ranging impact on culture. Only recently has there emerged a body of theory and research identifying specific psychological mechanisms that are responsive to the perceived threat of infectious diseases and that may play an important role in the construction of many different kinds of cross-cultural differences. With that context in mind, this chapter has three objectives.

First, we want to address the question of *why* disease prevalence might be expected to predict a broad range of cultural outcomes. To do so, we present a conceptual framework that draws explicitly on an evolutionary analysis. We discuss the negative fitness consequences of contracting infectious diseases, and we identify specific kinds of behavioral strategies that may limit exposure to infectious diseases. We also consider the fact that, in addition to conferring specific kinds of fitness benefits (i.e., reduced exposure to diseases), these same behavioral strategies may have specific costs as well. Through the logic of an evolutionary cost-benefit analysis, this framework implies that these behavioral strategies may vary in their functional utility, depending on the extent to which infectious diseases actually pose a prevalent threat in the immediate environment. This in turn implies a causal link between regional variability in disease prevalence and cross-cultural variability in attitudes and values that are relevant to those behavioral strategies.

Second, we want to address the question of whether disease prevalence does, in fact, predict important cross-cultural differences. We summarize a variety of ways in which the general theoretical framework can be applied to deduce hypotheses about specific kinds of cross-cultural differences that may result from regional variation in disease prevalence. In each case, we summarize empirical results supporting these predictions. It turns out that worldwide variation in disease prevalence predicts a remarkably wide range of cross-cultural differences, pertaining not only to overt cultural customs (e.g., food preparation) but also to many more subtle differences operating at a psychological level of analysis—including cross-cultural differences in basic personality traits (e.g., extraversion) and values (e.g., individualism versus collectivism). Furthermore, as a consequence of these effects on fundamental psychological variables, regional differences in disease prevalence may also lead to persistent differences operating at the societal level of analysis as well.

Third, we want to consider exactly *how* these interesting relationships might have emerged. We raise, and discuss, several important questions about the actual evolutionary mechanisms that might underlie links between regional variation in disease prevalence and cultural variation in human behavior. By doing so, we highlight some of the thorny conceptual issues that inevitably arise when applying an evolutionary perspective to human cultural variation—issues that can be resolved only with rigorous and sustained research efforts.

WHY THE PREVALENCE OF DISEASE CAN HAVE CULTURAL CONSEQUENCES

So why might cultural outcomes be influenced by the prevalence of infectious diseases in the local ecology? There are at least two different lines of reasoning.

Cultural Practices as Socially Constructed Defenses Against Disease

One reason is based on the premise that cultural practices can be promulgated as rational responses to the prospects and perils posed by immediate ecological circumstances. Coastal communities develop rituals and practices pertaining to boat building and fishing, for instance, whereas landlocked communities don't. Similarly, given the threat that infectious diseases have posed to human populations, some kinds of cultural practices may have been invented to serve as barriers to the transmission of those diseases. Such practices would be especially likely to be invented and sustained under ecological circumstances in which infectious diseases pose an especially substantial threat.

An example is provided by research documenting worldwide variation in the use of culinary spices (Sherman & Billing, 1999). Spices are natural antibiotics; they contain toxins that kill many of the potentially harmful bacteria that can be found in the food we eat. Thus the use of spices in the preparation of food can be very beneficial as a defense against bacterial infections. Of course, there may be costs as well. The cultivation of spices consumes resources that might otherwise be spent to obtain more nutritious foods. There may also be physiological costs associated with the ingestion of spices, as they do contain toxins. Consequently, the use of culinary spices would most likely emerge and persist as a cultural practice under ecological circumstances in which the benefits of this practice are especially likely to outweigh the costs: under circumstances in which there actually is a high likelihood of bacterial infestation in food. Sherman and Billing (1999) reasoned that the risk of bacterial infestation is fundamentally a product of ambient temperature (the hotter the ambient temperature, the more likely that foodstuffs will be contaminated by bacterial infestations). They proceeded to analyze the cuisines of dozens of countries worldwide and tested the hypothesis that in geographical regions that are especially hot (meteorologically speaking), the cuisines are spicily hot as well. This is indeed the case. Meals in Mexico and Ethiopia are spicier than those in Mongolia and Estonia, and these culinary differences aren't just random cultural quirks. They are part of a broader worldwide pattern of cross-cultural differences in food preparation. This pattern of cultural differences appears to be rooted, at least in part, in the differential prevalence of pathogens.

This particular relation between disease prevalence and cultural outcomes may not reflect an evolutionary process per se. More likely, it implies the operation of cultural transmission processes that are analogous to evolutionary processes (see chapters by Dutton and Heath and by Rozin in this volume). These cultural transmission processes certainly operate on cultural rituals and practices that—like food preparation—are explicitly imitated, taught, and learned. But it's not clear just how broadly they apply to many other, more subtle facets of culture. Therefore, it's important to consider an additional line of reasoning that links disease prevalence to cross-cultural differences. This line of reasoning explicitly draws on speculations about evolutionary adaptations that influence human cognition and their further implications for human culture.

The "Behavioral Immune System" and Its Cultural Implications

Infectious diseases pose a substantial threat to reproductive fitness. Examples abound. Tens of millions of European peoples succumbed to the so-called Black

Death during the Middle Ages, enormous numbers of aboriginal Americans died from bacterial diseases introduced by colonial Europeans, and so forth (Guerra, 1993; Lippi & Conti, 2002; McNeill, 1976). And these are just recent examples. Infectious diseases have been important agents of disability and mortality within human (and prehuman) populations for far longer than that (Wolfe, Dunavan, & Diamond, 2007) and have exerted very strong selection pressures on these populations. Therefore, it's no surprise that humans (like many other species) are characterized by an extraordinarily sophisticated suite of mechanisms—the immune system—that evolved the capacity to identify and defend against harmful pathogens when those pathogens come into contact with our bodies. Of course, we pay a price whenever we actually use our immune system (e.g., fever and other debilitating symptoms; the consumption of caloric resources). Therefore it's also no surprise that humans (and many other species) are equipped with a "behavioral immune system" that serves as a first crude line of defense against potentially harmful pathogens.

This system operates by facilitating the behavioral avoidance of pathogens and the conspecifics that carry them (Kiesecker, Skelly, Beard, & Preisser, 1999; Schaller, 2006; Schaller & Duncan, 2007). In humans, this system is sensitive to threats (including people) in the immediate environment that appear, superficially, to represent some sort of contagion risk. When these threats are perceived, aversive emotions and cognitions are automatically activated, facilitating avoidant behavioral reactions. Of course, there may be functional costs as well as benefits associated with the activation of these aversive reactions. Consequently, the behavioral immune system is likely to have evolved in such a way as to be flexibly engaged, depending on additional information indicating the relative ratio of benefits to costs in the immediate environment. Aversive reactions to potential contagion risks are expected to be especially profound when additional information indicates that an individual is especially vulnerable to disease (Schaller & Duncan, 2007).

Several recent lines of research have applied this conceptual framework toward a deeper understanding of specific interpersonal aversions and intergroup prejudices (Park, Faulkner, & Schaller, 2003; Park, Schaller, & Crandall, 2007; Schaller & Duncan, 2007). One line of work has focused on xenophobia and ethnocentrism. Historically, contact with foreign peoples may have posed an increased contagion risk. (There are at least two reasons for this risk: Foreign peoples may introduce novel pathogens; foreign peoples may also violate local customs—such as those pertaining to food preparation—that serve as barriers to disease transmission.) Consequently, the behavioral immune system may be hypersensitive to superficial cues connoting foreignness, and the perception of foreign peoples may trigger aversive emotional, cognitive, and behavioral responses. Moreover, these xenophobic responses are likely to be variable across persons and contexts, depending on the extent to which perceivers are vulnerable (or merely perceive themselves to be vulnerable) to the transmission of infectious diseases. Recent research reveals that this is the case. People who are more chronically worried about disease show stronger negative responses to subjectively foreign (but not familiar) ethnic groups (Faulkner, Schaller, Park, & Duncan, 2004; Navarrete & Fessler, 2006). Xenophobic and ethnocentric responses are also exaggerated among people whose immunological defenses are temporarily compromised (Navarrete, Fessler, & Eng,

2007) and are amplified by contextual cues that make the risk of disease temporarily salient (Faulkner et al., 2004).

An analogous phenomenon might exist at a cultural level of analysis. If aversive responses to foreigners are affected by differences in individuals' perceived vulnerability to infectious diseases, it's no stretch to suppose that xenophobia and ethnocentrism might also be affected by ecological factors that render some populations, more than others, to be chronically vulnerable to infectious diseases. Within geographical regions in which infectious diseases have historically been more prevalent, one might expect greater levels of xenophobia within the local population. In other words, worldwide variation in disease prevalence may predict cultural differences in xenophobic attitudes.

Is this actually so? To our knowledge, no empirical research has rigorously addressed the hypothesis, but a quick secondary analysis of data already in the scholarly literature suggests some provisional support. Recently (within the context of research described more fully later) we developed an index indicating the historical prevalence of infectious diseases in dozens of countries worldwide (Fincher, Thornhill, Murray, & Schaller, 2008; Schaller & Murray, 2008). This index can be used to predict country-level measures of xenophobia. One might employ results from an item, as one crude country-level indicator of xenophobia, on the World Values Survey (Inglehart et al., 2000) that assesses the percentage of people in each country who explicitly indicate that they wouldn't want "people of a different race" as their neighbors. Are the two variables related? Yes. Across a sample of 67 countries, the correlation is substantially positive and statistically significant ($r = .43, p < .001$). Ecological variability in disease prevalence does seem to predict cross-cultural variability in xenophobia.

Now, we don't want to make too much of this one empirical result. After all, it's just one correlation, and it reflects just a single indicator of cross-cultural differences in xenophobic attitudes. And we don't want to suggest that xenophobia is a defining cultural characteristic. We present this result here simply to illustrate a fundamental point: The evolutionary cost-benefit analysis that informs our understanding of the behavioral immune system not only implies specific ways in which individuals' psychological responses differ within different everyday contexts but also implies specific ways in which cultural outcomes may differ under different ecological circumstances. Many attitudes, values, and traits may dispose people to be either more or less vulnerable to infectious diseases. These dispositions are likely to be cross-culturally variable, contingent on the prevalence of infectious diseases within the local ecology.

DISEASE PREVALENCE PREDICTS MANY KINDS OF CROSS-CULTURAL DIFFERENCES

Exactly what kinds of characteristics might dispose people to be more vulnerable to infectious diseases? Conversely, what kinds of characteristics might help people avoid infectious diseases? At a cultural level of analysis, are these characteristics predicted by regional variability in the prevalence of disease? Here are some answers to those questions.

Mate Preferences and Mating Strategies

Many diseases may be transmitted via intimate physical contact of the sort associated with sexual behavior. Because of this, animals of various species are sensitive to superficial characteristics indicating the contagion risk posed by potential mates, and they selectively avoid mating with those who demonstrate these characteristics (e.g., Kavaliers, Colwell, Braun, & Choleris, 2003). Human mate preferences are also likely to be influenced by superficial physical traits connoting the presence of (or potential resistance to) contagious parasites. One such trait is physical (un)attractiveness. In many species, bilateral symmetry may be a cue indicating both the current absence of disease and a strong immune system in general (Møller, 1996). In humans, bilateral symmetry in facial features contributes to the subjective impression of physical attractiveness (Thornhill & Gangestad, 1999). Therefore, subjective impressions of physical attractiveness may serve as a heuristic cue indicating both the absence of and the resistance to infectious diseases. It follows that not only will physical attractiveness be a highly prized feature in a mate but it will be especially prized within cultures with historically high prevalence of infectious diseases. Exactly such a result was reported by Gangestad and Buss (1993; see also Gangestad, Haselton, & Buss, 2006). Disease prevalence strongly predicted cross-cultural differences in the value placed on the physical attractiveness of a potential mate.

In addition to hypotheses about cross-cultural differences in the characteristics that people value in a mate, we can also deduce additional hypotheses about mating behavior more broadly. Because sexual intimacy puts people at an increased risk of contracting infectious diseases, there may be fitness costs associated with any chronically incautious (e.g., promiscuous or unrestricted) approach to mating. These costs must be weighed against the potential fitness benefits that might accrue as a result of incautious mating strategies (among men especially, promiscuous mating strategies may produce more offspring). This cost-benefit ratio is likely to vary depending on the prevalence of infectious diseases in the immediate ecology: Within high-disease places, the costs associated with unrestricted mating strategies are most likely to outweigh the benefits. Consequently, in high-disease places, people are expected to exhibit a more cautious, restricted approach to sexual relations.

We recently tested this hypothesis (Schaller & Murray, 2008). To do so, we employed the results of a massive cross-cultural study reported by Schmitt (2005). Schmitt administered the *Sociosexual Orientation Inventory* (SOI; Simpson & Gangestad, 1991) to over 14,000 people living in 48 different countries. (Low SOI scores indicate a restricted approach to sexuality; higher SOI scores indicate a more unrestricted approach—including greater chronic interest in new sexual partners and greater comfort with casual sexual encounters.) Schmitt (2005) reported mean standardized SOI scores for each country, among other results. Separately, using methods modeled after previous investigations (e.g., Gangestad et al., 2006; Low, 1990), we computed an index indicating the historical prevalence of infectious diseases in each of these countries. Did regional variation in disease prevalence predict cross-cultural variation in the SOI? Yes. The correlation was negative: In places with a higher prevalence of disease, both men and women tend to report a more restricted approach to sexual relations.

Interestingly, this effect was substantially stronger for SOI scores for females than for SOI scores for males, and it was only on SOI scores for females that the effect remained statistically significant after statistically controlling for other country-level variables (e.g., GDP per capita, life expectancy, absolute latitude, mean annual temperature). This makes sense within the overall conceptual framework described earlier. Because of differential reproductive investment, the fitness benefits of an incautious approach to sexual relations are likely to be greater among men than among women. For men, these benefits may actually outstrip the costs even at relatively high levels of disease prevalence, and previous research indicates that this may indeed be the case (e.g., Low, 1990). Among women, however, the benefits of unrestricted sexuality behavior are lower and so may more readily be outweighed by the costs (disease transmission) when diseases are highly prevalent.

Fundamental Personality Traits

Behavior need not be sexual to increase the risk of disease transmission. Broader behavioral dispositions (i.e., those that are not specific to the domain of sexual behavior) can also affect the risk of exposure to disease-causing pathogens. Any interaction that places an individual in close proximity to other people may expose that individual to socially transmitted pathogens. Thus, a dispositional tendency toward gregariousness and extraversion may be associated with an enhanced risk of disease transmission (Hamrick, Cohen, & Rodriguez, 2002; Nettle, 2005)—particularly under ecological circumstances in which diseases are highly prevalent. This implies a predictable cross-cultural difference in dispositional extraversion: In regions characterized by high levels of diseases, there are likely to be generally lower levels of extraversion within the local population.

Only some diseases are transmitted through direct social interaction. Many others are transmitted indirectly as a result of contaminated water supplies, inexpertly prepared foods, and so forth. And, as we discussed earlier, many cultural norms serve as barriers against these forms of transmission (e.g., cultural practices that proscribe the placement of latrines near sources of drinking water or that prescribe the appropriate spicing of meats). To the extent that individuals violate these cultural conventions, they may expose themselves and others to infectious diseases—especially within regions in which the prevalence of disease is high. This has an implication for cross-cultural differences in dispositional tendencies toward curiosity, experimentation, and willingness to deviate from the status quo—the sorts of traits signaled by the fundamental personality trait that is sometimes called "openness to experience." In regions characterized by high levels of diseases, there are likely to be generally lower levels of openness within the local population.

Extraversion and openness are two of the "Big Five" personality traits—the dimensions of personality that are considered fundamental to an understanding of individual differences (John & Srivastava, 1999). It follows that variation along these same traits may be fundamental to the perception of overall dispositional differences between cultural populations. Several recent cross-cultural research projects have administered personality questionnaires to thousands of people in dozens of countries worldwide and as a result have documented cross-cultural differences along each of the Big Five personality traits, including extraversion and openness (McCrae, 2002; McCrae et al., 2005; Schmitt et al., 2007). Using the same index that we used in the aforementioned investigations, we found that the historical prevalence of diseases indeed predicts cross-cultural variation along the two hypothesized personality dimensions: Disease prevalence is negatively correlated with country-level mean scores on both extraversion and openness (Schaller & Murray, 2008).

Additional analyses ruled out various alternative explanations for these effects. A reverse causal explanation is rendered implausible by several considerations, including the fact that contemporary personality scores were more strongly predicted by our index of historical disease prevalence than by an analogous index based on contemporary epidemiological data. Additional analyses revealed that the relations between disease prevalence and personality traits persisted even when controlling for other variables (e.g., GDP per capita, life expectancy, absolute latitude, mean annual temperature), thus ruling out a variety of alternative causal explanations. The results suggest that cross-cultural differences in personality style may indeed result, in part, from regional differences in the prevalence of disease.

Individualism and Collectivism

Underlying many specific cross-cultural differences lies a fundamental distinction between individualistic and collectivistic value systems (Hofstede, 2001; Triandis, 1995). Indeed, it has been suggested that the individualism–collectivism dimension "may ultimately prove to be the most important dimension for capturing cultural variation" (Heine, 2008, p. 189). But why exactly are some cultures individualistic and others collectivistic? Disease prevalence may provide an important part of the answer to this question.

There are at least two reasons to suspect a causal link between disease prevalence and individualism and collectivism, each of which focuses on a specific component of the multifaceted systems of values to which the terms *individualism* and *collectivism* refer. First, compared to individualistic value systems, collectivistic value systems are defined in part by an especially sharp distinction between in-group and out-group and by the corollary tendency to be especially wary of out-group members (Gelfand, Bhawuk, Nishii, & Bechtold, 2004; Sagiv & Schwartz, 1995). The same deductive logic that implies a positive correlation between disease prevalence and xenophobia also implies a positive correlation between disease prevalence and collectivism. Second, collectivism is defined in part by a strong value placed on tradition and conformity, whereas individualism is characterized by a greater tolerance for (and encouragement of) deviation from the status quo (Oishi, Schimmack, Diener, & Suh, 1998). The same deductive logic that implies a negative correlation between disease prevalence and openness also implies a positive correlation between disease prevalence and collectivism. Thus, for multiple reasons, one might expect that worldwide variation in disease prevalence will predict cultural variation along the individualism–collectivism dimension.

This relationship is exactly what the empirical data show (Fincher et al., 2008). Across multiple country-level indices of individualism and/or collectivism (Gelfand et al., 2004; Hofstede, 2001; Kashima & Kashima, 1998; Suh, Diener, Oishi, & Triandis, 1998), our index of disease prevalence correlated positively with collectivism and negatively with individualism. These correlations were exceptionally strong.

And, just as with differences in personality style, these effects remained statistically significant even after controlling for a variety of other country-level variables (e.g., GDP per capita). In fact, our results indicate that a substantial part of the often-observed relation between GDP and individualism is attributable more directly to pathogen prevalence than to economic development. It appears that the scholarly literature on individualism and collectivism may have overestimated economic influences while underestimating the causal influence of pathogens.

The Political, Civil, and Spiritual Fabric of Society

Because cultural value systems (of the sort implicated by individualism and collectivism) have profound implications for how individuals think, feel, and behave in relation to one another, they inevitably have implications that manifest in the systematic workings of societal institutions—governments, economies, legal systems, and so forth. Within any culture that prizes tradition and conformity, certain kinds of civil liberties and individual rights (e.g., freedom of speech) may be perceived to pose a greater threat to the common good. Consequently, authoritarian governmental regimes may be more likely to emerge and persist, and individual liberties and rights may be systematically repressed by legal and/or political means (Conway, Sexton, & Tweed, 2006). It follows, therefore, that disease prevalence may predict legislated restrictions on individual civil liberties and political rights. And, more generally, disease prevalence may place limits on the democratization of societies. New research provides support for these hypotheses (Thornhill, Fincher, & Aran, 2009).

Conformity to cultural traditions is facilitated not only by legislated repression of dissent but also by the kinder, gentler sort of interpersonal interactions through which people learn to adopt, and reproduce anew, those cultural traditions. Much of this socialization and acculturation occurs during childhood, within the context of familial interactions. But the family is by no means the only social context through which cultural indoctrination occurs. Religion also serves this community function (Atran & Norenzayan, 2004; see also the chapter in this volume by Shariff, Norenzayan, & Henrich). Indeed, by invoking the specter of omniscient and powerful supernatural beings, religious teachings provide an especially potent set of mechanisms through which individuals learn to uphold cultural traditions and to fear the consequences of violating those traditions. One might expect, therefore, that when diseases pose an especially substantial problem (and ritualized behavioral practices provide a more consequential buffer against that problem), religiosity will be more prevalent as well. Again, preliminary empirical results indicate that this is the case (Fincher & Thornhill, 2008).

We should be careful not to overinterpret findings such as these. They are, after all, just correlations, and they reflect just a few of the many variables that must be assessed to rigorously document relations between disease prevalence and societal outcomes. Still, these findings are consistent with—and contribute to—the increasingly plausible conclusion that a considerable chunk of cross-cultural variability, across many different domains of social life, results from regional variability in the prevalence of infectious diseases.

GENES, DEVELOPMENT, COGNITION, AND CULTURE

Even if we allow ourselves to draw that conclusion, we are still a long way from truly understanding the actual mechanisms through which disease prevalence influences cultural outcomes. There are different kinds of mechanisms (operating over different timescales) that might plausibly account for a causal relationship between disease prevalence and cultural norms.

One possible mechanism is natural selection, through which different genes are selectively favored under different ecological circumstances. Consider, for instance, cross-cultural differences in fundamental personality traits. Well-established research shows that these traits are heritable (e.g., Jang, Livesley, & Vemon, 1996), and research has begun to identify specific genetic markers associated with specific personality traits (e.g., Savitz & Ramesar, 2004; Stein, Schork, & Gelernter, 2004). Also clear is that infectious diseases can pose a powerful selective force on gene frequencies within different human populations (Sabeti et al., 2002; Williamson et al., 2007). It is plausible that among populations living in chronically disease-ridden regions of the world, natural selection has favored alleles that are probabilistically associated with introversion rather than extraversion and with dispositional caution rather than curiosity and openness to experience. Moreover, if we assume that there is some heritable genetic contribution to other attitudes and dispositions (e.g., dispositional tendencies toward collectivism), a natural selection process can be extended to explain cross-cultural variability along those characteristics too.

This is by no means the only explanation. A causal link between disease and culture might also operate through an entirely different kind of evolved mechanism. Genes associated with specific traits and dispositions may be widespread across all human populations but may be differentially expressed depending on the prevalence of infectious diseases within the local ecology. Phenotypic differences often emerge from the differential expression of common genes; a gene's expression often depends on input from the immediate environment, and this context-contingent phenotypic plasticity is often adaptive (Carroll, 2005; Ridley, 2003). Consider personality traits again. Because these traits may confer either fitness costs or benefits, and these costs and benefits vary under different circumstances (Nettle, 2007), the expression of genes associated with these traits may have evolved to be sensitive to informational inputs signaling the prevalence of infectious diseases. The consequence of such an evolutionary process is that even genetically identical populations may, under different ecological circumstances, have different overall levels of extraversion, openness, and other traits.

This sort of phenomenon is sometimes called "evoked culture" (Gangestad et al., 2006; Tooby & Cosmides, 1992), a label emphasizing the important point that cultural differences may reflect universal human capacities that are differentially evoked under different ecological circumstances. Of course, even if we suggest that cultural differences might be evoked as the result of context-contingent gene expression, we have still just barely begun to elucidate the mechanisms through which this occurs (Schaller, 2006). To more fully elaborate on this phenomenon, one must articulate specific developmental and psychological processes. For instance, one might hypothesize that in high-disease environments, genes associated with introversion may be expressed early in individuals' development, with the phenotypic result that these individuals develop personalities characterized by a chronic tendency toward introversion. Or, somewhat differently, one might hypothesize that evolved developmental processes produce individuals who aren't chronically introverted or extraverted per se but are instead chronically sensitive to disease-relevant inputs from the environment; when those inputs are received, these individuals respond by being temporarily introverted or extraverted, depending on the nature of those inputs.

It is also possible that disease prevalence may produce these many cross-cultural differences through mechanisms that have nothing directly to do with genes at all. Some of these effects might be explained by the kind of social construction and cultural transmission processes that, earlier, we illustrated with cross-cultural differences in the use of culinary spices. In regions characterized by persistently high levels of disease, cultural learning and transmission processes may selectively sustain values that advise against extraversion, openness, individualism, and other potentially costly dispositional tendencies. These kinds of cultural processes don't render "real" evolutionary processes irrelevant. (After all, cultural transmission processes depend on evolved cognitive and behavioral capabilities.) They do, however, relegate these evolutionary processes to a role that is somewhat more distant from, and less specific to, observed cultural differences.

These different kinds of explanatory mechanisms are conceptually independent and are not mutually exclusive. Even if specific genes associated with specific traits are differentially pervasive across different human populations, the actual expression of those genes may still vary depending on immediate ecological circumstances. And even if there are evolved genetic mechanisms contributing directly to cultural differences in personality, cultural transmission processes may play a role as well. Each of these possible mechanisms has unique empirical implications. Many different kinds of research programs may prove useful in testing these implications. With advances in genomics, the coming years are likely to reveal many specific genetic markers for personality traits, attitudes, and values. This may allow direct tests of the possibility (which at this point is purely hypothetical) that infectious diseases have exerted selection pressures on the genetic bases of these traits, attitudes, and values. Similarly, advances in genetics and developmental biology may allow more direct tests of the possibility that cross-cultural differences result, in part, from the differential expression of these genes. Other research strategies can be pursued immediately. Useful insights might be gained by the careful longitudinal study of the behavioral dispositions of immigrants and their offspring. (Perhaps especially informative will be studies of immigrants who relocate from low-disease regions to places characterized by a much higher prevalence of pathogens.) It might also be informative to conduct analog studies on nonhuman animals, in which the developmental context (e.g., the prevalence of infectious diseases in laboratory populations) might be carefully manipulated so as to test for possible consequences on dispositional tendencies. And, of course, some of the hypothesized implications might be tested with experimental methodologies of the sort that have been used to document other consequences of the behavioral immune system (e.g., Navarrete et al., 2007; Park et al., 2003, 2007). Variations on these methods might be used to test whether the temporary salience of (or vulnerability to) infectious diseases affects the activation of collectivistic values into working memory or the temporary inhibition of extroverted behavioral tendencies. Other variations might test whether a temporary feeling of vulnerability influences the transmission and/or learning of specific values, traits, or dispositional tendencies.

The take-home message is this: It has become increasingly apparent through the application of an evolutionary perspective on human culture that important cross-cultural differences may result from regional differences in the prevalence of infectious diseases. But the exact causal mechanisms remain unclear. Future research must rise to this empirical challenge if we wish to truly understand the evolutionary origins of cross-cultural differences.

ACKNOWLEDGMENTS

This research was supported in part by research grants from the University of British Columbia Hampton Fund and the Social Sciences and Humanities Research Council of Canada.

REFERENCES

Atran, S., & Norenzayan, A. (2004). Religion's evolutionary landscape: Counterintuition, commitment, compassion, communion. *Behavioral and Brain Sciences*, *27*, 713–770.

Carroll, S. B. (2005). *Endless forms most beautiful*. New York: Norton.

Conway, L. G., III, Sexton, S. M., & Tweed, R. G. (2006). Collectivism and governmentally initiated restrictions: A cross-sectional and longitudinal analysis across nations and within a nation. *Journal of Cross-Cultural Psychology*, *37*, 20–41.

Faulkner, J., Schaller, M., Park, J. H., & Duncan, L. A. (2004). Evolved disease-avoidance mechanisms and contemporary xenophobic attitudes. *Group Processes and Intergroup Behavior*, *7*, 333–353.

Fincher, C. L., & Thornhill, R. (2008, February). *Infectious disease prevalence and the worldwide distribution of religiosity and religions*. Paper presented at the Evolutionary Psychology Pre-Conference, Society for Personality and Social Psychology, Albuquerque, NM.

Fincher, C. L., Thornhill, R., Murray, D. R., & Schaller, M. (2008). Pathogen prevalence predicts human cross-cultural variability in individualism/collectivism. *Proceedings of the Royal Society B: Biological Sciences*, *275*, 1279–1285.

Gangestad, S. W., & Buss, D. M. (1993). Pathogen prevalence and human mate preferences. *Ethology and Sociobiology*, *14*, 89–96.

Gangestad, S. W., Haselton, M. G., & Buss, D. M. (2006). Evolutionary foundations of cultural variation: Evoked culture and mate preferences. *Psychological Inquiry*, *17*, 75–95.

Gelfand, M. J., Bhawuk, D. P. S., Nishii, L. H., & Bechtold, D. J. (2004). Individualism and collectivism. In R. J. House, P. J. Hanges, M. Javidan, P. W. Dorfman, & V. Gupta (Eds.), *Culture, leadership, and organizations: The GLOBE study of 62 societies* (pp. 437–512). Thousand Oaks, CA: Sage.

Guerra, F. (1993). The European–American exchange. *History and Philosophy of the Life Sciences*, *15*, 313–327.

Hamrick, N., Cohen, S., & Rodriguez, M. S. (2002). Being popular can be healthy or unhealthy: Stress, social network diversity, and incidence of upper respiratory infection. *Health Psychology*, *21*, 294–298.

Heine, S. J. (2008). *Cultural psychology*. New York: Norton.

Hofstede, G. (2001). *Culture's consequences* (2nd ed.). Thousand Oaks, CA: Sage.

Inglehart, R., et al., (2000). *Codebook for World Values Survey*. Ann Arbor MI: Institute for Social Research.

Jang, K. L., Livesley, W. J., & Vemon, P. A. (1996). Heritability of the Big Five personality dimensions and their facets: A twin study. *Journal of Personality*, *64*, 577–592.

John, O. P., & Srivastava, S. (1999). The Big Five trait taxonomy: History, measurement, and theoretical perspective. In L. A. Pervin & O. P. John (Eds.), *Handbook of personality: Theory and research* (2nd ed., pp. 102–138). New York: Guilford.

Kashima, E. S., & Kashima, Y. (1998). Culture and language: The case of cultural dimensions and personal pronoun use. *Journal of Cross-Cultural Psychology*, *29*, 461–486.

Kavaliers, M., Colwell, D. D., Braun, W. J., & Choleris, E. (2003). Brief exposure to the odour of a parasitized male alters the subsequent mate odour responses of female mice. *Animal Behaviour*, *65*, 59–68.

Kiesecker, J. M., Skelly, D. K., Beard, K. H., & Preisser, E. (1999). Behavioral reduction of infection risk. *Proceeding of the National Academy of Sciences*, *96*, 9165–9168.

Lippi, D., & Conti, A. A. (2002). Plague, policy, saints and terrorists: A historical survey. *Journal of Infection*, *44*, 226–228.

Low, B. S. (1990). Marriage systems and pathogen stress in human societies. *American Zoologist*, *30*, 325–339.

McCrae, R. R. (2002). NEO-PI-R data from 36 cultures: Further intercultural comparisons. In R. R. McCrae & J. Allik (Eds.), *The five-factor model of personality across cultures* (pp. 105–126). New York: Kluwer Academic/Plenum.

McCrae, R. R., Terracciano, A., & 78 members of the Personality Profiles of Cultures Project. (2005). Universal features of personality traits from the observer's perspective: Data from 50 cultures. *Journal of Personality and Social Psychology*, *88*, 547–561.

McNeill, W. H. (1976). *Plagues and peoples*. New York: Anchor Books.

Møller, A. P. (1996). Parasitism and developmental instability of hosts: A review. *Oikos*, *77*, 189–196.

Navarrete, C. D., & Fessler, D. M. T. (2006). Disease avoidance and ethnocentrism: The effects of disease vulnerability and disgust sensitivity on intergroup attitudes. *Evolution and Human Behavior*, *27*, 270–282.

Navarrete, C. D., Fessler, D. M. T., & Eng, S. J. (2007). Elevated ethnocentrism in the first trimester of pregnancy. *Evolution and Human Behavior*, *28*, 60–65.

Nettle, D. (2005). An evolutionary approach to the extraversion continuum. *Evolution and Human Behavior*, *26*, 363–373.

Nettle, D. (2007). Individual differences. In R. I. M. Dunbar & L. Barrett (Eds.), *Oxford handbook of evolutionary psychology* (pp. 479–489). Oxford, UK: Oxford University Press.

Oishi, S., Schimmack, U., Diener, E., & Suh, E. M. (1998). The measurement of values and individualism–collectivism. *Personality and Social Psychology Bulletin*, *24*, 1177–1189.

Park, J. H., Faulkner, J., & Schaller, M. (2003). Evolved disease-avoidance processes and contemporary anti-social behavior: Prejudicial attitudes and avoidance of people with physical disabilities. *Journal of Nonverbal Behavior*, *27*, 65–87.

Park, J. H., Schaller, M., & Crandall, C. S. (2007). Pathogen-avoidance mechanisms and the stigmatization of obese people. *Evolution and Human Behavior*, *28*, 410–414.

Quinlan, R. J. (2007). Human parental effort and environmental risk. *Proceedings of the Royal Society B*, *274*, 121–125.

Ridley, M. (2003). *The agile gene*. New York: HarperCollins.

Sabeti, P. C., Reich, D. E., Higgins, J. M., et al. (2002). Detecting recent positive selection in the human genome from haplotype structure. *Nature*, *419*, 832–837.

Sagiv, L., & Schwartz, S. H. (1995). Value priorities and readiness for out-group contact. *Journal of Personality and Social Psychology*, *69*, 437–448.

Savitz, J. B., & Ramesar, R. S. (2004). Genetic variants implicated in personality: A review of the more promising candidates. *American Journal of Medical Genetics Part B: Neuropsychiatric Genetics, 131B*, 20–32.

Schaller, M. (2006). Parasites, behavioral defenses, and the social psychological mechanisms through which cultures are evoked. *Psychological Inquiry*, *17*, 96–101.

Schaller, M., & Duncan, L. A. (2007). The behavioral immune system: Its evolution and social psychological implications. In J. P. Forgas, M. G. Haselton, & W. von Hippel (Eds.), *Evolution and the social mind: Evolutionary psychology and social cognition* (pp. 293–307). New York: Psychology Press.

Schaller, M., & Murray, D. R. (2008). Pathogens, personality, and culture: Disease prevalence predicts worldwide variability in sociosexuality, extraversion, and openness to experience. *Journal of Personality and Social Psychology*, 95, 212–221.

Schmitt, D. P. (2005). Sociosexuality from Argentina to Zimbabwe: A 48-nation study of sex, culture, and strategies of human mating. *Behavioral and Brain Sciences*, *28*, 247–311.

Schmitt, D. P., Allik, J., McCrae, R. R., Benet-Martinez, V., et al. (2007). The geographic distribution of Big Five personality traits: Patterns and profiles of human self-description across 56 nations. *Journal of Cross-Cultural Psychology*, *38*, 173–212.

Sherman, P. W., & Billing, J. (1999). Darwinian gastronomy: Why we use spices. *BioScience*, *49*, 453–463.

Simpson, J. A., & Gangestad, S. W. (1991). Individual differences in sociosexuality: Evidence for convergent and discriminant validity. *Journal of Personality and Social Psychology*, *60*, 870–883.

Stein, M. B., Schork, N. J., & Gelernter, J. (2004). A polymorphism of the beta1-adrenergic receptor is associated with low extraversion. *Biological Psychiatry*, *56*, 217–224.

Suh, E., Diener, E., Oishi, S., & Triandis, H. C. (1998). The shifting basis of life satisfaction judgments across cultures: Emotions versus norms. *Journal of Personality and Social Psychology*, *74*, 482–493.

Thornhill, R., Fincher, C. L., & Aran, D. (2009). Parasites, democratization, and the liberalization of values across contemporary countries. *Biological Reviews*, *84*, 113-131.

Thornhill, R., & Gangestad, S. W. (1999). Facial attractiveness. *Trends in Cognitive Sciences*, *3*, 452–460.

Tooby, J., & Cosmides, L. (1992). The psychological foundations of culture. In J. H. Barkow, L. Cosmides, & J. Tooby (Eds.), *The adapted mind* (pp. 19–136). New York: Oxford University Press.

Triandis, H. C. (1995). *Individualism and collectivism*. Boulder, CO: Westview.

Williamson, S. H., Hubisz, M. J., Clark, A. G., Payseur, B. A., Bustamante, C. D., & Nielsen, R. (2007). Localizing recent adaptive evolution in the human genome. *PloS Genetics*, *3*(6), e90. doi:10.1371/journal.pgen.0030090.

Wolfe, N. D., Dunavan, C. P., & Diamond, J. (2007). Origins of major human infectious diseases. *Nature, 447*, 279-283.

17

Universal Mechanisms and Cultural Diversity

Replacing the Blank Slate With a Coloring Book

DOUGLAS T. KENRICK, SARYNINA NIEUWEBOER, and ABRAHAM P. BUUNK

The view of the mind as a Blank Slate, devoid of fixed instincts and subject only to the influence of experiences after birth, was eloquently championed by John B. Watson in 1925. That view, reinforced by reports of cultures seeming to violate most North American norms of conduct, strongly influenced 20th-century psychology (see Pinker, 2002, for a review). Experimental research on learning and cognition, though, did not support the Blank Slate view. It turned out that even so simple a process as classical conditioning is biased to favor associations of the sort regularly confronted by a given species. Rats, for example, whose ancestors often searched for food in the dark, more easily condition aversions to new tastes than to visual cues. Quail, on the other hand, whose ancestors searched for food in broad daylight, condition nausea more easily to visual than to gustatory cues (Wilcoxon, Dragoin, & Kral, 1971). Further problems for a Blank Slate view came from research revealing cross-culturally consistent behavior patterns rather than infinite malleability (e.g., Brown, 1991; Buss, 1989; Daly & Wilson, 1988; Kenrick & Keefe, 1992; Pinker, 1994).

To say that there is strong evidence against the Blank Slate view is not to suggest that thought and behavior can be understood instead as the product of fixed instincts or deterministic genes (Crawford & Anderson, 1989; Kenrick, 2006; Krebs, 2003). Indeed, alongside cultural regularities, there is variation within and between cultures in virtually all forms of complex behavior. Cross-cultural variations in behavior are sometimes misconstrued as prima facie evidence against an evolutionary perspective (e.g., Eagly & Wood, 1999). This is partly because evolutionary accounts presume some psychological universals, and there is often confusion about the meaning of the term *universal* (Norenzayan & Heine, 2005). Some universals are found at the surface level (all human groups recognize angry expressions, for example); others require an understanding of underlying motivations that may be expressed in different ways (all human groups have humor and moral outrage, but which verbal statements elicit laughter or horror depends partly on an understanding of local cultural norms).

Researchers adopting an evolutionary perspective view the mind as a set of evolved psychological systems selected to enhance individual fitness by solving particular adaptive problems our ancestors faced over evolutionary history. Those problems were not constant, and they varied not only across situations and individual life histories but also across ecological contexts. Hence, although evolutionary accounts presume psychological universals underlying many behaviors, they do not presume that overt behaviors will be invariant across cultural contexts (in the same way that bipedal locomotion or visual adaptation to darkness are invariant across cultural contexts). For example, spoken language is a human universal, despite the fact that people in different cultures use completely different words for the same concepts. Yet, the emergence of language depends on environmental inputs (other human beings reliably using patterned vocalizations to refer to objects and events in the child's world), but it is not solely determined by either the inputs (household pets hear the same patterned vocalizations but do not learn to imitate them) or the underlying neural architecture (children growing up in Bogotá, Colombia, learn very different words than children in British Columbia, Canada, although both organize them according to a universal human grammar; Pinker, 1994).

In this chapter, we consider several lines of research demonstrating phenotypic cultural variations that, on closer examination, demonstrate the operation of underlying human universals. We also consider how to understand the dynamic processes underlying the emergence of such variations from universals, and we suggest a more appropriate metaphor with which to replace the outmoded, yet still influential, Blank Slate.

AGE PREFERENCES IN MATES: A HUMAN UNIVERSAL?

In searching for mates, American men tend to seek relatively younger women, and American women tend to seek relatively older men (Kenrick & Keefe, 1992). The age discrepancy was originally explained in terms of sex-role norms of North American society, but several aspects of the evidence raise problems for this account. The preferences of adolescent males raise immediate questions. Although young men are highly attuned to sex roles, they blatantly violate the age preference norm. Instead, males in their teenage years express a strong attraction toward relatively older females. A 16-year-old boy asked to imagine an ideally attractive date tends to choose not an age-mate but a woman several years older than himself (Kenrick, Gabrielidis, Keefe, & Cornelius, 1996). Do teenage boys think they stand any chance of winning the affections of these older females? Not at all. Teenage boys are keenly aware that their attraction to older girls is not reciprocated. Men in their mid-20s are interested in women around their own age, but still their preferences indicate no normative pressures to prefer younger women—they are attracted to slightly older as well as slightly younger women (Kenrick & Keefe, 1992). It is not until men are over 30 that the supposed social norm is clearly demonstrated. Men in their 30s begin to express a clear preference for relatively younger partners—and this bias only becomes extreme in men over 40.

We have argued that men's age preferences are not ultimately driven by a normative compulsion to date and marry relatively younger women. Instead, men of all ages are attracted to female features that were associated with fertility in the ancestral past (Buunk, Dijkstra, Kenrick, & Warntjes, 2001; Kenrick & Keefe, 1992). Whereas women before the late teens have not fully matured, and women over 40 decrease in fertility, women during their 20s and early 30s manifest features associated with fertility, such as relatively rounded hips and youthful hair and skin. According to an evolutionary life history account, our male ancestors would have been selected for any inclinations to prefer females manifesting those characteristics. Thus, to say simply that men are attracted to "younger" women oversimplifies the picture and obscures the causal mechanism.

To say that women are generally attracted to older men is more descriptively accurate but may also miss the point. From an evolutionary life history perspective, women are attracted not to seniority but to status and wealth. Female attraction to status is presumed to stem from an intrinsic sex difference in the types of resources each sex provides to offspring (Geary, 2000). Males across human societies do not contribute physiological resources directly from their bodies to their offspring, as females do, but instead contribute indirect resources, such as food and shelter. The ability to provide such resources correlates with a man's status and material wealth. Because males continue to accumulate resources beyond their years of peak physical fitness and do not undergo menopause, a strong inclination to prefer young males would not have been selected for in our female ancestors.

Age-linked changes in fertility are universal, with puberty and menopause forming lower and upper limits, respectively, on fertility in all societies. If we are correct in presuming that male attraction toward women of a certain age is linked to evolved predispositions favoring the choice of fertile partners, then males the world over ought to become more attracted to relatively younger women as the males themselves age. Consistently, data from a large number of societies indicate a change in males' relative preferences as men age. Young men are interested in mates around their own age, and older men become progressively interested in partners who are relatively younger. So, for example, a typical 25-year-old American man might specify an interest in women between the ages of 20 and 30 (Kenrick & Keefe, 1992). A typical 40-year-old man might specify an interest in women between the ages of 25 and 35. Women tend not to change their preferences across the life span but to prefer men around their own age, regardless of their own age. So, for example, a typical 25-year-old American woman might specify an interest in a man between 24 and 30, whereas a typical 40-year-old woman might specify an interest in men between the ages of 39 and 45.

The same pattern of changes over the life span is found in singles ads from the Netherlands and Germany, in marital ads from India, and in marriage records from several American cities, Brazil, Africa, and a remote island in the Philippines (Harpending, 1992; Kenrick & Keefe, 1992; Otta, Queiroz, Campos, daSilva, & Silveira, 1998). Likewise, UN data from various locations around the world reveal teenage women are several times more likely to marry than teenage men, but the reverse is true for people over 50 (Kenrick & Keefe, 1992).

These data, in aggregate, strongly support the hypothesis of a universal male preference for women during their years of fertility (Kenrick & Keefe, 1992). A preference for women in the years of peak fertility is, of course, not the only factor influencing male mate choice but presumed to operate all else equal. Men also have countervailing attractions to similarity on age and other dimensions and are constrained by female choices (females who married much older males risked losing a partner during the years he was most needed to help raise children). As a result of female choice, not all older men can attract women in the years of maximal fertility (see Kenrick & Keefe, 1992, for detailed discussion of these issues). Older men who are not in a position to attract highly desirable (and highly selective) young women may nevertheless gain other benefits from marrying women closer to their own age. For example, men with mates live longer, and their wives may assist them in contributing to their grandchildren. Nevertheless, whether or not the attraction is mutual, older men across societies continue to be sexually attracted to younger women as sexual partners (Buunk et al., 2001; Kenrick & Keefe, 1992).

TIWI ABORIGINES AND DUTCH WIDOWS: TWO EXCEPTIONS TO THE RULE?

Although the human reproductive life span has varied historically and geographically with the availability of resources such as food, the basic sex difference has not likely changed noticeably during recorded history. If the evolutionary life history account were true, a mutual attraction between older men and relatively younger women ought to be universal across societies and historical time periods. Although our own research supported this reasoning, an anthropological case study of the Australian Tiwi seems at first glance to undermine that conclusion. Hart and Pillig (1960) reported, "Nearly every man in the tribe in the age group from thirty-two to thirty-seven was married to an elderly widow. … But very few of them had a resident *young* wife" (p. 25, emphasis in original).

Another possible exception emerged from an extensive data set we gathered from an immense historical archive. The city of Amsterdam has records of marriages going back for several centuries, and we examined 1,613 marriages from the 1600s, as well an additional 811 marriages from the 1700s and 1800s. The general pattern of the Dutch historical data corresponds with that found in other countries at other times. The data from all centuries revealed the expected drop in relative age of men's wives as the men aged. Also similar to the data from the 20th century, the data for age of women's husbands were relatively more flat across the life span. Although the vast majority of Dutch people got married during their 20s, and older men were, as usual, much more likely to get married than older women, the few older women who did get married included a small number of cases who married men who were much younger than themselves. For example, our sample from the 1700s included only two women above age 50, but both married men who were over a decade younger than themselves. So, at least in some cases, one can find men marrying women well past the years of fertility.

Why the variations from the statistical norm? A possible explanation for the Dutch data links to the importance of agriculture and commerce in the Netherlands. Younger men without property sometimes married older widows who owned farms or business. In this exchange, the older woman gained needed assistance in running the business; the younger man raised his income and his social status. From an evolutionary perspective, this arrangement could have served the younger man's interests if it increased his chances of later attracting younger females (via either legal marriage after the older woman died or extramarital liaisons). Many centuries earlier, the Muslim prophet Mohammed married a wealthy older widow during his 20s, helped manage her profitable business interests, and later married several younger wives after his status increased.

Considering such exchange processes sheds light on Tiwi marriage patterns. Although young Tiwi men marry older women, a closer examination reveals that they are, like men in other societies, strongly attracted to young women as sexual partners. Indeed, Hart and Pillig (1960) noted that a big problem in Tiwi society is keeping unmarried young men away from young women. A young man caught with the young wife of a patriarch could be punished by being expelled from the group or by being gored with a hunting spear. Besides these threats, several other features of Tiwi society allow older men to monopolize desirable young wives. For one, an older Tiwi father chooses husbands for his daughters and considers a daughter "an asset … to be invested in his own welfare" (Hart & Pillig, 1960, p. 15). The society is polygynous, and older men mostly betroth their young daughters to other patriarchs, who are in a position to reciprocate. Young men, with no daughters to offer, are thereby excluded from obtaining young wives.

The sanctions explain why younger men and women do not marry but not why young men and older widows marry. The explanation for that custom lies in several other features of the social system. All Tiwi females (but not all males) are required to be married. At birth, an infant girl is betrothed to an older man, and a widow must remarry on her husband's death. Older widows are less desirable to powerful older men, who frequently have numerous young wives. On the other side, a man under age 30 has insufficient social power and influence to induce a powerful patriarch to bestow a young daughter on him. By marrying a widow, a young man cements alliances with her relatives and gains rights in determining her daughters' later marriages if they become widowed early. Consequently, a young man's status within the group goes up after he marries an older widow, and he is subsequently more likely to acquire desirable younger wives (Hart & Pillig, 1960).

Men from the Mardudjara tribe, another Australian aborigine group, also reportedly prefer younger wives as sexual partners (Tonkinson, 1978). But young wives are hard to find, as girls are frequently betrothed to powerful older men. Again, a young man may marry a widow as his first wife, thereby increasing his status and social connections and increasing his chance of later attracting younger wives. The Mardudjara pattern is less pronounced than the Tiwi pattern because the Mardudjara do not require all girls and women to marry.

There is likely a wide range of individual variation among men and women and across societies in attraction toward relatively older versus younger partners. But any monogamous society in which young men were attracted exclusively to older women and in which only older women could find partners would not have lasted (given the constraints posed by female reproductive life history). Instead, the Tiwi pattern suggests a dynamic interaction in which new social norms emerge from the combined influences of evolved psychological mechanisms (male attraction toward women in the years of peak fertility, female attraction toward high-status men) and local social factors (a geriatric patriarchy that monopolizes younger women, in combination with a rule that all women must be married).

The Tiwi marriage patterns are difficult to explain from other simple accounts, such as Eagly and Wood's (1999) social power hypothesis. That hypothesis suggests that the sex difference in age preferences found in North American and European society can be explained in terms of the differences in social power between men and women and implies that women who gained more power should come to express preferences similar to those of men (i.e., for younger, more attractive partners). Aging Tiwi women marry much younger men, analogous to the "trophy wives" married by high status men in modern urban societies. But this is not associated with female power at all; Tiwi women live in a society that is in fact vastly less egalitarian than European and North American societies. Instead, Tiwi society is extremely patriarchal, with older men trading their daughters with one another and older women's remarriages being negotiated by their sons (Hart & Pillig, 1960). The society is, as noted, highly polygynous; when the older women's younger husbands get more power, they take on additional wives.

ECOLOGICAL VARIATIONS MAY DIFFERENTIALLY EVOKE CONTINGENT PROGRAMS

From the ethnocentric viewpoint of one's own society, cultural variations may seem arbitrary and illogical. As in the case of Tiwi marriages, however, apparently puzzling variation across cultures may be linked to common features of human nature. Numerous other examples can be found. For example, marriages across human societies range from strict monogamy through polygyny and polyandry, seeming to negate any easy generalizations about evolved human mating patterns. A deeper analysis, however, reveals the operation of broad evolutionary principles. To begin with, human mating bonds (found in all types of marriage) are unusual for mammals. Generally, male mammals contribute few direct resources to the offspring. Because of internal gestation and later nursing, most mammalian females can raise their young without the help of males. Pair bonds are more common in birds because females lay large eggs requiring continuous incubation and because helpless chicks must be protected while the parents search for food. Helpless offspring are a characteristic humans share with birds and with other mammals in which male parental investment has evolved (Geary, 2000).

Like humans, many bird species vary their mating patterns, and those variations are linked to ecological factors. Male lark buntings with poor territories may attract no females as mates, males with decent territories mate monogamously, and those with rich territories may attract multiple females (Pleszczynska & Hansell, 1980). Female birds mate polygynously only when male territories vary in quality, and females may be as attracted to the resource-rich territory as to the male who occupies it (Orians, 1969). Although human behavior is undoubtedly more flexible than avian behavior, some similar principles nevertheless apply to polygynous human societies. Multiple women are especially likely to marry the same man when several conditions converge: (a) a steep social hierarchy, (b) a generally rich environment allowing one family to accumulate wealth, and (c) occasional famines so the poor face risks of starvation (Crook & Crook, 1988). Under these circumstances, a woman joining a large wealthy family reaps benefits for her offspring (i.e., lower odds of starvation), even though she must share her husband.

Other cross-cultural variations also operate within general evolutionary guidelines (e.g., Gangestad, Haselton, & Buss, 2006). For example, although polygyny is common across different societies, polyandry is very rare. The evolutionary explanation of this bias (also found across animal species) implicates certainty of parental investment: A female mammal is certain whether a given offspring is hers (regardless of how many females share her mate); a male's parental certainty decreases with each new male with whom his partner mates. In the relatively rare cases when polyandry is found, however, it seems to reflect an interaction between evolved mechanisms and local environmental factors. For example, brothers in resource-poor Tibetan highlands sometimes marry the same woman. There is a net genetic payoff to combining their efforts: Monogamous Tibetan families have fewer surviving children than those in which brothers pool their resources (Crook & Crook, 1988). By sharing one wife, Tibetan brothers preserve their family estate and may actually increase the number of their genes passed to subsequent generations. Because brothers share roughly 50% of genes, children share some of each brother's genes, regardless of who the father was. If a Tibetan family produces all girls, the polyandrous pattern switches to a polygynous one, in which several sisters marry one man, again preserving the family estate. Thus, alongside the flexibility and variability of human social norms, polyandrous and polygynous marriages nevertheless reflect trade-offs following general rules of evolutionary economics.

Cultural variations in norms outside the realm of mating behavior also illustrate a flexible interplay of evolved mechanisms and local ecology. Nisbett and his colleagues (Cohen & Nisbett, 1994; Nisbett, 1993) demonstrated a number of links between the "culture of honor" and violent behavior. Consistent with evolutionary analyses of the function of violence, the culture of honor applies more to men than to women (Wilson & Daly, 1985). Honor-motivated violence is linked to competition for status, which has greater reproductive implications for males than for females. Males in the southern United States, where the culture of honor is prominent, advocate relatively more violent responses to a number of threats. Beyond this, however, cultural variations in honor-linked violence stand alongside substantial cross-cultural regularities. For example, the absolute levels of violence approved for particular affronts (such as attacking one's child, breaking into one's home, or disagreeing with one's views) are quite similar among males from different geographic regions. The commonalities may reflect more fundamental similarities across human groups. Furthermore, the differences between societies, rather than being random, may themselves reflect common ecological problems faced by different groups, such as the association between pastoral lifestyles and cultures of honor (Nisbett, 1993).

Fiske (1992) reviewed data in other domains of social life suggesting that across a wide range of cultures, different rules of exchange apply between family members, friends, and strangers. Family members, for example, generally use communal sharing rules (everyone takes what they need from a common pool). Friendships, on the other hand, are more likely to operate according to sharing rules (everyone gets an equal share), whereas exchanges between strangers tend to follow market rules (you get what you pay for). Whether a society is generally labeled as "individualistic" or "collectivistic," close kin consistently use different exchange rules than do strangers (Fiske, Kitayama, Markus, & Nisbett, 1998). At the same time, cultures vary in the extent to which members are expected to rely on the different rule systems.

These examples suggest that the human mind is surely flexible, but it is hardly a Blank Slate when it comes to absorbing, and constructing, cultural practices. Instead, the mind might better be conceived as a Coloring Book (Kenrick, Becker, Butner, Li, & Maner, 2003). From this view, all human brains come equipped with general outlines, but the particular "colors" drawn inside those outlines depend on cultural norms attuned to local physical and social ecology. At the end of this chapter, we return to consider the advantages of replacing the Blank Slate idea with this more vividly hued metaphor.

INDIVIDUAL DECISION-RULES AND THE EMERGENCE OF CULTURAL NORMS

To say that human cultures are constructed on a foundation of evolved psychological mechanisms does not imply invariance at either the individual or the cultural level (Cohen, 2001; Norenzayan & Heine, 2005). Modern evolutionary theorists conceptualize evolved psychological mechanisms not as inflexible fixed action patterns but more as conditional decision-rules (Cosmides & Tooby, 1992; Kenrick, Li, & Butner, 2003). How do rules in individual heads lead to the emergence of shared cultural norms? Insights from dynamical systems models can help answer that question.

There has been a traditional gulf between psychologists, who focus on individuals, and sociologists, who focus on groups. Evolutionary approaches to psychology have been especially focused on the individual. An individual human being, like any other living organism, is viewed from this perspective as a collection of psychological mechanisms designed to promote the replication of his or her genes. Evolutionary psychologists, like other psychologists, have repeatedly acknowledged that individual psychological mechanisms operate in "interaction" with the social environment (e.g., Crawford & Anderson, 1989). Thus far, however, there has been little research bridging the gap between individualistic psychological mechanisms and group-level processes such as the formation of social norms.

Two recent conceptual insights may help bridge that gap (Kenrick et al., 2002). One comes from research on complex dynamic systems (e.g., Latané & L'Herrou, 1996; Nowak & Vallacher, 1998). That research suggests that initially random interactions between interconnected units (such as neurons in a brain or people in a group) tend to move toward self-organized patterns. The other insight comes from the integration of ideas in cognitive science and evolutionary psychology. Rather than viewing organisms as generally adaptive "fitness-maximizing" machines, the emerging view focuses on specific cognitive mechanisms that embody conditional decision-rules. These contingency-sensitive mechanisms are presumably designed to serve fundamental motivations associated with key problems regularly confronted by our human ancestors (Bugental, 2000; Buss, 2004; Cosmides & Tooby, 1992; Schaller, Park, & Kenrick, 2007). According to that view, there are qualitatively different decision-rules associated with different problem domains, and individuals differ in decision-rules as a function of adaptive and random variation. Because humans are social organisms, decision mechanisms within individuals do not specify rigidly programmed behaviors but instead unfold in dynamic interplay with decision mechanisms of other people in their social network.

Imagine a neighborhood made up of individual people who interacted preferentially with their nearest neighbors and who had to decide whether to adopt a cooperative or a competitive stance in interacting with those immediate neighbors. Imagine further that each individual was motivated not to be in a local minority; to be cooperative when the majority of neighbors were competitive would mean getting exploited, and to be competitive when the majority of neighbors were cooperative would mean eliciting retaliation from otherwise agreeable neighbors. Because individual neighbors would initially shift in different directions depending merely on the random placement of peaceful and competitive neighbors, who would themselves sometimes be shifting their strategies in opposite directions, changes in dynamic influence networks are cognitively challenging to track. Computer simulations can provide a conceptually useful model of such processes, however. Such simulations suggest that regardless of a wide variety of random initial conditions and independent of many other factors, such neighborhoods will tend to "self-organize." That is, any inclinations toward conformity will tend to produce clusters of cooperators and clusters of competitors.

Small random factors play an important role in the emergence of local norms—sometimes a whole neighborhood will become cooperative or competitive just because one key person leans toward one of these strategies. Another key finding of such simulations is this: Different local ecologies may induce a person to act very differently even if his or her underlying trait, or decision-rule, does not vary (Kenrick, Li, & Butner, 2003). Individual differences can radically change community-level outcomes, depending on initial conditions. Nevertheless, neighborhoods typically move toward normative consistency, and changing particular individuals has little impact after the system organizes itself. New immigrants, using their own decision-rules, will generally end up acting as locals do. There are numerous benefits of matching the most common local behavior in human groups, and Henrich and Boyd (1998) demonstrated how natural selection often favors such conformity.

Spatial self-organization is ubiquitous in dynamical systems, but it is only one type of dynamic emergent process. Dynamic processes are found at all levels, from interactions of different motives within one individual to the formation of status hierarchies over the life span and the coevolution of traits in males and females over evolutionary time (Kenrick et al., 2002). Indeed, natural selection is itself an elegant example of dynamically self-organizing processes. A great deal of understanding about cultural norms will follow as social scientists increasingly combine insights from evolutionary psychology, which has led to a great number of discoveries regarding human decision-rules, and dynamical systems theory, which has much to say about the emergence of complexity from simple rules (Kenrick et al., 2002; Kennick, Li, & Butner, 2003).

Returning to the Tiwi, status hierarchies, social regulations, and mating decisions in males and females in that society are dynamically interconnected in ways that produced a stable normative structure different from that found in most other societies. One need not assume young Tiwi men use different mating decision-rules than men in other societies. Like younger men in other societies, they are attracted to women their own age. The powerful patriarchs' own attraction to younger women, however, led to a system in which older males monopolize all younger females and punish liaisons between young men and women. Because decision-rules are intrinsically contingent on environmental inputs, young Tiwi males do not simply rigidly persist in seeking younger wives but instead move to an alternative strategy: mate with an older widow and elevate their status to that of an older male with the means to play the game. Thus, Tiwi society at one level demonstrates several general universals expected in species with differential parental investment (females pairing with high-status males and males vigorously competing for status), as well as a particular feature expected in a species within which females terminate reproduction with menopause (older males desiring younger mates). Because older Tiwi males have been unusually successful in monopolizing social power and fertile young females, there are emergent consequences on other features of Tiwi social structure, which must reach a new dynamic equilibrium between opposing motivations in different individuals.

Tooby and Cosmides (1992) suggested an interesting analogy of a jukebox to represent the interactions between genetic and environmental factors. On that view, ecological factors evoke different contingent programs. That model has limitations in explaining Tiwi social organization, which seems unlikely to have automatically emerged from a preselected set of genetically programmed options. Instead, a combination of several genetic influences and several ecological and historical factors has led to a unique social organization. At the same time, such interactions cannot be understood on the Blank Slate metaphor, because the form of the emergent cultural pattern must be made compatible with strong predispositions (e.g., attraction toward fertile females, and male competition for power). Considering evolved psychological mechanisms in terms of emergent social dynamics bridges evolutionary and sociocultural approaches and raises a host of empirical questions about how cultural diversity emerges from particular mechanisms interacting with particular environments.

UNIVERSAL PROBLEMS, LOCAL SOLUTIONS

Success in human groups involves solving different sets of problems in different domains of social life (Buss, 2004; Kenrick, Li, & Butner, 2003). Because our ancestors reliably confronted certain problems, human cognitive mechanisms may favor particular decision-rules in each domain. For instance, because human parents form pair bonds and share a heavy investment in their joint offspring, jealousy involving mates' reproductive behaviors is likely more intense than jealousy involving friends' sharing tools or food (Buunk, Massar, & Dijkstra, 2007). Human parental investment also predisposes strong nepotistic favoritism and protective feelings toward one's own offspring, different from those felt regarding friends who are nonkin (Kenrick, Sundie, & Kurzban, 2008). There are probably a limited set of centrally important social domains—collections of distinct problems and opportunities that humans had to face in dealing with potential friends, enemies, potential mates, and offspring (Kenrick, Li, & Butner, 2003; Schaller et al., 2007).

As a consequence of these types of different decision-rules, different social geometries emerge within each domain (see Kenrick, Li, & Butner, 2003). For example, jealousy and parental investment combine to make parental pair bonds the prevalent arrangement, even in societies that permit polygyny. Aggregations of more than two have advantages for networks based on cooperative sharing, but too many close friends can become unworkable (think of a dinner at a conference that expands from 4 people to 20, or imagine trying to arrange a picnic with 60 people as opposed to 10). Assuming that humans everywhere share certain basic underlying motivations, these geometries may lead to general similarities in social organization across different societies (cf. Fiske, 2000). All around the world, these factors lead everyday coalitional groups to be loosely overlapping (as a friend of my brother becomes my friend), status competitions to be hierarchical (there are fewer positions at the top than at the bottom), intergroup conflict to inspire the formation of larger groups (via safety in numbers) with clear barriers (an enemy of my in-group member is also my enemy), and so on (see Kenrick, Li, et al., 2003, for further discussion).

Despite broad similarities across societies, universal psychological mechanisms do not imply phenotypically rigid social patterns. Variations in physical or social ecology lead to local variations in rules about how to choose mates or share resources. Besides ecological variation and individual differences, some phenotypic variation in norms is expected from random spatial factors. Despite local variation, though, common human motivations should lead to a limited class of recurrent solutions, each bearing the imprint of human nature (as in the correlations between local resources, social hierarchies, and marriage patterns). In other words, universals are expected at the level of underlying motivation rather than particular overt behaviors.

Such analyses make it clear why "genetic determinism" is an outmoded notion. Psychological mechanisms always involve environmentally contingent decision-rules. Research on individual psychology, elucidated by evolutionary analyses of other species and other societies, can inform us about how those decision mechanisms operate. Dynamical models can help us understand how such mechanisms interact to lead to emergence of self-organized cultures. The ultimate form of social self-organization depends on the specific decision-rules used by individuals. Because those rules vary for different domains of social life, models of emergent social dynamics will be more useful if they take content seriously.

A truly cross-cultural psychology is essential to understanding the common problems human beings confront (Norenzayan & Heine, 2005; Schaller & Crandall, 2004). Cross-cultural approaches will be most useful if they move beyond culturally relativistic (and ultimately ethnocentric) accounts of how "we" are different from "them." At the same time, we ought not expect absolute universals at the level of phenotypic behaviors. The Tiwi case suggests the usefulness of searching for common motivations underlying overtly different cultural norms. Incorporating insights from the study of complex dynamical systems can help elucidate how simple motivations unfold into diverse cultural norms. This broader perspective can help avoid two extremes: a nihilistic cultural relativism that fails to account for recurring patterns in human behavior, and a simplistic genetic determinism that fails to account for phenotypic variation across individuals and societies.

BLANK SLATES, JUKEBOXES, COMPUTERS, AND COLORING BOOKS

Few psychologists today would argue for a completely Blank Slate. Yet there are remaining controversies about how many lines are predrawn in the Coloring Book and whether the slate is blank in some cognitive and behavioral domains. Some accounts acknowledge evolved differences between the sexes in parenting capacity and physical strength, for example, but posit that most or all psychological differences, including mate preferences, flow from social roles constructed around those differences (e.g., Eagly & Wood, 1999). Others acknowledge differences in aggression or sexuality but do not place these differences in evolutionary perspective (e.g., Hyde, 2005). Explicitly evolutionary models all include a presumption that the slate is not blank, but they vary regarding the number of discrete mechanisms and whether these are seen as hierarchically organized into a smaller number of general motivational–cognitive systems (Cosmides & Tooby, 1992; Kenrick, Sadalla, & Keefe, 1998; Kurzban & Aktipis, 2006). Thus, most of the empirical questions about the degree and nature of constraints on human preferences and decisions are yet to be answered.

But despite the general agreement that the Blank Slate is an outdated metaphor, it remains a simple and cognitively compelling image. We ought to replace it with an image that is equally straightforward and understandable. Tooby and Cosmides's jukebox metaphor is an interesting alternative. Compared with a Blank Slate, a jukebox is pleasingly interactive—its outputs are determined not solely by what's inside or by environmental inputs alone (pressing F6 will not result in a tune unless there's a record inside). As we noted, however, many cultural norms are not straightforward and automatic consequences of pushing a particular set of preordained buttons.

A Coloring Book, like a jukebox, also includes an explicit consideration of what's inside (the predrawn lines that suggest a giraffe versus a zebra versus a rocket ship) and environmental inputs (the young artist wielding the crayons). But the Coloring Book metaphor has a few advantages. For one, a Coloring Book leaves a lot more room for flexible and potentially unpredicted outcomes—one child might choose to color his giraffe purple and green instead of tan and brown. At the same time, a Coloring Book paradoxically allows for more built-in constraints, alongside its flexibility. More than the buttons on a jukebox, the predrawn outlines in a Coloring Book strongly solicit particular inputs from the environment (most children coloring a giraffe will be inspired to search for tan, brown, and yellow rather than purple, blue, and green). Thus, although a Coloring Book can, in one sense, be colored in an infinite number of ways, in another sense it is not really completely passive because the outlines tend to strongly suggest (though not determine) particular palettes of inputs to be used on the different pages.

MUST WE BE METAPHORICAL?

Why seek any new metaphor at all? Why not just replace the Blank Slate with a more realistic picture—billions of neurons arranged in an interconnected network of parallel information-processing mechanisms, all simultaneously and dynamically interacting with an array of diverse auditory, tactile, visual, gustatory, and internal biochemical inputs? We argue that metaphors are essential as cognitive tools for beginning students and scientists alike. Metaphors are critical to help students as well as scientists think about topics that are too complex for any human mind, which is limited to thinking about seven bits, plus or minus two, of information at any given time. Even philosophers and cognitive neuroscience researchers probably have to strain to really think about the brain as a set of several billion neurons divided into *n* number of physiological components responding to *y* number of external inputs about current temperature, balance, social structure, and conversational content in light of past temperature, balance, social structure, and conversational content, as well as numerous internal inputs from testosterone, adrenaline, and dopamine levels in various concentrations at various locations in the central and autonomic nervous system and how all of that is influenced by particular sets of early experiences and genetic blueprints that unfolded in particular nutritional contexts, and so on. For students, even the preceding sentence probably computes merely as booming, buzzing confusion. Instead, we need simplified images of how the whole system functions, and for this we need metaphors. As it turns out, people, scientists included, regularly and spontaneously think in terms of metaphors all the time (Lakoff, 1987).

Besides the jukebox, there are other more complicated technological metaphors that have been offered over the years. During the 19th century, some scientists conceptualized the mind as a steam engine. During the early 20th century, a telephone switchboard seemed a better image, and during the latter half of the 20th century, the mind was upgraded to a digital computer. As Pinker (2002) noted, the computer metaphor was in some ways a modernized variant of the Blank Slate—envisioning the mind as an infinitely programmable set of identical "on–off" switches.

One could imagine a more complicated and modernized digital computer, with parallel central processing units, but a metaphor is, after all, merely a thinking aid, not the thing itself. The Coloring Book metaphor does not pretend to be an actual representation of the human brain, but it does provides a straightforward contrast to the Blank Slate, conceptually extending this old and powerful metaphor to encompass interactions in a way that is easy to visualize. Indeed, the Coloring Book actually incorporates the old metaphor, but it inspires us to think about the mind as having some built-in outlines as well as a great deal of blank space to be filled by environmental inputs.

An additional advantage of a Coloring Book metaphor is that, unlike a Blank Slate, a Coloring Book includes not just one page but many. Thinking about the different pages in a Coloring Book can provide a mnemonically accessible way to think about the otherwise difficult notion of domain specificity. Just as there are different patterns of predrawn lines on different pages of a Coloring Book so there are likely different patterns of constraint in different domains of decision making. Language learning, avoidance of poisonous foods, fear conditioning, and mate preference develop according to different dynamic constraints and consequently involve different palettes of inputs from which they are likely to be colored (e.g., Kenrick, Keefe, Bryan, Barr, & Brown, 1995; Öhman & Mineka, 2001; Pinker, 1994; Sherry & Schacter, 1987). We expect future research will reveal that few important domains of human psychology are rigidly colored in at birth. At the same time, only a very blurry image is likely to emerge unless we better discern the adaptive guidelines provided by human nature.

REFERENCES

Brown, D. E. (1991). *Human universals*. New York: McGraw-Hill.

Bugental, D. B. (2000). Acquisition of the algorithms of social life: A domain-based approach. *Psychological Bulletin*, *126*, 187–219.

Buss, D. M. (1989). Sex differences in human mate preference: Evolutionary hypothesis tested in 37 cultures. *Behavioral and Brain Sciences*, *12*, 1–49.

Buss, D. M. (2004). *Evolutionary psychology: The new science of mind* (2nd ed.). Boston: Allyn & Bacon.

Buunk, B. P., Dijkstra, P., Kenrick, D. T., & Warntjes, A. (2001). Age preferences for mates as related to gender, own age, and involvement level. *Evolution and Human Behavior*, *22*, 241–250.

Buunk, A. P., Massar, K., & Dijkstra, P. (2007). A social cognitive evolutionary approach to jealousy: The automatic evaluation of one's romantic rivals. In J. P. Forgas, M. G. Haselton, & W. von Hippel (Eds.), *Evolution and the social mind* (pp. 213–228). New York: Psychology Press.

Cohen, D. (2001). Cultural variation: Considerations and implications. *Psychological Bulletin*, *127*, 451–471.

Cohen, D., & Nisbett, R. E. (1994). Self-protection and the culture of honor: Explaining southern violence. *Personality and Social Psychology Bulletin*, *20*, 551–567.

Cosmides, L., & Tooby, J. (1992). Cognitive adaptations for social exchange. In J. H. Barkow, L. Cosmides, & J. Tooby (Eds.), *The adapted mind* (pp. 163–228). New York: Oxford University Press.

Crawford, C. B., & Anderson, J. L. (1989). Sociobiology: An environmentalist discipline. *American Psychologist*, *44*, 1449–1459.

Crook, J. H., & Crook, S. J. (1988). Tibetan polyandry: Problems of adaptation and fitness. In L. Betzig, M. Borgerhoff Mulder, & P. Turke (Eds.), *Human reproductive behavior: A Darwinian perspective* (pp. 97–114). Cambridge, UK: Cambridge University Press.

Daly, M., & Wilson, M. (1988). *Homicide.* Hawthorne, NY: Aldine de Gruyter.

Eagly, A. H., & Wood, W. (1999). The origins of sex differences in human behavior: Evolved predispositions versus social roles. *American Psychologist*, *54*, 408–423.

Fiske, A. P. (1992). The four elementary forms of sociality: Framework for a unified theory of social relations. *Psychological Review*, *99*, 689–723.

Fiske, A. P. (2000). Complementarity theory: Why human social capacities evolved to require cultural complements. *Personality and Social Psychology Review*, *4*, 76–94.

Fiske, A. P., Kitayama, S., Markus, H. R., & Nisbett, R. E. (1998). The cultural matrix of social psychology. In D. T. Gilbert, S. T. Fiske, & G. Lindzey (Eds.), *Handbook of social psychology* (4th ed., pp. 915–982). New York: McGraw-Hill/Oxford University Press.

Gangestad, S. W., Haselton, M. G., & Buss, D. M. (2006). Evolutionary foundations of cultural variation: Evoked culture and mate preferences. *Psychological Inquiry*, *17*, 75–94.

Geary, D. C. (2000). Evolution and the proximate expression of human paternal investment. *Psychological Bulletin*, *126*, 55–77.

Harpending, H. (1992). Age differences between mates in Southern African pastoralists. *Behavioral and Brain Sciences*, *15*, 102–103.

Hart, C. W. M., & Pillig, A. R. (1960). *The Tiwi of North Australia.* New York: Holt.

Henrich, J., & Boyd, R. (1998). The evolution of conformist transmission and the emergence of between-group differences. *Evolution and Human Behavior*, *19*, 215–241.

Hyde, J. S. (2005). The gender similarities hypothesis. *American Psychologist*, *60*, 581–592.

Kenrick, D. T. (2006). Evolutionary psychology: Resistance is futile. *Psychological Inquiry*, *17*, 101–109.

Kenrick, D. T., Becker, D. V., Butner, J., Li, N. P., & Maner, J. K. (2003). Evolutionary cognitive science: Adding what and why to how the mind works. In K. Sterelney & J. Fitness (Eds.), *From mating to mentality: Evaluating evolutionary psychology* (pp. 13–38). New York: Psychology Press.

Kenrick, D. T., Gabrielidis, C., Keefe, R. C., & Cornelius, J. (1996). Adolescents' age preferences for dating partners: An evolutionary model of life-history strategies. *Child Development*, *67*, 1499–1511.

Kenrick, D. T., & Keefe, R. C. (1992). Age preferences in mates reflect sex differences in human reproductive strategies. *Behavioral and Brain Sciences*, *15*, 75–133.

Kenrick, D. T., Keefe, R. C., Bryan, A., Barr, A., & Brown, S. (1995). Age preferences and mate choice among homosexuals and heterosexuals: A case for modular psychological mechanisms. *Journal of Personality and Social Psychology*, *69*, 1166–1172.

Kenrick, D. T., Li, N. P., & Butner, J. (2003). Dynamical evolutionary psychology: Individual decision-rules and emergent social norms. *Psychological Review*, *110*, 3–29.

Kenrick, D. T., Maner, J. K., Butner, J., Li, N. P., Becker, D. V., & Schaller, M. (2002). Dynamical evolutionary psychology: Mapping the domains of the new interactionist paradigm. *Personality and Social Psychology Review*, *6*, 347–356.

Kenrick, D. T., Sadalla, E. K., & Keefe, R. C. (1998). Evolutionary cognitive psychology. In C. Crawford & D. L. Krebs (Eds.), *Handbook of evolutionary psychology* (pp. 485–514). Hillsdale, NJ: Lawrence Erlbaum.

Kenrick, D. T., Sundie, J. M., & Kurzban, R. (2008). Cooperation and conflict between kith, kin, and strangers: Game theory by domains. In C. Crawford & D. Krebs (Eds.), *Foundations of evolutionary psychology* (pp. 353–370). New York: Lawrence Erlbaum Associates.

Krebs, D. L. (2003). Fictions and facts about evolutionary approaches to human behavior. *Psychological Bulletin, 129*, 842–847.

Kurzban, R., & Aktipis, C. A. (2006). Modular minds, multiple motives. In M. Schaller, J. A. Simpson, & D. T. Kenrick (Eds.), *Evolution and social psychology* (pp. 39–54). New York: Psychology Press.

Lakoff, G. (1987). *Women, fire, and dangerous things.* Chicago: University of Chicago Press.

Latané, B., & L'Herrou, T. (1996). Spatial clustering in the conformity game: Dynamic social impact in electronic groups. *Journal of Personality and Social Psychology, 70*, 1218–1230.

Nisbett, R. E. (1993). Violence and U.S. regional culture. *American Psychologist, 48*, 441–449.

Norenzayan, A., & Heine, S. J. (2005). Psychological universals: What are they and how can we know? *Psychological Bulletin, 131*, 763–784.

Nowak, A., & Vallacher, R. R. (1998). *Dynamical social psychology.* New York: Guilford.

Öhman, A., & Mineka, S. (2001). Fears, phobias, and preparedness: Toward an evolved module of fear and fear learning. *Psychological Review, 108*, 483–522.

Orians, G. H. (1969). On the evolution of mating systems in birds and mammals. *American Naturalist, 103*, 589–603.

Otta, E., Queiroz, R. D. S., Campos, L. D. S., daSilva, M. W. D., & Silveira, M. T. (1998). Age differences between spouses in a Brazilian marriage sample. *Evolution and Human Behavior, 20*, 99–103.

Pinker, S. (1994). *The language instinct.* New York: William Morrow.

Pinker, S. (2002). *The Blank Slate.* New York: Viking Penguin.

Pleszczynska, W. K., & Hansell, R. I. C. (1980). Polygyny and decision theory: Testing of a model in lark buntings (*Calamospiza melanocorys*). *American Naturalist, 116*, 821–830.

Schaller, M., & Crandall, C. S. (2004). *The psychological foundations of culture.* Mahwah, NJ: Lawrence Erlbaum.

Schaller, M., Park, J. H., & Kenrick, D. T. (2007). Human evolution and social cognition. In R. I. M. Dunbar & L. Barrett (Eds.), *Oxford handbook of evolutionary psychology* (pp. 491–504). Oxford, UK: Oxford University Press.

Sherry, D. F., & Schacter, D. L. (1987). The evolution of multiple memory systems. *Psychological Review, 94*, 439–454.

Tonkinson, R. (1978). *The Mardudjara Aborigines: Living the dream in Australia's desert.* New York: Holt, Rinehart & Winston.

Tooby, J., & Cosmides, L. (1992). The psychological foundations of culture. In J. H. Barkow, L. Cosmides, & J. Tooby (Eds.), *The adapted mind: Evolutionary psychology and the generation of culture* (pp. 19–136). New York: Oxford University Press.

Wilcoxon, H. C., Dragoin, W. B., & Kral, P. A. (1971). Illness-induced aversions in rat and quail: Relative salience of visual and gustatory cues. *Science, 171*, 826–828.

Wilson, M., & Daly, M. (1985). Competitiveness, risk taking, and violence: The young male syndrome. *Ethology and Sociobiology, 6*, 59–73.

Subject Index

L

M

N

O

P

Author Index

Note: Page numbers with "n" after them indicate a note. Page number in **bold** refer to figures or tables.

H

T

For Product Safety Concerns and Information please contact our EU representative GPSR@taylorandfrancis.com
Taylor & Francis Verlag GmbH, Kaufingerstraße 24, 80331 München, Germany

www.ingramcontent.com/pod-product-compliance
Lightning Source LLC
LaVergne TN
LVHW011117300926
849160LV00008B/412